AF787858

Selected Topics in Plasma Physics

Edited by Sukhmander Singh

Published in London, United Kingdom

IntechOpen

Supporting open minds since 2005

Selected Topics in Plasma Physics
http://dx.doi.org/10.5772/intechopen.78866
Edited by Sukhmander Singh

Contributors
Eduard Rostomyan, Theodore Anderson, Chinmoy Mallick, M. Bandyopadhyay, R Kumar, Sukhmander Singh, Shahid Ali, Yas Al-Hadeethi, Prasanta Chatterjee, Laxmikanta Mandi, Kaushik Roy

© The Editor(s) and the Author(s) 2020
The rights of the editor(s) and the author(s) have been asserted in accordance with the Copyright, Designs and Patents Act 1988. All rights to the book as a whole are reserved by INTECHOPEN LIMITED. The book as a whole (compilation) cannot be reproduced, distributed or used for commercial or non-commercial purposes without INTECHOPEN LIMITED's written permission. Enquiries concerning the use of the book should be directed to INTECHOPEN LIMITED rights and permissions department (permissions@intechopen.com).
Violations are liable to prosecution under the governing Copyright Law.

Individual chapters of this publication are distributed under the terms of the Creative Commons Attribution 3.0 Unported License which permits commercial use, distribution and reproduction of the individual chapters, provided the original author(s) and source publication are appropriately acknowledged. If so indicated, certain images may not be included under the Creative Commons license. In such cases users will need to obtain permission from the license holder to reproduce the material. More details and guidelines concerning content reuse and adaptation can be found at http://www.intechopen.com/copyright-policy.html.

Notice
Statements and opinions expressed in the chapters are these of the individual contributors and not necessarily those of the editors or publisher. No responsibility is accepted for the accuracy of information contained in the published chapters. The publisher assumes no responsibility for any damage or injury to persons or property arising out of the use of any materials, instructions, methods or ideas contained in the book.

First published in London, United Kingdom, 2020 by IntechOpen
IntechOpen is the global imprint of INTECHOPEN LIMITED, registered in England and Wales, registration number: 11086078, 5 Princes Gate Court, London, SW7 2QJ, United Kingdom
Printed in Croatia

British Library Cataloguing-in-Publication Data
A catalogue record for this book is available from the British Library

Additional hard and PDF copies can be obtained from orders@intechopen.com

Selected Topics in Plasma Physics
Edited by Sukhmander Singh
p. cm.
Print ISBN 978-1-83962-678-4
Online ISBN 978-1-83962-679-1
eBook (PDF) ISBN 978-1-83962-680-7

For EU product safety concerns:
IN TECH d.o.o., Prolaz Marije Krucifikse Kozulić 3, 51000 Rijeka, Croatia, info@intechopen.com or visit our website at intechopen.com.

We are IntechOpen, the world's leading publisher of Open Access books

Built by scientists, for scientists

5,100+
Open access books available

126,000+
International authors and editors

145M+
Downloads

156
Countries delivered to

Our authors are among the
Top 1%
most cited scientists

12.2%
Contributors from top 500 universities

BOOK CITATION INDEX
CLARIVATE ANALYTICS
INDEXED

WEB OF SCIENCE™

Selection of our books indexed in the Book Citation Index in Web of Science™ Core Collection (BKCI)

Interested in publishing with us?
Contact book.department@intechopen.com

Numbers displayed above are based on latest data collected.
For more information visit www.intechopen.com

Meet the editor

Dr. Sukhmander Singh is currently an assistant professor in the Department of Physics at Central University of Rajasthan, Ajmer, India. Previously, he was assistant professor at Motilal Nehru College, University of Delhi. He obtained his MSc (Physics) from Jawaharlal Nehru University New Delhi and PhD (Plasma Physics) from IIT Delhi, New Delhi, India. He has published numerous research papers and conference proceedings in journals of international repute as well as contributed many book chapters. He is an active reviewer for many international journals. His areas of interest include theory and simulation of plasma waves and instabilities.

Contents

Preface

Almost 99 percent of matter in the universe exists in a plasma state, as in the form of the Sun, stars, lightning, welding arcs, interplanetary and interstellar space, and in neon and fluorescent tubes. Plasma is an electrically conducting medium in which there are roughly equal numbers of neutral and positively and negatively charged particles.

This book introduces topics of plasma physics and presents related applied research. Several chapters deal with basic concepts in plasma physics, nonequilibrium plasma modeling, space plasma applications, and plasma diagnostics. The book also gives an overview of the linear and nonlinear aspects of plasma physics. As such, it is a useful resource for many young researchers and students in the field.

The first chapter provides an introduction to plasma physics, occurrence of plasma in universe, types of plasma, plasma quasi-neutrality, plasma frequency, Debye shielding, and waves in plasma. It also discusses stable and unstable plasma oscillations with the help of two fluids, also called plasma hydrodynamics.

Chapter 2 discusses the technological application of plasma physics in the field of electric propulsion. These devices have much greater exhaust velocities, longer lifetime, and greater thrust density than chemical propulsion devices. The chapter discusses different types of plasma propulsion devices by providing mathematical formulations. It also presents the current status of the research on Hall thrusters.

Chapter 3 discusses the interaction of microwave electric fields with plasma. It also gives the dispersion relations of electrostatic and electromagnetic waves. The spatio-temporal evolution pattern of microwave-radiated plasma parameters highlight the role of these electric fields in power coupling processes.

Chapter 4 deals with the design of smart plasma antennas to achieve high gain and reconfigurable beam width without physically moving the antenna. It also reports on the generation of microwaves and their reflection from plasma.

Chapter 5 explains the principles of the diagnostic technique for a test charge in Lorentzian dusty plasma with the help of space-time Fourier transformations. Finally, Chapter 6 deals with the theory of nonlinear waves in plasma to obtain the Korteweg-de Vries (KDV) equation by using the Reductive Perturbation Technique.

We are grateful to our authors Rostomyan, Mallick, Bandyopadhyay, R. Kumar, Theodore Anderson, S. Ali, Y. Al-Hadeethi, Laxmikanta Mandi, Kaushik Roy, and Prasanta Chatterjee. I especially thank Mr. Anil Joshi, who always gave me good advice and helped on several occasions. Special thanks to my parents, sisters, grandfather, and wife Rajvinder Kaur for her cooperation in my life and for sharing the responsibility of family affairs so that I could spend my time for this book. I gratefully acknowledge the technical support provided by the publisher IntechOpen for bringing out this book in a very short time.

Dr. Sukhmander Singh
Assistant Professor,
Department of Physics,
Central University of Rajasthan,
Ajmer, Kishangarh, India

Section 1

Introduction to Plasma State

Chapter 1

Introduction: Plasma Parameters and Simplest Models

Eduard Vardges Rostomyan

Abstract

Plasma is ionized gas (partially or fully). Overwhelming majority of matter in the universe is in plasma state (stars, Sun, etc.). Basic parameters of plasma state are given briefly as well as classification of plasma types: classic-quantum, ideal-nonideal, etc. Differences between plasma and neutral gas are presented. Plasma properties are determined by long distance electrostatic forces. If spatial dimensions of a system of charged particles exceed the so-called Debye radius, the system may be considered as plasma, that is, a medium with qualitatively new properties. The expressions for Debye radius for classical and quantum plasma are carried out. Basic principles of plasma description are presented. It is shown that plasma is a subject to specific electrostatic (or Langmuir) oscillations and instabilities. Simplest plasma models are given briefly: the model of "test" particle and model of two (electron and ion) fluids. As an example, Buneman instability is presented along with qualitative analysis of its complicate dispersion relation. Such analysis is typical in plasma theory. It allows to easily obtain the growth rate.

Keywords: plasma, quasi-neutrality, Langmuir frequency, Debye length, simplest plasma models, unstable plasma oscillations

1. Introduction

Everyone knows the three states of matter: solids, liquids, and gases. Plasma is often called the fourth state of matter, bear in mind that with increasing temperature, the following transitions take place: solids-liquids-gases-plasma. Under the last transition, atoms lose electrons. Plasma consists (along with neutral atoms) of charged particles: electrons and positively charged ions (single and/or multiple ionized). This definition of plasma is far from complete. The complete definition of plasma is, in fact, impossible. It must cover a very wide range of phenomena in a wide variety of conditions.

Plasma is very common in the universe. Most of the substance in it (more than 99%) is in plasma state. Media consisting of ionized atoms is found almost everywhere. The upper layers of the Earth's and stellar atmospheres, interstellar medium, etc. actually are in plasma state. Stellar plasma is another widespread example. In the plasma of stars, in particular the Sun, reactions of the synthesis of light elements, the so-called thermonuclear reactions, provide a huge release of energy and plasma heating. Currently, scientists from many countries around the world are studying the possibility of creating such a high-temperature plasma in terrestrial conditions, setting the task of implementing controlled thermonuclear fusion and providing humanity with an inexhaustible supply of energy.

There are two fundamentally different approaches to the implementation of controlled thermonuclear fusion. The first approach is obtaining the reactions in the so-called magnetic traps. Hot plasma should not come in contact with the walls of the chamber, as this will lead to its actual destruction. The confinement of the plasma by a magnetic field, in theory, should prevent the contact of the plasma with the walls of the chamber, since the magnetic field bends the trajectory of charged particles. However, despite tremendous efforts, the task of plasma confining has not been completely solved yet. Special configurations of magnetic field, magnetic traps help only partially. Plasma is an unstable medium in which small perturbations increase and destroy its given state. Instabilities are intrinsic for plasma. It turned out that any nonequilibrium initial distribution of particles is unstable. Below we show how instability follows from general electrodynamical consideration and give an example.

The second approach is the very quick heating of plasma up to thermonuclear temperatures. The reaction itself and energy removal also take place quickly. These processes recur very fast, and confinement of such plasma is not needed. This approach is called inertial fusion. It was proposed when very fast heating of plasma becomes possible by using intense laser beams and/or high-current relativistic electron beams.

The development of these studies is associated with the rebirth of the concept of plasma, which arises upon the investigations of gas-discharge processes. The processes in gas-discharge plasma have also been intensively studied. The studies were associated with the development of the needs of classical and quantum electronics, for which gas-discharge appliances play an important role. Finally, solid-state plasma should be noted: electron plasma of metals and electron–hole plasma of semiconductors.

The listed series can be continued almost unlimitedly, speaking about plasma in magnetohydrodynamic and thermionic converters of thermal energy into electrical energy, about plasma in solutions of electrolytes, etc. However, the above examples are sufficient to make sure the extremely wide prevalence of plasma in nature and the importance of studying its properties.

A vast literature has been grown to describe plasma state (see, e.g., [1–7] and many others). Our further presentation is based on the principles of plasma electrodynamics considering plasma as a continuous medium with a large number of free charged carriers.

2. Plasma parameters

Plasma is an ionized gas consisting of free electrons and various types of ions and neutrals. First of all, it is necessary to know the charge e_α and concentration n_α of plasma components (here the index takes different values corresponding to the types of particles in the plasma, $\alpha = e$ for electrons, $\alpha = i_1, i_2 \dots$ for ions of various types, and $\alpha = n$ for neutrals). All plasma particles are in chaotic motion, but full thermodynamic equilibrium is absent. Usually each component has its own temperature T_α, which is also necessary to know.

In solid-state and semiconductor plasma, the conception of temperature should be given more accurately. If the Fermi energy $E_{F\alpha}$ of α-type particles exceeds their thermal energy

$$E_{F\alpha} = \frac{\left(3\pi^2\right)^{2/3}\hbar^2 n_\alpha^{2/3}}{2m_\alpha} >> T_\alpha \tag{1}$$

(here m_α is the mass of the particle, and $\hbar$ is the Planck constant), quantum effects should be accounted. In this case the Maxwell's distribution function does not describe the behavior of charged particles. It should be described by Fermi distribution function, and the $E_{F\alpha}$ (1) plays the role of temperature. In this case plasma is *degenerate*.

If neutrals are absent, plasma is *fully ionized*. In the opposite case, plasma is *partially ionized*, and it is necessary to know the level of plasma ionization. This is the ratio of neutrals' density to the density of charged particles (or to the full density of plasma).

An important characteristic peculiarity of plasma state is a very wide range of values of these (and other) parameters. For example, plasma in some stars (white dwarfs) has a density of 10^{25}–10^{26} cm^{-3}, but in the interstellar space, plasma has a density of 1–10 cm^{-3}. The ratio is 10^{26}. The ratio of other parameter values is a bit less. This leads to important consequences. In the example above, different approaches may be required to describe the plasma inside the stars and in the interstellar space. The most interesting cases will be mentioned below.

An important condition for plasma existence is its quasi-neutrality. The condition of quasi-neutrality has the form

$$\sum_{\alpha} e_\alpha n_\alpha \approx 0 \tag{2}$$

where the summation is made over all types of charged particles $\alpha = e, i_1, i_2, i_3 \ldots$ When it is violated, strong electric fields arise, which restore plasma quasi-neutrality. Violations of quasi-neutrality are possible only in spatial and temporal scales, small in comparison with the characteristic plasma scales. The temporal characteristic scale of plasma is determined by its proper oscillations, but the spatial scale is determined by the length of plasma shielding (Debye length; see below).

2.1 Plasma oscillations

2.1.1 Langmuir frequency

Plasma, as a medium with a large number of free charged particles, is a subject to oscillations. Consider in detail the oscillations of uniform electron plasma. Ions are heavy (immobile) and serve for charge neutralization. Let a small displacement of an electron layer relative to the ions take place (see **Figure 1**). We denote the displacement vector by **X**. The density of the uncompensated electron charge at the displacement **X** may be found from the continuity equation:

$$\rho = \nabla n_e e \mathbf{X} = n_e e \nabla \mathbf{X} \tag{3}$$

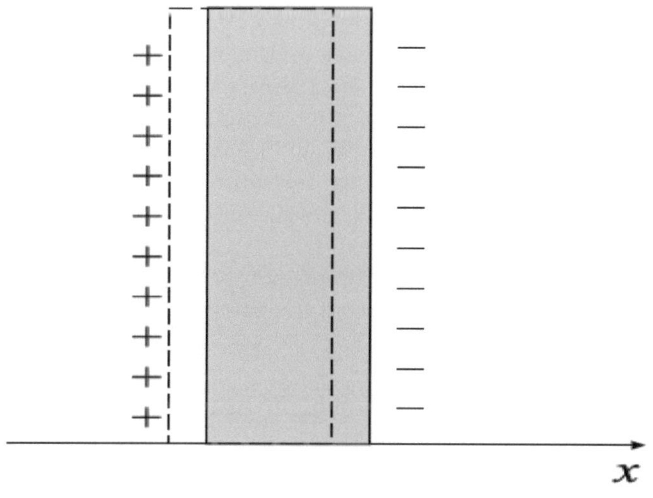

Figure 1.
Oscillations of the electron layer.

This charge creates an electric field **E**, the value of which may be determined from Poisson's equation:

$$\nabla \cdot \mathbf{E} = 4\pi\rho = 4\pi n_e |e| \nabla \cdot \mathbf{X} \quad (4)$$

Hence, we can write (given that for $\mathbf{X} = 0$, we have $\mathbf{E} = 0$)

$$\mathbf{E} = 4\pi n_e e \mathbf{X} \quad (5)$$

Thus, the field **E** is parallel to the displacement of electrons and acts on each electron with a force

$$\mathbf{F} = -e\mathbf{E} = -4\pi n_e e^2 \mathbf{X} \quad (6)$$

tending to return the electron to its original equilibrium position. As a result, we have the equation of motion of an electron in the form

$$m\frac{d^2\mathbf{X}}{dt^2} = -e\mathbf{E} = -4\pi n_e e^2 \mathbf{X} \quad (7)$$

This equation describes the oscillations of plasma electrons near the equilibrium position ($\mathbf{X} = 0$) with a frequency

$$\omega = \omega_{Le} = \sqrt{\frac{4\pi n_e e^2}{m}}, \quad (8)$$

which is known as the electron Langmuir frequency. If one uses MKS units, the expression for electron Langmuir frequency is

$$\omega_{Le} = \sqrt{\frac{n_e e^2}{\varepsilon_0 m}} \quad (9)$$

where ε_0 is the dielectric permittivity of vacuum. Violations of plasma quasi-neutrality are possible only on a temporal scale, small in comparison with time $\tau \sim 1/\omega_{Le}$.

2.2 Gas parameter. Debye length

The behavior of an ionized gas is determined by long-distance electrostatic forces. These forces significantly influence on the plasma behavior and, actually, determine its parameters. First of all it is necessary to find out under what conditions a system of electrostatically interacting particles can be considered as a gas. The main peculiarity of a gas is the following: its particles interact during very small time intervals only (during collisions); the rest time every particle moves independently on others. At distances exceeding the size of the gas molecules, there is no interaction (its potential is equal to zero). Or, in other words, the potential energy of a particle is much lesser than their kinetic energy. In this case, the ratio of the distance, at which the interaction between the particles is significant to the average distance between particles, is small [8]:

$$\Lambda_G = \frac{a}{\langle r \rangle} \approx a n^{1/3} \ll 1 \quad (10)$$

(here a is the molecule size, $\langle r \rangle$ is the average distance between particles, and n is the density). The condition (10) also holds for the interaction of electrons with neutrals and ions with neutrals. However, if we consider long-distance interaction between charge particles, the parameter Λ_G (gas-like parameter) requires rethinking. Its physical meaning becomes slightly different. The gas approximation is valid if the energy of the interaction between particles $U(\langle r \rangle)$ is smaller than the average thermal energy T of the particles itself, i.e.

$$U(\langle r \rangle) \approx \frac{e^2}{\langle r \rangle} \ll T \tag{11}$$

In other words, the following parameter, determining the plasma state

$$\Lambda_G \to \Lambda_P = \frac{U(\langle r \rangle)}{T} \approx \frac{e^2/\langle r \rangle}{T} \approx \frac{e^2 n^{1/3}}{T} \ll 1 \tag{12}$$

must be small. The first condition (for a neutral gas) means that in a sphere with a radius equal to the radius of interaction, there are few particles. The meaning of a similar condition for plasma is the opposite. To prove this we, first of all, determine the interaction radius in plasma. For the determination we consider in detail the potential of a test particle in plasma. Let a particle with a charge q be placed at the point $r = 0$. We intend to find its potential φ from Poisson's equation. Assuming, for simplicity, that the charge of the single type of singly charged ions is not changed by the test particle (i.e., $e_i = -e$), we can write Poisson's equation in the form

$$\Delta\varphi = -4\pi q\delta(r) - 4\pi e\left[n_e \exp\frac{-e\varphi}{T_e} - n_i \exp\frac{e\varphi}{T_i}\right] \tag{13}$$

where Δ is the Laplace operator and $\delta(r)$ is the Dirac function. Assuming that $|e\varphi| \ll T_e, T_i$, we find

$$\varphi(r) = \frac{q}{r}\exp\left(-\frac{r}{r_D}\right) \tag{14}$$

where

$$r_D = \left\{\sum_\alpha \left(\mathrm{v}_{\mathrm{T}\alpha}^2/\omega_{\mathrm{L}\alpha}^2\right)^{-1}\right\}^{-\frac{1}{2}} \tag{15}$$

is the so-called Debye radius. It shows the distance, in which the Coulomb forces are acting in plasma. Outside of the Debye radius, the interaction between charged particles is exponentially small and may be neglected. Comparative characteristics of the two curves are given in **Figure 2**. Curve (a) presents Debye potential, and curve (b) presents the vacuum potential $\sim 1/r$.

The electrostatic forces are, in fact, shielded. Now we can compare the average distance between charged particles with the Debye radius and make sure that the number of particles in the Debye sphere is large. For a simple case of plasma with singly charged ions and $T_e \approx T_i \approx T$, we have

$$r_D n^{1/3} = n^{1/3}\sqrt{\frac{T_e T_i}{4\pi\left(e^2 n_e T_i + e_i^2 n_i T_e\right)}} \approx \frac{\sqrt{T}}{e n^{1/6}} = \sqrt{\frac{1}{\Lambda_P}} \gg 1 \tag{16}$$

This condition is essentially the opposite of analogous condition for gas (10). In a gas, the particles generally do not interact. The interaction takes place only at very

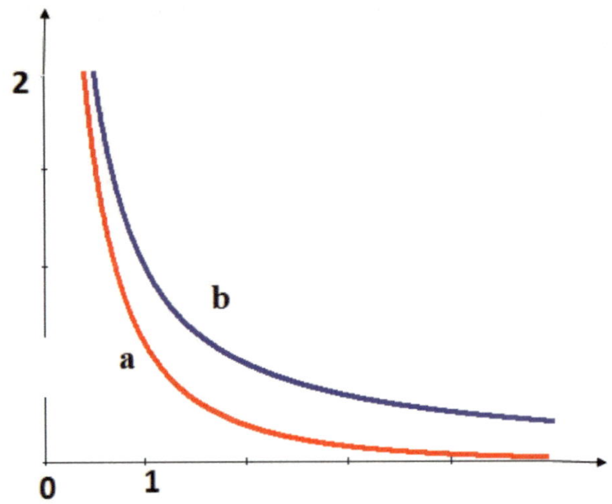

Figure 2.
Comparative characteristic of the two curves. Curve (a) presents Debye potential, and curve (b) presents the vacuum potential.

short intervals during collisions. In plasma, on the contrary, particles experience an interaction almost always. But, at the same time, the interaction is weak. It does not outrage their movement.

Debye radius, in particular, for electrons is

$$r_{De} = \frac{v_{Te}}{\omega_{Le}} = \sqrt{\frac{T_e}{4\pi n_e e^2}} \tag{17}$$

For the quasi-neutrality of plasma, it is necessary that its characteristic dimensions L be much larger than the Debye radius $L >> r_D$. Moreover, under this condition a system of charged particles can be considered as plasma, i.e., a material medium with qualitatively new properties. Otherwise, it is a simple collection of individual charged particles, to which vacuum electrodynamics is applicable.

2.3 Degenerate plasma

It remains to determine the gas-like parameter for degenerate plasma as well as to answer the question of whether is there in quantum plasma Debye screening. For this we first recall that the expression for average energy of the particles, which is valid both for classical and quantum cases, may be written in the following form:

$$\langle E \rangle = \begin{cases} T & \text{if} \quad T > E_F \\ E_F & \text{if} \quad E_F > T \end{cases} \tag{18}$$

i.e., in the quantum case, the average energy of the state is equal to Fermi energy E_F (see (1)). The gas-like parameter for degenerate plasma may be obtained if we replace $T \to E_F$ in the expression (12). It becomes

$$\Lambda_P^{(D)} = \frac{e^2 n^{1/3}}{E_F} << 1 \tag{19}$$

Now we show that in quantum (degenerate) plasma of metals, shielding of electrostatic field also takes place and derives an expression for characteristic length for the Debye radius in degenerate plasma.

The energy of free electrons is $p^2/2m$. In the presence of a field with the potential $\Phi(r)$, it is $p^2/2m + \Phi(r)$. As a result, Fermi particles become distributed uniformly in the spherical layer between $p_{\min} = \sqrt{2me\Phi}$ and $p_{\max} = \sqrt{p^2 + 2me\Phi}$. Given this circumstance, one can find the expression for the electron density:

$$n_e = n_{0e}\left(1 + \frac{e\Phi}{E_{Fe}}\right)^{3/2}, \tag{20}$$

where n_{0e} is the density in the absence of a field (which coincides with the density of the neutralizing ion background). Now it is not difficult to write Poisson's equation for the potential of the test particle placed at the point $r = 0$, and the charge of which is q. The equation is

$$\Delta\Phi = -4\pi q\delta(\mathbf{r}) - 4\pi e n_{0e}\left[\left(1 + \frac{e\Phi}{E_{Fe}}\right)^{3/2} - 1\right] \tag{21}$$

The solution of this equation in the limit of weak fields, $|e\Phi| \ll E_{Fe}$, gives the shielded Coulomb potential:

$$\Phi(r) = \frac{q}{r}\exp\left(-\frac{r}{r_{De}}\right) \tag{22}$$

with a Debye radius r_{De}, the expression for which is

$$r_{De} = \sqrt{\frac{E_{Fe}}{6\pi e^2 n_{0e}}} \tag{23}$$

The gas parameter for plasma (12) is similar to condition (10) for neutral gas in the following sense. Both of these conditions are fulfilled better for fewer densities of plasma and neutral gas. The better the gas condition is satisfied, the more ideal is plasma. For degenerate plasma (in which particles need quantum description), in the contrary, the gas condition depends on density inversely, i.e., with the increase in density, the ideality becomes better (see (19). As $E_F \sim n^{2/3}$, it turns out that with the increase in density, the average energy of Coulomb interaction increases slower. As a result $\Lambda_G^{(D)} \sim n^{-1/3}$. So, the denser the degenerate metal component, the better gas condition is fulfilled for it.

The diagram below presents the areas of the charge carriers' degeneracy and areas of the gas approximation applicability.

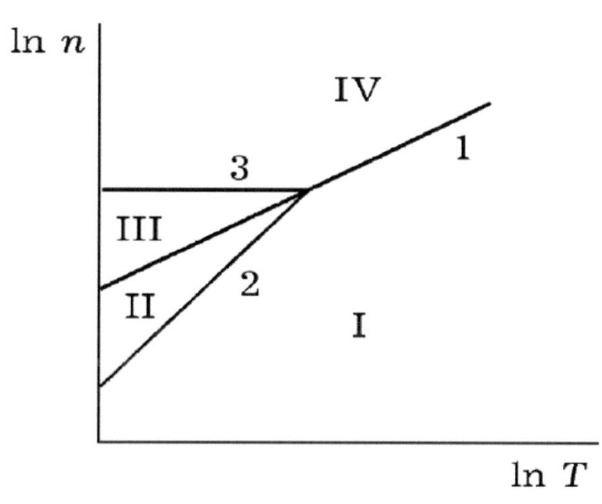

The degeneracy condition for electron plasma has the form $E_F > T$ (E_F; see (1)). In the diagram $\ln n$ vs $\ln T$, this condition gives line 1 dividing the region of the degenerate plasma from the nondegenerate (classical) state. The condition for the applicability of the gas approximation in a nondegenerate state is $\Lambda_P = \frac{e^2 n^{1/3}}{T} \ll 1$. In the same diagram, the condition $\Lambda_P = 1$ gives line 2. In the degenerate state, the condition for applicability of gas approximation is $\Lambda_P^{(D)} = e^2 n^{1/3}/E_F \ll 1$, in which E_F does not depend on T. In these conditions $\Lambda_P^{(D)} = 1$ gives line 3 passing through the point, where lines 1 ($E_F = T$) and 2 ($\Lambda_P = 1$) intersect. Therefore, region I is the region of a nondegenerate plasma with weak interaction (the gas approximation is applicable). Region II is the region in which the plasma is nondegenerate with strong interaction, i.e., classic fluid. In region III, the plasma is degenerate with strong interaction, i.e., quantum fluid. Both in region II and region III, the gas approximation is not applicable. Finally, region IV of the parameter variations characterizes degenerate plasma with a weak interaction (gas approximation is applicable).

In conclusion, we give an estimate of the applicability conditions for the gas approximation (10) and (12) for various plasmas. First of all, we note that the size of atoms and molecules is of order $a \sim 10^{-7}$–10^{-8} cm and the condition for gas approximation (10) is satisfied up to $n < 10^{21}$–10^{22} cm^{-3}, i.e., in gases at normal temperature up to a pressure of hundred atmospheres. It is obvious that in gas plasma both in the ionosphere and in the laboratory, this condition is fulfilled perfectly, with a large margin.

A somewhat different situation holds for the condition of gas approximation in plasma (12). In the ionosphere plasma, where $n_e \sim 10^6$–10^7 cm^{-3} and $T_e \approx 10^4$ K, we have $\Lambda_P \approx 10^{-4} \ll 1$, i.e., the condition is well satisfied. In ordinary gas-discharge fluorescent lamps, as well as in discharges used in laboratory experiments, where $n_e \approx 10^{10}$–10^{14} cm^{-3} and $T_e \sim 10^4$–10^5 K, the value of $\Lambda_P \ll 1$. However, at the discharge in dense gases used in the light sources for laser pumping, as a rule, $n_e \ll 10^{18}$–10^{19} cm^{-3}, and $T_e \ll 1$–10 eV. Herewith $\Lambda_P \approx 0.1$–0.5, which indicates a significant violation of the applicability condition for the gas approximation and a significant manifestation of the properties of non-ideal plasma or, as one says, liquid effects.

In a thermonuclear plasma, in facilities with magnetic confinement, $n_e \approx 10^{14}$–10^{15} cm^{-3}, and $T_e \approx T_i \approx 10^8$ K. As a result, we have $\Lambda_P \approx 10^{-5} \ll 1$, i.e., the ideality of plasma is guaranteed. However, in the inertial thermonuclear reactors, where experimentators strive to obtain plasma with $n_e \sim n_i \sim 10^{24}$–$10^{25}$ cm^{-3} at the temperature of $T \approx 10^8$ K, it turns out that $\Lambda_P \geq 0.01$ and may be even more. This, apparently, will require a consideration of weakly non-ideal plasma, especially in conditions of pollution (the presence of multiply charged ions).

Finally, a brief summary on ideality of plasma in solids is presented. Even in good conductors, such as copper, where $n_e \approx 5 \cdot 10^{22}$ cm^{-3} and $E_F \sim 1$ eV, we have $\Lambda_P \equiv \Lambda_p^{(D)} \sim 1$, i.e., plasma of metals is always non-ideal, and it is more correct to consider it as electron liquid. Nevertheless, it turns out that the application of the gas approximation to metals leads to good results from the point of view of the comparison with experiments. As for the electron–hole plasma of semiconductors, they are not degenerate at normal temperature, due to the small density of the carriers. For this reason, the condition of gas approximation (19) is well satisfied. Exceptions can occur only at very low temperatures.

3. Plasma description

3.1 Self-consistent approach

The main feature of plasma and plasma-like media, such as gas plasma, plasma of metals, semimetals, and semiconductors, is the presence of a large number of free charge carriers. Here we present general principles of their description as a continuous media. The term "plasma-like media" was first introduced in [6] (see also [7]), the authors of which understood such seemingly different states of matter as an ionized gas, or actually plasma; metals, semiconductors, and even molecular colloidal crystals and electrolytes may be described based on similar principles – the principles of electrodynamics of plasma-like media. In this section we briefly formulate these principles.

A self-consistent interaction of the electromagnetic field and charge carriers takes place in plasma-like media. Field equations are the equations of Maxwell in which the current and the charge must be represented by a sum over all carriers (charged particles) in the plasma:

$$\begin{aligned} &\nabla \times \mathbf{B} = \frac{1}{c}\frac{\partial \mathbf{E}}{\partial t} + \frac{4\pi}{c}\mathbf{j} = \frac{1}{c}\frac{\partial \mathbf{E}}{\partial t} + \frac{4\pi}{c}\sum_{\alpha} n_\alpha \mathbf{v}_\alpha e_\alpha \\ &\nabla \times \mathbf{E} = -\frac{1}{c}\frac{\partial}{\partial t}\mathbf{B} \\ &\nabla \mathbf{E} = 4\pi\rho = 4\pi\sum n_\alpha e_\alpha; \quad \nabla \mathbf{B} = 0 \end{aligned} \tag{24}$$

where $\mathbf{E}$ is the electric field strength, $\mathbf{B}$ is the magnetic induction, and n_α, e_α, and $\mathbf{v}_\alpha$ are the density, charge, and the velocity of α-th carrier $\alpha = e, i_1, i_2, i_3 \ldots$ ..

The equations for fields are written in this form (i.e., in terms of $\mathbf{E}$ and $\mathbf{B}$), because these quantities have direct physical meaning: they determine the Lorentz force $\mathbf{F}_n$ that acts on n-th carrier of α type (it may be an electron or ion of arbitrary type):

$$\mathbf{F}_n = e_\alpha\left\{\mathbf{E} + \frac{1}{c}[\mathbf{v}_n \times \mathbf{B}]\right\} \tag{25}$$

According to charge conservation law, the continuity equation must be satisfied for electrons and all types of ions, i.e., for $\alpha = e, i_1, i_2, \ldots$ The continuity equation is

$$\frac{\partial n_\alpha}{\partial t} + \nabla(n_\alpha \mathbf{v}_\alpha) = 0 \tag{26}$$

Here and in consideration below, we do not take into account the processes of ionization and recombination.

Eqs. (24)–(26) describe the self-consistent interaction between electromagnetic fields and statistically large numbers of charged particles (plasma). According to the set, electromagnetic fields determine the motion of charged particles. In its turn, the same electromagnetic fields are induced by moving plasma particles.

If the plasma (or plasma-like media) are in external fields (electric $\mathbf{E}_0$ and/or magnetic $\mathbf{B}_0$), the equations must be written in somewhat other form. External fields should be singled out. If the external fields are excited by external current $\mathbf{j}_0$ and charge ρ_0 densities, the set (24) should be written as

$$\begin{aligned} &\nabla \times \mathbf{B} = \frac{1}{c}\frac{\partial \mathbf{E}}{\partial t} + \frac{4\pi}{c}(\mathbf{j}_0 + \mathbf{j}) = \frac{1}{c}\frac{\partial \mathbf{E}}{\partial t} + \frac{4\pi}{c}\left\{\mathbf{j}_0 + \sum_{\alpha} n_\alpha \mathbf{v}_\alpha e_\alpha\right\} \\ &\nabla \mathbf{E} = 4\pi(\rho + \rho_0) = 4\pi\sum n_\alpha e_\alpha + 4\pi\rho_0 \\ &\nabla \mathbf{B} = 0; \qquad \nabla \times \mathbf{E} = -\frac{1}{c}\frac{\partial}{\partial t}\mathbf{B} \end{aligned} \tag{27}$$

The external current $\mathbf{j}_0$ and charge density ρ_0 do not depend on the processes in plasma. Their values, along with $\mathbf{E}_0$ and $\mathbf{B}_0$, satisfy Maxwell's equations. These fields also influence on the motion of plasma particles. The expression for Lorentz force looks like (25)

$$\mathbf{F}_n = e_\alpha \left\{ \mathbf{E} + \frac{1}{c} [\mathbf{v}_n \times \mathbf{B}] \right\} \tag{28}$$

However, here fields **E** and **B** are induced by external charge and current also.

An important conclusion follows from the sets (24) and (27). Only one additional vector quantity appears in the field equations – the current in the plasma **j** (the charge density may be expressed in terms of **j** by solving the continuity equation):

$$\mathbf{j} = \mathbf{j}(\mathbf{E}) = \sum_\alpha n_\alpha e_\alpha \mathbf{v}_\alpha \tag{29}$$

In this expression $\mathbf{v} = \mathbf{v}(\mathbf{E})$. The expression (29) shows that the current, induced in plasma, depends on velocities $\mathbf{v}_\alpha$ which are found independently (e.g., from equations of motion).[1] In the following, we will consider linear phenomena only. This implies linear dependence $\mathbf{j}(\mathbf{E})$, which is true, if the fields are comparatively small. It is to the point to note that in spite of our (and most other) consideration, nonlinear effects reveal themselves, first of all, in plasma and plasma-like media. Under linear consideration in isotropic media (media with no preferred directions), we actually have proportional dependence of the plasma current on the electric field or

$$\mathbf{j} = \sigma \mathbf{E} \tag{30}$$

This dependence represents Ohm's law for plasma, and σ is plasma conductivity for the considered case of isotropic plasma. However, if the plasma is in the external field, it loses the isotropy, and the relationship between **j** and **E** becomes much more complicated.

3.2 Waves in plasma and plasma-like media

The most important solutions of formulated set of equations are the solutions in the form of traveling waves, i.e., the solutions that depend on space coordinate **r** and time t as

$$\sim \exp(-i\omega t + i\mathbf{kr}) \tag{31}$$

Solutions of this type are the simplest solutions. In this case the initial equations may be essentially simplified. In (24) derivations may be replaced by multiplication:

$$\frac{\partial}{\partial t} \to -i\omega \quad \nabla \equiv \frac{\partial}{\partial \mathbf{r}} \to i\mathbf{k} \tag{32}$$

[1] The expression for current **j** in plasma depends on the model, which is chosen for plasma description (see examples below). If one explores the most complete kinetic consideration, the expression for the plasma current changes and becomes

$$\mathbf{j} = \sum_\alpha e_\alpha \int \mathbf{v} d\mathbf{v} f_\alpha(\mathbf{v})$$

where $f_\alpha(\mathbf{v})$ is the distribution function for α-type plasma particles.

And the initial set reduces to a set of linear algebraic equations:

$$\begin{aligned} &[\mathbf{kE}] = \frac{\omega}{c}\mathbf{B}; && i[\mathbf{kB}] = \frac{4\pi}{c}\mathbf{j} - i\frac{\omega}{c}\mathbf{E} \\ &i\mathbf{kE} = 4\pi\rho; && \mathbf{kB} = 0 \end{aligned} \tag{33}$$

Any other, more complicate, solution of the initial equations in linear theory can be presented as a superposition of the simplest solutions with various amplitudes. As follows from the general principle of electrodynamics, this superposition is also a solution of the initial set. This emphasizes the importance of consideration of the solutions in the form of traveling waves.

The motions and continuity equations also may be reduced to algebraic form. Here we present the reduced form of continuity equation only as the motion equations for plasma particles depend on the model, chosen for plasma description (see below):

$$\omega\rho = \mathbf{kj} \tag{34}$$

So, the initial equations (consisting of Maxwell's, continuity, and motion equations) reduce to linear algebraic set. The condition for the existence of nonzero solutions of the set is called *dispersion relation*. It, actually, presents a certain relation between frequency ω and components of wave vector $\mathbf{k}$. This relation helps determine ω for the given $\mathbf{k}$ and, vice versa, to determine $\mathbf{k}$ (one of its components) if ω and other components of $\mathbf{k}$ are given. These statements of the problem are called initial and boundary problems accordingly. These statements are widely used in plasma physics. Herewith in plasma many cases are encountered, in which solution of the initial problem gives complex frequency $\omega = \omega' + i\omega''$ for real $\mathbf{k}$. In this case the real part of the frequency $\mathrm{Re}\,\omega = \omega'$ shows the frequency of the wave, but the imaginary part shows (depending on its sign) either the increasing of the wave's amplitude if $\omega'' > 0$ or decreasing if $\omega'' < 0$ in accordance to

$$\exp(-i\omega t) = \exp(\omega'' t)\exp\{-i\omega' t + i\mathbf{kr}\} \tag{35}$$

The solution of the boundary problem should be interpreted in the same manner. If the solution of the problem gives a complex component of the wave vector $\mathbf{k}$, its imaginary part shows either amplification of the given wave in a given direction or its quenching.

3.3 Electrostatic waves in plasma

Plasma is a medium, where propagation of specific, electrostatic (or plasma) waves is possible. These waves have no oscillating magnetic field. These waves are also called space charge or Langmuir waves. In the waves electric field is parallel to its propagation direction $\mathbf{k}\|\mathbf{E}$. Oscillations of plasma particles also are parallel to the propagation direction, i.e., the waves are purely longitudinal. These waves play the most important role in plasma and strongly influence on its stability, much more than the usual electromagnetic waves (propagation of which in plasma is also possible).

The explicit expression for plasma conductivity σ helps to obtain the dispersion relation for longitudinal waves. When we consider solutions of Maxwell's equations in the form of traveling waves, the conductivity σ depends on frequency ω and wave vector $\mathbf{k}$, i.e., $\sigma \equiv \sigma(\omega, \mathbf{k})$. The dispersion relation for electrostatic waves in plasma can be expressed in terms of $\sigma(\omega, \mathbf{k})$. It has the following form:

$$1 + i\frac{4\pi}{\omega}\sigma(\omega, \mathbf{k}) \equiv \varepsilon(\omega, \mathbf{k}) = 0 \tag{36}$$

where $\varepsilon(\omega, \mathbf{k})$ is the well-known dielectric permittivity of the given media.

Propagation of usual electromagnetic waves in plasma is also possible. The dispersion relation for this case is

$$k^2 = \frac{\omega^2}{c^2}\varepsilon(\omega, \mathbf{k}) \tag{37}$$

In a particular case of vacuum, this expression gives propagation of usual vacuum electromagnetic waves. For vacuum $\varepsilon(\omega, \mathbf{k}) \equiv 1$, and the dispersion relation (37) takes a familiar look, $\omega = kc$.

For further development of the properties of plasma-like media, it is necessary to specify the plasma models.

4. The simplest plasma models

Each model of plasma specifies how its particles interact with the electromagnetic field, as well as specifies the behavior of plasma particles inside and between plasma species. Here we consider the simplest models only, leaving aside the most rigorous kinetic consideration. On examples of simple models, we show how the models work as well as some of their advantages and disadvantages.

4.1 One-particle model

We begin with the model of one, "average" (or test) particle. In this model particles interact via electromagnetic field, and the interaction, in fact, is weak (it weakly perturbs the motion of the particles). Collisions inside and between species are also taken into account. This model describes the oscillatory properties of gas-discharge and ionosphere plasma well enough. In particular, the model was successfully used for the description of radio-frequency wave propagation through the ionosphere [9].

The initial set of equations in the model of "average" particle includes Newton equations for the "average" electron and for "average" ion along with equations for electromagnetic field and continuity equation:

$$\begin{aligned} \frac{d}{dt}\mathbf{v}_e &= \frac{e}{m}\left\{\mathbf{E} + \frac{1}{c}[\mathbf{v}_e \times \mathbf{B}]\right\} - \nu_{en}\mathbf{v}_e - \nu_{ei}(\mathbf{v}_e - \mathbf{v}_i) \\ \frac{d}{dt}\mathbf{v}_i &= \frac{e_i}{M}\left\{\mathbf{E} + \frac{1}{c}[\mathbf{v}_i \times \mathbf{B}]\right\} - \nu_{in}\mathbf{v}_i - \nu_{ie}(\mathbf{v}_i - \mathbf{v}_e) \end{aligned} \tag{38}$$

Here $\mathbf{v}_e$ and $\mathbf{v}_i$ are the velocities of electrons and ions; m and M are their masses; ν_{en}, ν_{ei} and ν_{in}, ν_{ie} are the frequencies of their collisions, which determine the friction forces inhibiting their motion; ν_{en} is the frequency of collisions of electrons with neutral atoms (molecules) and ν_{ei} with ions, respectively; and for ions, this is ν_{in} and ν_{ie}. According to Newton's third law, $m\nu_{ei} = M\nu_{ie}$. A similar system of equations is also used to describe the dynamics of solid-state plasma, but in this case the meaning of the collision frequencies differs from the abovementioned. The frequencies, actually, are the inverse lifetimes of electrons and holes, respectively.

First we show how this model works on the simplest example of plasma oscillations, as well as how easy the Langmuir frequency follows from this model. Consider pure electron plasma. Ions are heavy and immobile. They serve only for neutralization of electrons' charge. Upon derivation of equations, which describe plasma oscillations, one should recall that we consider linear plasma phenomena. This means that the equations should be linearized, i.e., we consider small perturbations of physical quantities from their basic (equilibrium) state. For example, the density of electrons is considered as $n_e = n_{e0} + n_e'$ where $n_e' \ll n_{e0}$, and in the resulting expressions we retain the first-order terms only and neglect the terms of second-order smallness (multiplication of the first-order terms). We also take into account that the Langmuir oscillations are potential and use Poisson's equation instead of a full set of Maxwell's equations. For one more simplification, we consider one-dimensional case: let electrons oscillate along the z axis. All this leads to the following: the initial equations (motion, Poisson's, and continuity) are reduced to simple form presented below:

$$\frac{\partial}{\partial t}\mathrm{v}_e{}' = \frac{eE}{m} - \nu_{ei}\mathrm{v}_e{}'; \quad \frac{\partial}{\partial t}n_e' = -n_{0e}\frac{\partial}{\partial z}\mathrm{v}_e{}'; \quad \frac{\partial}{\partial z}E = 4\pi e n_e' \tag{39}$$

Here $\mathrm{v}_e{}'$ is the velocity of electrons, e and m are their charge and mass, ν_{ei} is the frequency of electron-ion collisions, and t is the time.

For the solutions of (39) that depend on z and t in the form $\exp(-i\omega t + ikz)$, we have the equations

$$-i(\omega + i\nu_e)\mathrm{v}_e{}' = \frac{eE}{m}; \quad \omega n_e' - ikn_{0e}\mathrm{v}_e{}' = 0; \quad ikE = 4\pi e n_e' \tag{40}$$

Thus, we arrive to the set of simple algebraic equations, from which the following expression for plasma current results

$$j = en_e{}'\mathrm{v}_e{}' = i\frac{e^2 n_{0e}}{m}E \equiv \sigma E, \quad \sigma = i\frac{e^2 n_{0e}}{m} \tag{41}$$

From (41) one can easily obtain the corresponding expression for dielectric permittivity as well as the dispersion relation, which is

$$\varepsilon = 1 - \frac{\omega_p^2}{\omega(\omega + i\nu_{ei})} = 0 \tag{42}$$

If $\nu_{ei} \ll \omega$ (realizes in most cases), the relation (42) leads to $\omega = \pm\omega_p$, i.e., we have free plasma oscillations. If one takes into account plasma collisions, he obtains small negative imaginary correction to the frequency: $\omega_p \to \omega_p - i\nu_e/2$. This shows the decay of the oscillations. The decay takes place as a result of collisions.

Obtained results on plasma oscillations and their decay coincide to experimental data. In fact, the model of "average" particle describes plasma well in the considered range of frequencies. Namely this model was used by Langmuir to describe the oscillatory properties of gas-discharge plasma. Also the model was successfully used especially for describing the propagation of radio-frequency waves through the ionosphere [9]. Thereby, the model of "average" particle is justified for high-frequency range. However, in the opposite limit of low frequencies, this model does not lead to reasonable results. That is why, new, more complicate models have been explored.

4.2 Two-fluid hydrodynamic: relative electron-ion motion

The idea to consider plasma as a system consisting of electron and ion fluids arose long time ago. In this model plasma species are described by hydrodynamic equation and interact through the electromagnetic field and through the collisions. The interaction leads to various effects. In particular instability can follow from the interaction. The general theory of plasma instabilities shows that instability is a result of thermodynamically nonequilibrium initial distribution of plasma components.

The initial equations describing electron and ion fluids and their interaction are somewhat more complex than the previous case of "average" particle. They contain additional terms that follow from classical hydrodynamics:

$$\frac{d}{dt}\mathbf{v}_e \equiv \left[\frac{\partial}{\partial t} + (\mathbf{v}_e\nabla)\right]\mathbf{v}_e = -\frac{\nabla(n_eT_e)}{mn_e} + \frac{e}{m}\left\{\mathbf{E} + \frac{1}{c}[\mathbf{v}_e \times \mathbf{B}]\right\} - \nu_{en}\mathbf{v}_e - \nu_{ei}(\mathbf{v}_e - \mathbf{v}_i)$$
$$\frac{d}{dt}\mathbf{v}_i \equiv \left[\frac{\partial}{\partial t} + (\mathbf{v}_i\nabla)\right]\mathbf{v}_i = -\frac{\nabla(n_iT_i)}{Mn_i} + \frac{e_i}{M}\left\{\mathbf{E} + \frac{1}{c}[\mathbf{v}_i \times \mathbf{B}]\right\} - \nu_{in}\mathbf{v}_i - \nu_{ie}(\mathbf{v}_i - \mathbf{v}_e) \tag{43}$$

Here m and M are electron and ion mass, T_e, T_i, and n_e, n_i are their temperatures and densities. Other denotations coincide to the previous case of "average" particle above.

The equations describing the electron and ion fluid motion (43) should be supplemented by Maxwell's and continuity equations. Equations for T_e and T_i (energy balance equations or equations for heat) are also needed. Specific forms of these equations depend on the physical meaning of the problem, which is considered. For simplicity, it may be assumed as $T_e, T_i = const$. These assumptions greatly simplify further analysis.

Here we briefly consider a simple example of two-fluid model. In order to show the role of ions and how this role can lead to instability, we consider a case in which electron fluid moves relative to ions in rest. Let $\mathbf{u}$ be the constant velocity of moving electrons. Neutrals are absent. The initial equations of electron and ion fluids (43) and Maxwell's and continuity equations after linearization are reduced to the following set of equations (we assume $T_e, T_i = 0$ and consider the potential oscillation in one-dimensional system and choose the z axis along $\mathbf{u}$):

$$\begin{aligned}
&\left(\frac{\partial}{\partial t} + u\frac{\partial}{\partial z}\right)\mathrm{v}_e{}' = \frac{eE}{m} - \nu_{ei}\mathrm{v}_e{}'; \quad \frac{\partial}{\partial t}\mathrm{v}_i{}' = \frac{e_iE}{M} \\
&\left(\frac{\partial}{\partial t} + u\frac{\partial}{\partial z}\right)n_e{}' = -n_{0e}\frac{\partial}{\partial z}\mathrm{v}_e{}'; \quad \frac{\partial}{\partial t}n_i{}' = -n_{0i}\frac{\partial}{\partial z}\mathrm{v}_i{}' \\
&\frac{\partial}{\partial z}E = 4\pi(en_e{}' + e_in_i{}')
\end{aligned} \tag{44}$$

Here $\mathrm{v}_e{}'$ and $\mathrm{v}_i{}'$ are the perturbations of electron and ion fluid velocities accordingly, n_e' and n_i' are the perturbations of their densities, n_{0e} and n_{0i} are their unperturbed densities, ν_{ei} is the frequency of electron-ion collisions, and E is the electric field. We look for solutions of the set (44) in the form of waves propagating along the z axis $\sim \exp(-i\omega t + ikz)$. In this case the equations (44) are reduced to algebraic set. If one performs further steps (determination of induced plasma current, finding plasma conductivity and dielectric permittivity) by analogy to the previous case, he arrives to the expression for dielectric permittivity of considered system, which, in this case, consists of three terms:

$$\varepsilon = 1 + \Delta\varepsilon_e + \Delta\varepsilon_i \tag{45}$$

Here the second and third terms are electron and ion contributions in the dielectric permittivity:

$$\Delta\varepsilon_e = -\frac{\omega_{Le}^2}{(\omega - ku)(\omega - ku + i\nu_{ei})}; \quad \Delta\varepsilon_i = -\frac{\omega_{Li}^2}{\omega^2} \tag{46}$$

but the first term is, in fact, the vacuum unit. In (46) $\omega_{Le} = \sqrt{4\pi e^2 n_{0e}/m}$ and $\omega_{Li} = \sqrt{4\pi e_i^2 n_{0i}/M}$ are the electron and ion Langmuir frequencies. The dispersion relation of the considered system

$$1 - \frac{\omega_{Le}^2}{(\omega - ku)(\omega - ku + i\nu_{ei})} - \frac{\omega_{Li}^2}{\omega^2} = 0 \tag{47}$$

is an algebraic equation of the fourth order. Despite the solutions of the fourth-order equations are known (see any reference book on mathematics, e.g., [10]), they are not suitable for our purposes. Very cumbersome expressions with many radicals cannot give physical information, and we choose another way of finding the solutions of (47) and their analysis. The way presented below is typical for plasma theory.

For the analysis we first consider the case of the ion's absence, i.e., if there are streaming electrons only. The dispersion relation is of the second order, with the solution

$$\omega_{\pm} = ku \pm \omega_{Le} \tag{48}$$

This solution represents the waves in the electron stream: fast (+) and slow (−) beam waves. An important fact is that the energy of the slow beam wave is negative [11]. This means that for excitation of the slow wave in the electron beam, one should withdraw energy, but not put energy into the beam. This circumstance plays an important role in theory of streaming instabilities. It lies on the basis of explanation of physical meaning of the most well-known plasma instabilities, beam-plasma instability, as well as the meaning of dissipative beam instabilities. In the last case, dissipation leads not to quenching of the oscillations (as one expects) but to their amplification. The matter is in the following: dissipation serves as a channel for energy withdrawal for excitation of the beam negative-energy wave.

Continuing the analysis of the dispersion relation (47), one can easily see its important peculiarity, which appears as a result of the inequality $\omega_{Li} \ll \omega_{Le}$; ions play a role under small ω only:

$$\omega_{Li} \gg \omega \to 0 \tag{49}$$

In the opposite case, the contribution of the ions in (47) results in small corrections to the roots that describe the proper oscillations of streaming electron (48). However, if the condition (49) holds, two additional roots of (46) (from a total of four) are approximately equal:

$$\omega = \pm \frac{\omega_{Li}}{\sqrt{1 - \frac{\omega_{Le}^2}{k^2 u^2}}} \tag{50}$$

This expression shows that the instability (caused by the presence of ions) exists if $ku < \omega_{Le}$. In this case the imaginary part of the root (growth rate of unstable oscillations) attains its maximum if $ku \approx \omega_{Le}$. The instability in this case is called resonance instability. If the two conditions ($\omega \to 0$ and $ku \approx \omega_{Le}$) are realized, one can at once calculate the roots of the dispersion relation (47), if he rewrites it in the form

$$\omega \left(\frac{\partial(\Delta \varepsilon_e)}{\partial \omega} \right)_{\omega \to 0} = \frac{\omega_{Li}^2}{\omega^2} \tag{51}$$

From this expression the growth rate of resonant instability follows. It is

$$\delta_{Bn} = \frac{\sqrt{3}}{2^{4/3}} \omega_{Le} \left(Z^2 \frac{m}{M} \right)^{1/3} \tag{52}$$

In the given case, the growth rate is of the same order as the frequency of the unstable oscillations in plasma. This instability was first discovered by Buneman [12]. It (as well as other low-frequency instabilities) plays an important role in many scenarios in space physics and geophysics. The physical essence of this instability lies in the fact that the proper space charge oscillations of moving electrons (beam slow wave) in the frame, associated with the ions in rest due to the Doppler effect, experience red shift, and this greatly reduced frequency becomes close to the proper frequency of ions. Actually the instability is due to the resonance of the negative-energy wave with the ion oscillations.

The models of plasma that have been considered above are very simplified. In spite of this, we have seen that the models describe some phenomena in plasma well enough. For more detailed consideration, one should use the most complete kinetic consideration.

Author details

Eduard Vardges Rostomyan
Institute of Radiophysics and Electronics, Armenian National Academy of Sciences, Armenia

*Address all correspondence to: eduard_rostomyan@mail.ru

IntechOpen

© 2020 The Author(s). Licensee IntechOpen. This chapter is distributed under the terms of the Creative Commons Attribution License (http://creativecommons.org/licenses/by/3.0), which permits unrestricted use, distribution, and reproduction in any medium, provided the original work is properly cited. cc BY

References

[1] Fitzpatric R. The Plasma Physics: An Introduction. New York: CRC Press, Taylor & Francis; 2014

[2] Chen FF. Introduction to Plasma Physics and Controlled Fusion. 2nd ed. New York and London: Plenum Press; 1984

[3] Dendy RO. Plasma Physics. Cambridge, UK: Cambridge University Press; 1995

[4] Bellan PM. Fundamentals of Plasma Physics. Cambridge, UK: Cambridge University Press; 2006

[5] Bittencort JA. Fundamentals of Plasma Physics. NY: Springer; 2004

[6] Silin VP, Rukhadze AA. Electromagnetic Properties of Plasma and Plasma-Like Media (in Russian). Moscow: Atomizdat; 1961

[7] Rukhadze AA. Principles of Plasma Electrodynamics. Berlin: Springer-Verlag; 1984

[8] Lifshits EM, Pipaevskii LP. Physical Kinetics. Amsterdam: Elsevier; 1995

[9] Ginzburg VL. Propagation of Electromagnetic Waves in Plasma (in Russian). Moscow: Nauka; 1967

[10] Korn G, Korn T. Mathematical Handbook. New York: McGraw-Hill Book Company; 1968

[11] Briggs RJ. Electron Stream Interaction with Plasma. Cambridge, Massachusetts: M.I.T. Press; 1964

[12] Buneman O. Instability, turbulence, and conductivity in current-carrying plasma. Physical Review Letters. 1958;**1**:8

Section 2

Plasma Applications in Space Technology

Chapter 2

Hall Thruster: An Electric Propulsion through Plasmas

Sukhmander Singh

Abstract

The chapter discussed the technological application of plasma physics in space science. The plasma technology is using laser-plasma fusion, inertial fusion, Terahertz wave generation and welding of metals. In this chapter, the application of plasma physics in the field of electric propulsion and types has been discussed. These devices have much higher exhaust velocities, longer life time, high thrust density than chemical propulsion devices and useful for space missions with regard to the spacecraft station keeping, rephrasing and orbit topping applications. The mathematical relation has been derived to obtain the performance parameters of the propulsion devices.

Keywords: electric propulsion, Hall thruster, impulse, exhaust velocity

1. Overview of propulsion devices and rocket equation

Electric propulsion (EP) devices use electric power to produce thrust. These devices have much higher exhaust velocities than chemical propulsion devices. Therefore, EP devices require much less propellant mass than chemical systems for a given space task. Here first we have overview of different thrusters and their basic mechanism based on type of propellant used to get the thrust.

The motion of any propulsion devices is given by Newton's 3rd Law of action and equal, opposite reaction which forms the basis for the motivation for the study of electric propulsion. The rocket equation states that a device can accelerate to a desired final velocity by reaction against an expelled propellant stream [1].

Consider a rocket of mass m, which expels an infinitely small unit of fuel dm at an exhaust velocity $\vec{U}_{ex}$. The exhaust velocity $\vec{U}_{ex}$ is almost constant and it is a fixed property of the propellant [2]. Conservation of linear momentum requires that the spacecraft experience a small change in velocity $d\vec{v}$, such that

$$m\frac{d\vec{v}}{dt} + \vec{U}_{ex}\frac{dm}{dt} = 0 \tag{1}$$

Integrating by setting appropriate limits in mass and velocity yields

$$\int_{v_i}^{v_f} \frac{d\vec{v}}{\vec{U}_{ex}} + \vec{U}_{ex}\int_{m_i}^{m_f} \frac{dm}{m} = 0 \tag{2}$$

After simplification, we get

$$\ln\left(\frac{m_f}{m_i}\right) = -\frac{v_f - v_i}{\vec{U}_{ex}} = -\frac{\Delta\vec{v}}{\vec{U}_{ex}} \tag{3}$$

The above rocket equation provides the relationship between the mission velocity and the mass of propellant $m_p = m_i - m_f$ required for a given mission. It is clear that a higher $d\vec{v}$ demands more propellant. Unfortunately, the mass ratio cannot be increased so much to avoid payloads problems in space mission, therefore for a given mass fraction, the exhaust velocity $\vec{U}_{ex}$ of the propellant needs to be the order of $d\vec{v}$ and the higher the propellant exit velocity, the less propellant mass is required.

2. Thrust, impulse and efficiency

The performance of thrusters is usually characterized by a number of parameters. A first quantity relevant to thruster performance is the thrust T, which is the total force undergone by the rocket. The specific impulse is used to compare the efficiencies of different type of propulsion systems [2]. The performance parameter is the specific impulse I_{sp}, defined below

$$I_{sp} = \frac{T}{\dot{m}_p g} \tag{4}$$

Here $\dot{m}_p$ is the mass flow rate and g is the acceleration due to gravity. The specific impulse has the dimension of time and is a measure for the effective lifetime of the thruster, when lifting its own propellant from the earth's surface.

For the case of a constant mass flow rate the thrust is also constant as

$$T = \dot{m}_p \vec{U}_{ex}, \tag{5}$$

and the specific impulse simplifies to

$$I_{sp} = \frac{\vec{U}_{ex}}{g} \tag{6}$$

Finally, the rocket equation turned into

$$\frac{m_f}{m_i} = e^{\frac{-\Delta\vec{v}}{g I_{sp}}} \tag{7}$$

The rocket equation is equally applicable to all type of propulsion systems. Therefore high specific impulse related to better efficiency for a propellant. Based on the acceleration of gases for propulsion, electrical thrusters have been classified into three main categories.

2.1 Electrothermal thrusters

In electrothermal thrusters, the hot gas is expanded through a nozzle without ionizing it. When it is being passed through a thin nozzle, the thermal energy of gas gets converted into kinetic energy and produce a thrust.

2.2 Electromagnetic thrusters

In electromagnetic thrusters an inert gas is used as a propellant and it is ionized by heating to produce plasma. Then these ionized gas (charged particles) are acceleration by electromagnetic force to generate thrust.

2.3 Electrostatic thrusters

In electrostatic thrusters only ions are accelerated by applying direct electric field at the exit side of the thruster to produce thrust.

3. Hall thruster operation

Hall effect thrusters (HETs) were originally developed in United States and Russia about 60 years ago, and the first working devices were reported in U.S. in the early 1960s. Now a days, most of the countries using the Hall thruster technology in their space mission. Unlike chemicals and electric rockets, the propulsive thrust in a Hall thruster is achieved by an ionized inert gas (Xenon) which has high atomic number and low ionization potential. For this Xenon is mostly used. In a Hall thruster, the propellant is ionized and then accelerated by electrostatic forces.

Figure 1 shows the internal parts of a plasma Hall thruster. Generally, the discharge channel is cylindrical shape made up with metallic material. The magnetic field of the order of 150 Gauss is applied to produce closed drift of electrons inside the channel. The applied magnetic field which is strong enough so that the electrons get magnetized, i.e. they are able to gyrate within the discharge channel, but the ions remain unaffected due to their Larmor radius much larger than the dimension of the thruster. Thus the electrons remain effectively trapped in azimuthally $\vec{E} \times \vec{B}$ drifts around the annular channel and slowly diffuse towards the anode. This azimuthal drift current of the electrons is referred to as the Hall current. The propellant enters from the left side of the channel via anode and gets ionized through hollow cathode of the device. The electric field of strength $\sim$1000 V/m gets

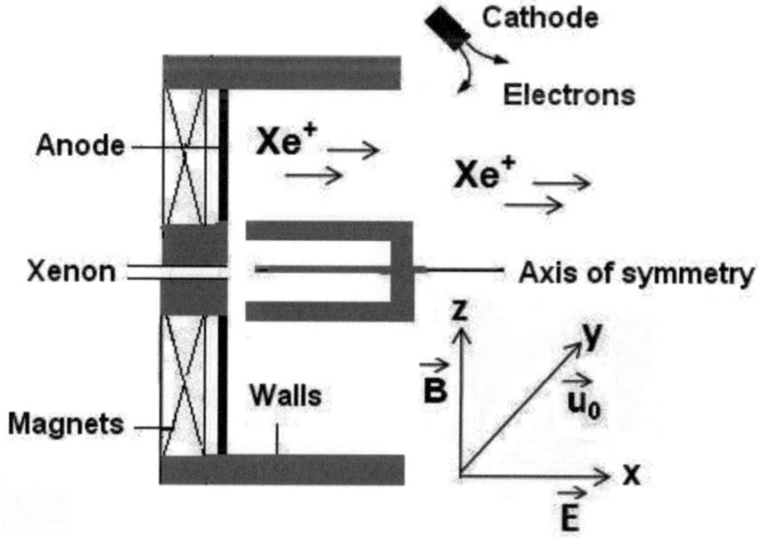

Figure 1.
Schematic diagram of a typical Hall plasma thruster.

generated inside the discharge channel along the axial direction of the device [3]. In addition, these kind of devices have implication in partially ionized plasmas (tokamaks), in ionosphere (base of the solar photosphere), in protoplanetary discs, circum nuclear discs in active galactic nuclei and neutron stars. Hall thruster has high thrust resolution, it is being used for the adjustment of the location of the satellite onboard.

4. Spacecraft issues

The first issue is that the divergence angle of these devices is about 60°, which relatively large and cause problem related to erosion of the channel walls and outer surfaces of the thruster. The erosion of the walls decreases the lifetime of the device. This channel usually has a length of the order of centimeters. In addition, densities in the channel are typically in the range between 10^{17} and 10^{18} m^{-3} for the plasma, and 10^{18} and 10^{20} m^{-3} for the neutral gas [4]. The plasma in a Hall thruster does not stay uniform and an inhomogeneous plasma immersed in the external electric and magnetic fields is not in the thermodynamically equilibrium state, this deviation in general is a source of plasma instabilities. The amplitudes of the waves and instabilities are attributed by the density scale lengths of plasma and magnetic field and other parameters. These waves/oscillation and instabilities may affect the efficiency of the device, hence forth research on studies on oscillation/instabilities always attracted the investigators.

5. Types of Hall plasma thruster

Two types of Hall thrusters have been developed: a thruster with closed electron drift and extended acceleration zone or stationary plasma thruster and a thruster with a very short acceleration channel or thruster with anode layer.

In **Table 1**, typical values of some of the pertinent properties are listed at the thruster exit for the SPT-100.

5.1 Dielectric wall thruster or stationary plasma thruster

Such thrusters have a wall made up of dielectric of boron nitride or silicon carbide and extended channel compared to its width. The role of the wall is that the collisions of the electrons and ions with the wall generate low energy secondary electrons. These secondary electrons keep tending the electron temperature low in the discharge plasma. By reducing the discharge electron energy, a smooth and continuous variation in plasma potential between the anode and the cathode is

Property	Value	Property	Value
Inner diameter	60 mm	Neutral velocity	~300 m/s
Outer diameter	100 mm	Electron temperature	5–10 eV
Plasma density	$\sim 10^{17}/m^3$	Ion temperature	1–5 eV
Neutral density	$\sim 10^{18}/m^3$	Neutral temperature	0.9 eV
Ion velocity	$\sim 10^4$ m/s	Debye length	$\sim 10^{-5}$ m
Collision mean free	~1 m		

Table 1.
Typical plasma parameters for Hall Thrusters.

obtained. Since the dielectric walls are not conductive, charge builds up along the length of the acceleration channel that leads to a variable potential profile along its length.

5.2 Thruster with anode layer

Thruster with anode layer also developed in Russia has a narrow acceleration zone associated with the narrow electric field region near the anode. This geometry considerably shortens the electric field region in the channel, where the ion acceleration occurs. However, this configuration does not change the basic ion generation or acceleration method. The channel wall made up of conductor, which is usually also a part of the magnetic circuit, is biased negatively (usually cathode potential) to repel electrons in the ionization region and to reduce electron-power losses. This reduces the loss caused by the ion and electron collisions with the walls. Since the walls are conductive, a constant potential (same as that of the cathode) is observed along the entire wall. Very high electron temperatures, i.e. more than 50 eV, are typically observed in such thrusters [1].

6. Review of status of current research and development in the subject

The range of the oscillations lies from few kHz to MHz in the acceleration channel of the thrusters and has been given in **Table 2**. Rayleigh-Taylor (RT) instability takes place when a lighter fluid supports a heavy fluid. The plasma in the Hall thruster possesses Rayleigh-Taylor instability, resistive instability, transit time instability, electromagnetic instability and sheath instabilities [5–11]. These systems are rampant with plasma instabilities and fluctuations, many of which are responsible for performance, driving electron transport across magnetic field lines and contributing to propellant ionization. Over the last decade several studies have been carried out with HET to characterize the low frequency azimuthal and axial oscillations and optimizing magnetic field profile for a wide range of operating conditions for better efficiency and performance. Singh and Malik [10, 11], investigated that temperature of the ion and drift velocity profiles of the electron modifies the conditions for Rayleigh type instability under the effects of thermal motions of ions.

The plasma resistivity induces resistive instabilities (electrostatic and electromagnetic) [6–9] associated with azimuthal and axial directions. High-frequency (1–10 MHz) instabilities have been studied in the Hall-effect thruster [6–9], where it was found that these instabilities have the highest level near the thruster exit plane. These oscillations in the Hall thruster determine the efficiency of the system and may affect the divergence of the ion beam and electron transport across the

Range (kHz)	Type	Driving mechanism
10–20	Loop or circuit oscillations	Magnetic field, discharge voltage and electron wall collision frequency
5–25	Rotating spokes	Ionization process
20–60	Azimuthal modes or drift instability	Gradient of density and magnetic field
70–500	Transient time oscillations	Plasma density gradient and low ionization
0.5–5 MHz	Azimuthal waves	Drift velocity of plasma species

Table 2.
Range and classification of oscillations in a Hall Thrusters.

magnetic field. Smolyakov et al. reported that sheath instabilities has a vital role in anomalous transport phenomena in Hall plasma thruster [12, 13].

7. Ion stream speed study for electrostatic thruster

Let us consider a Hall thruster having potential difference between anode and virtual cathode is Φ Volt and ions density (mass) n (M). The mass flow of propellant of ions of mass M through an area A is given by $\dot{m}_p = nMA\vec{U}_{ex}$.

The thrust is also constant as

$$T = \dot{m}_p\vec{U}_{ex}, \tag{8}$$

Substituting the value of mass flow rate, the thrust per unit area

$$\frac{T}{A} = \left(nM\vec{U}_{ex}\right)\vec{U}_{ex} = \frac{J_i}{q}M\vec{U}_{ex} \tag{9}$$

where J_i the current density of ions.

From the definition of work energy theorem, that the kinetic energy of each ion should equal to the work done in moving the charge across a potential drop. That is

$$\frac{1}{2}M\vec{U}^2_{ex} = q\Phi \tag{10}$$

Or

$$\vec{U}_{ex} = I_{sp}g = \sqrt{\frac{2q\Phi}{M}} \tag{11}$$

Thus the specific impulse or exhaust velocity of the ions depends on the potential drop developed across the anode and cathode and to the mass of the ion.

8. Conclusions

The $\vec{E} \times \vec{B}$ configurations of fields are used to confine electrons, increasing the electron residence time and allowing ionization and plasma sustainment. The magnetron sputtering used in material science for ion implantation is also based on the same $\vec{E} \times \vec{B}$ drift. The primary concern of the study to enhance the lifetime and performance of the Hall thruster by studying the instabilities that takes place in the channel and optimization of profile of the magnetic field which is the main parameter in respect to the erosion of the channel walls.

Acknowledgements

The University Grants Commission (UGC), New Delhi, India is thankfully acknowledged for providing the startup Grant (No. F. 30-356/2017/BSR).

Author details

Sukhmander Singh
Plasma Waves and Electric Propulsion Laboratory, Department of Physics, Central University of Rajasthan, Ajmer, Kishangarh, India

*Address all correspondence to: sukhmandersingh@curaj.ac.in

IntechOpen

© 2020 The Author(s). Licensee IntechOpen. This chapter is distributed under the terms of the Creative Commons Attribution License (http://creativecommons.org/licenses/by/3.0), which permits unrestricted use, distribution, and reproduction in any medium, provided the original work is properly cited.

References

[1] Kaufman HR. Technology of closed-drift thrusters. AIAA Journal. 2012; **23**(1):78-86. DOI: 10.2514/3.8874

[2] Goebel DM, Katz I. Fundamentals of Electric Propulsion: Ion and Hall Thrusters. New York: Wiley; 2008

[3] Jahn RG. Physics of Electric Propulsion. New York: McGraw-Hill; 1968

[4] Martinez-Sanchez M, Pollard JE. Spacecraft electric propulsion: An overview. Journal Propulsion Power. 1998;**14**(5):688-699

[5] Choueiri EY. Plasma oscillations in hall thrusters. Physics of Plasmas. 2001; **8**(4):1411-1426

[6] Singh S, Malik HK, Nishida Y. High frequency electromagnetic resistive instability in a hall thruster under the effect of ionization. Physics of Plasmas. 2013;**20**:102-109 (1–7)

[7] Singh S, Malik HK. Growth of low frequency electrostatic and electromagnetic instabilities in a hall thruster. IEEE Transactions on Plasma Science. 2011;**39**:1910-1918

[8] Singh S, Malik HK. Resistive instabilities in a hall thruster under the presence of collisions and thermal motion of electrons. The Open Plasma Physics Journal. 2011;**4**:16-23

[9] Malik HK, Singh S. Resistive instability in a hall plasma discharge under ionization effect. Physics of Plasmas. 2013;**20**:052115 (1–8)

[10] Singh S, Malik HK. Role of ionization and electron drift velocity profile to Rayleigh instability in a hall thruster plasma: Cutoff frequency of oscillations. Journal of Applied Physics. 2012;**112**:013307 (1–7)

[11] Malik HK, Singh S. Conditions and growth rate of Rayleigh instability in a hall thruster under the effect of ion temperature. Physical Review E. 2011; **83**:036406 (1–8)

[12] Nejoh Y, Yamamura Y. Simulation of the collisional sheath structure near the outer wall of the hall thruster. Physics of Plasmas. 2005;**12**:033506

[13] Smolyakov AI, Frias W, Kaganovich ID, Raitses Y. Sheath-induced instabilities in plasmas with $E \times B$ drift. Physical Review Letters. 2013;**111**:115002

Section 3

Microwave Plasma Interactions

Chapter 3

Evolution of Microwave Electric Field on Power Coupling to Plasma during Ignition Phase

Chinmoy Mallick, Mainak Bandyopadhyay and Rajesh Kumar

Abstract

During the gas ignition process, the plasma and the microwave electric fields are evolved with time together in the plasma volume. The spatio-temporal evolution pattern of microwave-radiated plasma parameters is reported here, highlighting the role of these electric fields on power coupling processes. Evolutions of electric field and so power coupling processes are calculated using the finite element method (FEM). It is observed that the main power coupling mechanism is electron cyclotron resonance (ECR) method; however, with the evolution of plasma, the mode shifts from ECR to off-ECR-type heating with time. Off-ECR heating in the form of upper hybrid resonance (UHR) method, electrostatic (ES) ion acoustic wave heating method is important heating mechanisms during highly dense plasma condition, when density is above critical density for launched frequency, 2.45 GHz. The conclusions on the shifting of heating mechanisms are also drawn based on the 3D maps of spatio-temporal plasma density and hot electron temperature evolution.

Keywords: microwave plasma simulation, COMSOL multi-physics, magnetized plasma evolution, ECR, off-resonance, ion source, electrostatic heating, electric field evolution, hot electron temperature, experimental validation

1. Introduction

Gaseous particles are ionized to bring them in the form of plasma through the various heating techniques. One of the popular heating techniques is the injection of high frequency microwaves (MW) to a cylindrical cavity that has comparable dimension to the injected MW wavelength. The MW plasma generated by the continuous or pulse feeding of the MW is used in the applications of industrial and accelerator fields for the material science and nuclear applications, respectively. In both of the feeding cases, the plasma is basically produced due to the power absorption by the electrons from the space-time dependent electric field of the MW. The spatio-temporal dynamics and also the steady-state behaviors of the plasma are governed by the ways the MW are coupled to the plasma sustained inside a cavity. The behavioral pattern of the electric field during the plasma evolution can help us to comment on the different MW coupling ways/mechanisms that are involved in the formation of plasma particles and their confinement scenarios. By mastering the basic concepts on those different coupling mechanisms, the coupling efficiency and so the performance of that particular plasma source can be optimized. Performance optimization for this kind of plasma source is indispensable as these are involved in various kinds of applications as mentioned above. One of the important plasma devices is the microwave ion sources that are operated in continuous as well as pulse mode to extract the ion beam during the transient and steady state periods of plasma loading conditions [1–5]. The beam qualities are influenced by the MW coupling mechanisms as they are involved in deciding the plasma parameters during the extraction of a particular instant of the plasma evolution time. Several studies have already reported the electric field evolution during few 10 s of microsecond range when the plasma density was increasing in the very similar plasma device. The electric field was dropped by about more than 50% within a span of few microseconds after the MW launch (t = 0 s) into the cavity [6–11].

Since the electric fields can affect the different power coupling mechanisms during the gas ignition moment (ns to μs), the spatio-temporal plasma parameters are influenced significantly especially in the low pressure regime. Many researchers have used the kinetic models like PIC/MCC or even the hybrid fluid/PIC to obtain more precise results in the MW plasma discharge. But they failed to estimate the hot electron dynamics efficiently in lower pressure condition, as these models demand intensive computational hardware due to its particle approach. Therefore, the current chapter presents the electric field evolution and its impact on the plasma parameter build-up during low pressure plasma state. Here, the model used is based on the finite element method (FEM) that gives more appropriate results for the transient plasma parameters through fluid modeling approach and time-dependent, partial differential equation solver (TDPDE) using fewer computer resources [11]. The different MW-plasma coupling mechanisms (ECR, UHR and electric field polarity reversal associated with ES wave heating) during the plasma density evolution after the MW launch (t = 0 s) can be understood from the behaviors of electric fields.

2. Basic theory of microwave plasma interaction

The study on the propagation and interactions of the microwave with the plasma is important to optimize the performance of any plasma devices like the microwave ion sources. The microwave propagation in the plasma is affected by the dielectric prosperities of the plasma medium. The dielectric property, i.e., the permittivity or the refractive index of the plasma, depends on the external magnetic field distribution that is used to confine the plasma particles and also the electrostatic fields that are present in the plasma. Therefore, the microwave while propagating in different directions within the plasma encounters different values of the refractive index as well as the permittivity that makes the magnetized plasma to be anisotropic and inhomogeneous, respectively. To generate a high plasma density, which is one of the primary requirements in some microwave plasma devices, viz., the microwave discharge ion source (MDIS) or electron cyclotron resonance ion sources (ECRIS), an optimum coupling of the microwave energy through the different interaction mechanisms to the plasma medium is necessary. The microwave propagation and the coupling mechanisms are also influenced by the boundary conditions present in the plasma devices and the geometrical shape of the plasma device. In most cases, the dimension of the plasma reactor used for the purpose of ion sources lie in the comparable range of the launched microwave wavelength. This means the microwave electromagnetic field propagation within the ion source reactor (or cavity) is guided by the boundary conditions and the geometrical shape of the ion source cavity. So, the microwave electromagnetic field coupling to the plasma is affected if the cavity geometry is perturbed. Due to the modification of the cavity geometry, the resonating properties of the cavity resonator are no longer dominated by the fundamental cavity resonant mode. The cavity can resonate with some additional resonating frequencies including the fundamental one. The additional resonating frequencies can lie near to the fundamental one. Due to this reason, if the microwave is launched to the modified cavity, the total microwave field is shared among the cavity resonant modes including the fundamental one and contributes to the power coupling to the plasma.

From the electromagnetic theory of a resonant cavity, only particular cavity resonant modes can exist having fixed frequencies that are given by [12, 13]

$$f = \frac{1}{2\pi}\sqrt{\frac{r^2\pi^2}{l^2} + h^2} \tag{1}$$

where r, l and h are the integer, the length of the plasma cavity and the eigenvalues of the cavity, respectively. The eigenvalues are obtained from the solution of the equations gives by:

$$\nabla_t^2 H_{z0} + h^2 H_{z0} = 0 \tag{2}$$

$$\nabla_t^2 E_{z0} + h^2 E_{z0} \tag{3}$$

A theoretical calculation for the cavity resonant modes from the eigenvalue equations for the electromagnetic field [12, 13] is performed from the empty and completely closed cavity. By considering a simplest cylindrical cavity of radius r and the length 'd' that is filled with a dielectric constant ε_r and the relative permittivity μ_r, the resonant frequencies allowed by the cavity that can be determined from the eigenvalue equations are shown below from the TE and TM modes:

$$f_{nml}^{TE} = \frac{c}{2\pi\sqrt{\mu_r\varepsilon_r}}\sqrt{\left(\frac{P'_{nm}}{r}\right)^2 + \left(\frac{l\pi}{d}\right)^2} \tag{4}$$

$$f_{nml}^{TE} = \frac{c}{2\pi\sqrt{\mu_r\varepsilon_r}}\sqrt{\left(\frac{P_{nm}}{r}\right)^2 + \left(\frac{l\pi}{d}\right)^2} \tag{5}$$

Here, P_{nm} and P'_{nm} are, respectively, the zeros of order m of the Bessel functions of order n and its first derivative. The indices represent the three-dimensional electromagnetic field patterns for each cavity resonant mode.

2.1 Microwave propagation in plasmas

In case of magnetized plasma, the microwave propagation is influenced by the plasma particle dynamics. As the plasma particle dynamics are represented by the particle velocity, the thermal velocity of the plasma particles should be considered when the microwave propagation in plasma is discussed. If the thermal velocity of the plasma particles is negligible with respect to the phase velocity of the microwave, i.e., $v_{thermal} \ll v_{phase}$, this approximation is useful for determining the dispersion of the microwave in a magnetized plasma. This approximation also known as 'cold plasma' is not applicable to the locations where the microwave encounters resonance in the magnetized plasma. On the other hand, if the temperature of the plasma particles is high enough that makes the velocity of the particle to be relativistic (also called as 'warm plasma'), the influence of the plasma particles cannot be neglected while determining the microwave propagation. In that case, the damping of the microwave field is greatly influenced by the plasma particles.

For the un-magnetized plasma case in which the plasma is considered to be isotropic and the condition, $v_{thermal} \ll v_{phase}$ is satisfied, the plasma fluid equation is written as:

$$\frac{\partial \tilde{v}}{\partial t} = q\tilde{E} \tag{6}$$

If the microwave electric field ($\tilde{E}$) and the velocity ($\tilde{v}$) are assumed to be varying with $e^{i\omega t}$, the plasma dielectric constant can be easily shown as:

$$\epsilon = \epsilon_0 \left(1 - \frac{\omega_{pe}^2}{\omega^2}\right) \tag{7}$$

where the ω_{pe} is the electron plasma frequency that is also obtained from the electron density by the relation:

$$\omega_{pe} = \sqrt{\frac{n_e e^2}{\epsilon_0 m_e}} \tag{8}$$

where m_e, e and ϵ_0 are the electron mass, charge and electrical permittivity in vacuum condition, respectively. So the electrical permittivity in plasma is affected by the plasma density. As the electrical permittivity has to be positive in ideal case, the plasma density has to be lower than the corresponding microwave frequency that is launched externally to energize the plasma particles. Therefore, in case of homogeneous and isotropic plasma, it is not possible to raise the plasma density beyond certain level that is known as the critical density, which is written as $n_{critical} = \omega_{pe}^2 \left(\frac{\epsilon_0 m_e}{n_e e^2}\right)$. The density below and above the critical density is referred to be underdense and overdense plasma. So the critical density is the main limitation in the un-magnetized plasma for the microwave propagation. The dispersion for the wave in magnetized plasma is written as:

$$\omega^2 - \omega_{pe}^2 = k^2 c^2 \tag{9}$$

In the overdense plasma, if $\omega < \omega_{pe}$, the propagation k-vector $'k'$ becomes imaginary that means the complete reflection of the incoming microwave from the plasma. Therefore, if the microwave propagation is assumed to be propagating in the x-direction in the overdense plasma, the electric field of the microwave can be written as:

$$\tilde{E} = \tilde{E}_0 e^{(ik.\tilde{r} - \omega t)} = \tilde{E}_0 e^{(ikx - \omega t)} = e^{\frac{x}{\delta}} e^{-i\omega t} \tag{10}$$

So, before encountering the overdense plasma, the electric field becomes an evanescent wave as its magnitude decays exponentially within a distance of approximately the skin depth value; $(\delta) = c/\left(\sqrt{\omega_{pe}^2} - \omega^2\right)$.

Under the externally applied magnetic field, the dielectric constant for the anisotropic plasma becomes a tensor quantity. It means the microwave propagation becomes dependent on the plasma dielectric properties while propagating in various directions with respect to the externally applied magnetic field. If the magnetic field is oriented axially ($\tilde{B} = B_0\tilde{z}$) in a cylindrical plasma chamber, the plasma can act as a dielectric with current $\tilde{J}$ and also the Maxwell's equations can be rewritten as:

$$\tilde{\nabla} \times \tilde{B} = \mu_0 \left(\tilde{J} + \varepsilon_0 \frac{\partial \tilde{E}}{\partial t} \right) = \mu_0 \frac{\partial \tilde{D}}{\partial t} \tag{11}$$

Now if the plasma motion follows the $e^{i\omega t}$ dependence and the plasma conductivity tensor is related with the current as $\bar{\bar{\sigma}}. \tilde{E} = \tilde{J}$, the relation between $\tilde{D}$ and $\tilde{E}$ is obtained as:

$$\tilde{D} = \varepsilon_0 \left(\bar{\bar{I}} + \frac{i}{\varepsilon_0 \omega} \bar{\bar{\sigma}} \right) \tilde{E} \tag{12}$$

where the symbol $\bar{\bar{I}}$ represents an identity tensor. In short, the parameter $\tilde{D}$ can also be written as $\tilde{D} = \varepsilon_0 \bar{\bar{\epsilon_r}} \tilde{E}$, where the effective dielectric tensor $\bar{\bar{\epsilon}}$ is represented as $\epsilon_0 \left(\bar{\bar{I}} + \frac{i}{\varepsilon_0 \omega} \bar{\bar{\sigma}} \right)$. The plasma conductivity tensor is evaluated from the fluid plasma approach. Therefore, the fluid equation in case of 'cold plasma' approximation is rewritten by neglecting the collisional and pressure term and the magnetic field term is considered as:

$$m_i \frac{\partial \tilde{v}}{\partial t} = e\left(\tilde{E} + \tilde{v} \times \tilde{B} \right) \tag{13}$$

The solution to this equation brings out the relation between the velocity ($\tilde{v}$) and the electric field of the microwave. As $\tilde{J} = ne\tilde{v}$, the final expression for the effective dielectric tensor $\bar{\bar{\epsilon}}$ can be shown in a determinant form as:

$$\bar{\bar{\epsilon}} = \epsilon_0 \begin{vmatrix} S & -iD & 0 \\ iD & S & 0 \\ 0 & 0 & P \end{vmatrix} \equiv \epsilon_0 \bar{\bar{\epsilon_r}} \tag{14}$$

where S, D and P can be written in terms of three different kinds of frequencies: $S = \left[1 - \frac{\omega_{pe}^2}{\omega^2} \left(\frac{1}{1-\omega_c} \right) \right]$, $D = [-\frac{\omega_{pe}^2}{\omega^2} \frac{\omega_c}{\omega} (\frac{1}{1-\omega_c})]$ and $P = 1 - \frac{\omega_{pe}^2}{\omega^2}$.

The wave equation can be derived from the Maxwell's curl equation for the electric field by following the standard procedure that shows

$$\tilde{\nabla} \times \tilde{\nabla} \times \tilde{E} = -\frac{\partial \tilde{\nabla} \times \tilde{B}}{\partial t} = -\frac{1}{c^2} \bar{\bar{\epsilon}} \cdot \frac{\partial^2 \tilde{E}}{\partial t^2} \tag{15}$$

The above equation is rewritten by assuming the $e^{i\tilde{k}.\tilde{r}}$ dependence of the electric field as:

$$\tilde{k} \times \tilde{k} \times \tilde{E} + \frac{\omega^2}{c^2} \bar{\bar{\epsilon}} . \tilde{E} = 0 \tag{16}$$

The equation can be represented in matrix form by assuming the angle between the wave vector and magnetic field to be 'θ' and denoting the vector $\tilde{N} = \frac{c}{\omega} \tilde{k}$.

$$\begin{bmatrix} S - N^2 \cos^2\theta & -iD & N^2 cos\theta sin\theta \\ iD & S - N^2 & 0 \\ N^2 cos\theta sin\theta & 0 & P - N^2 \sin^2\theta \end{bmatrix} \begin{bmatrix} E_x \\ E_y \\ E_z \end{bmatrix} = 0 \tag{17}$$

The solution to these equations can exist if the determinant is zero. This condition brings out the dispersion relation of the microwave in the plasma. By making the determinant to be zero, two solutions are obtained that are written as:

$$N_{O,X}^2 = \frac{2x(1-x)}{2(1-x) - y^2 sin^2\theta \pm \sqrt{y^4 sin^4\theta \pm 4y^2(1-x)^2 cos^2\theta}} \tag{18}$$

where the symbols '*O*' and '*X*' correspond to the ordinary microwave and extraordinary microwave, respectively. The notations x and y in Eq. (18) represent the electron density scale and magnetic field scale respectively and are denoted as: $x = \frac{\omega_{pe}^2}{\omega^2} \propto n_e$ and $y = \frac{\omega_c}{\omega} \propto B_0$.

From the solutions, the microwave propagation and damping properties can be explained considering the values of the electron density, magnetic field and the angle of propagation. By the definition, the refractive index (N) has to be positive for enabling the microwave to propagate in plasma. The two different values for the refractive index $N = 0$ and $N = \infty$ imply the cut-off and the resonance conditions for the microwave, respectively. From Eq. (18), three different types of cut-offs can be obtained by putting $N = 0$ that depends on the plasma parameters. Three cut-offs are: (1) $\omega = \omega_{pe}$; this cut-off is also known as 'O' cut-off; (2) $\omega = \omega_R = 0.5\sqrt{\omega_c^2 + 4\omega_{pe}^2} + \omega_c$; this cut-off is also called upper or R cut-off frequency that occurs both above ω_{pe} and ω_c; and (3) $\omega = \omega_L = 0.5\sqrt{\omega_c^2 + 4\omega_{pe}^2} - \omega_c$; it is also called the lower cut-off or L cut-off for the microwave. Based on the local plasma density values and magnetic field values in the plasma volume, "ordinary waves" have 'O' cut-off region, and extra-ordinary waves have two cut-offs, R cut-off, and L cut-off regions, from where corresponding waves are reflected back. To satisfy resonance condition ($N = \infty$), it is found from Eq. (18) that the term $Pcos^2\theta + S sin^2\theta$ has to be zero. It gives the resonance angle in terms of the plasma parameters that are given by:

$$cos^2\theta = \frac{x + y^2 - 1}{xy^2} \tag{19}$$

It is clear that the wave can resonate at the resonance angle (θ) that is decided by the angle of microwave propagation with respect to the magnetic field. From Eq. (18), it is also confirmed that only the extraordinary wave can have the resonances. From Eq. (19), the resonance angles for the extraordinary wave are determined by the plasma density and the magnetic field. For example, if the resonance angle is set to be zero, it is verified that the resonance takes place at $\omega = \omega_{RF} = \omega_c$. This means the microwave matches the larmour frequency of the electrons that are gyrating around the magnetic field lines. It is also known as electron cyclotron resonance (ECR). Again, for the resonance angle, $\theta = 90°$, another type of resonance can occur when $x + y^2 - 1 = 0$, and this condition yields the resonance frequency to be $\omega = \omega_{RF} = \sqrt{\omega_c^2 + \omega_{pe}^2}$. This type of resonance is known as upper hybrid resonance (UHR) applicable to the wave (e.g., extraordinary wave) that is propagating perpendicularly with respect to the magnetic field. The resonance of the ordinary wave can also be understood from Eq. (18). It can be seen from one of the solutions obtained from Eq. (19) that another resonance can exist in the region where $x, y \geq 1$. This is the forbidden region for the ordinary wave, which is the 'O' cut-off region. Therefore, the ordinary-type microwave can reach this resonance only by tunneling through this cut-off region.

To visualize the cut-offs and resonances for the different types of microwaves in a better way, the dispersion plots are shown in a single diagram, also known as Clemmow-Mullaly-Allis (CMA) diagram. **Figure 1** is applicable to the 'cold plasma' approximation case as discussed before. In the CMA diagram, the x- and y-axes represent the plasma density and the magnetic field, respectively. It means that for a given launch microwave frequency, a point on the diagram dictates the experimental situation that is characterized by a particular plasma density and the magnetic field. The diagram is divided into two regions: underdense and overdense plasma regions that are based upon the launch microwave frequency. The boundaries shown in the diagram correspond to the cut-off and resonances for the different types of microwaves.

The polarization of the electric field of the microwaves also plays important role on the damping of microwave power into the plasma. From Eq. (17), the relation between the x- and y-components of the electric field can be found out to understand the polarization in the plane perpendicular to the magnetic field. The relation is obtained from Eq. (17) as:

$$\frac{iE_x}{E_y} = \frac{N^2 - S}{D} \tag{20}$$

Eq. (20) indicates that the waves become circularly polarized and linearly polarized when the cut-off condition ($N^2 = 0$) and the resonance condition ($N^2 = \infty$) are satisfied respectively. The waves are circularly and linearly polarized, respectively. However, it is confirmed that the ordinary waves are left hand polarized and the extraordinary waves are right hand polarized. The electric field of the right hand polarized waves and the electrons gyrating across the magnetic rotate in the same direction. Therefore, if the microwave-launched frequency matches the electron gyration frequency, electrons observe a constant electric field at the resonance location and are resonantly accelerated by the electric field. It is known as electron

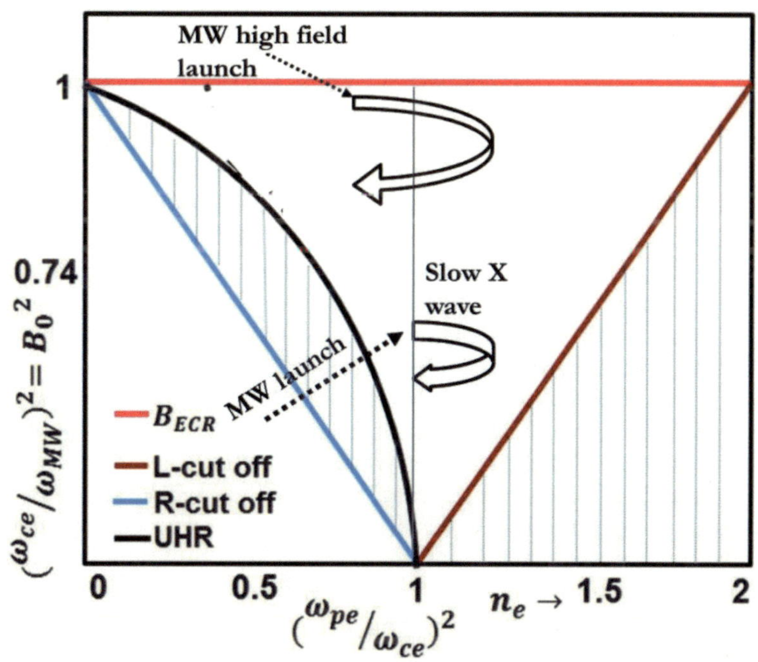

Figure 1.
The CMA diagram shows the propagation of microwave launched from high and low magnetic field side. The arrow bend implies the cut-off region and the mode conversion region near the upper hybrid resonance location.

cyclotron resonance heating (ECR). It is to be noted that only in the ECR heating mechanism the energy is transferred directly from the microwave to the gaseous electrons. As the electrons gain sufficient energy by resonance, they ionize the neutral gases and thus form plasma. There are also other mechanisms for energy transfer from the microwave to the plasma. But the direct energy transfer is not possible in these cases. In fact, the energy is shared among the microwave and the plasma oscillation modes by means of collisional absorption or non-linear phenomenon. The plasma oscillation modes are excited by the '*O*' and '*X*' modes of microwaves. A detailed description of the energy damping mechanisms is discussed below.

2.2 Electron cyclotron resonance (ECR)

As discussed above, the ECR heating is an attractive tool in direct energy transferring to the plasma specifically in the ion source applications. The working principle of ECR mechanism is based on the frequency matching condition in which the microwave frequency with same polarization matches with electron cyclotron motion. It looks pretty much simpler in a qualitative approximation. But in case of quantification, it appears to be a non-deterministic method. It means along with the frequency, the phase difference between the microwave and electron motion also plays important role in energy transferring through ECR mechanism. Under frequency matching condition, if the phase difference between the microwave and the electron motion is in the same phase, then the electrons are accelerated by the microwave electric field. On the other hand, if it is 180° out of phase, the electrons are decelerated. Practically, the temporal phase difference between the microwave and the electron motion is a random phenomenon. So, it becomes necessary to take an average energy gain temporally of the electrons for several microwave periods. It is demonstrated by several groups worldwide that the average temporal energy gain of the electrons has positive value if the averaging calculation is performed irrespective of the phase difference between the microwave and the electron cyclotron motion. It has been proved that the net energy gain is related non-linearly to the microwave electric field ($\propto E^2$). In ECR heating-based plasma, the collisions in plasma can play important role on the confinement of the plasma particles, thermalizing the plasma particles, and also on the ionization rate. The collision properties of plasma are determined by two physical parameters: mean free path ($\lambda_{mean} = \frac{1}{n\sigma}$) and collision frequency (? = $n\overline{\sigma v}$). Here, $\overline{\sigma v}$ is the product between cross-section and the velocity of plasma particles that is determined from the velocity distribution function (Maxwellian type). The particle confinement time is inversely proportional to the collision frequency. A contradiction arises from these two relations. The plasma can have a temperature even in collision-less case and follow the Maxwellian distribution. In collision-less case, the particle does not collide with each other and so there is no thermalization. But in ECR case, the kinetic energy of electrons increases, but there would be low temperature of the plasma. It suggests that there must be other mechanisms that are involved for the solution of this contradiction.

In most of the ECR ion source plasmas, the plasma is collision-less. But in other ion sources operating at microwave discharge, where the temperature of the plasma is comparatively lower and the pressure is also higher, a complete collision-less approximation is not valid. The other heating mechanisms that are involved in the collisional absorption condition must be taken into account. Considering the collisional term in the equation of motion,

$$m_e \frac{\partial \tilde{v}}{\partial t} = e\left(\tilde{E} + \tilde{v} \times \tilde{B}\right) - m_e \omega_c \tilde{v} \tag{21}$$

Here, ω_c denotes the collision frequency. Then the dispersion relation is modified as:

$$N_{O,X}^2 = 1 - \frac{2x(1 + iZ - x)}{2(1 + iZ)(1 - x) - y^2 sin^2\theta \pm \sqrt{y^4 sin^4\theta \pm 4y^2(1 + iZ - x)^2 cos^2\theta}} \tag{22}$$

Here, $Z = \frac{\omega_c}{\omega}$. It can be observed that there is collisional term present in the modified dispersion equation of the wave propagation. This collisional term limits the attainable energy during the ECR heating unlike the collision-less case, as discussed before.

2.3 Wave propagation in warm plasma condition

Although the cold plasma approximation is not valid for the wave dispersion in the region very close to the cut-off, the effect of warm plasma condition cannot be ignored. This is because the wavelength in the latter case is not negligible compared to the scale length of the plasma parameters [14]. At the resonance, where the refractive index is infinity, the wavelength becomes equivalent to the electron larmor radius. So, the finite larmor parameter effect is not negligible and has to be considered. The larmor radius can be written as $ŕ = 0.5(k_\perp^2 v_{th}^2)/\omega_c^2$ where $k_\perp$ denotes the perpendicular component of the wave vector with respect to the magnetic field (Z-direction). For the non-relativistic case assuming the Maxwellian distribution function, the distance of frequency from the nth cyclotron harmonic resonance in terms of the Doppler shift unit is written as $\xi_n = (\omega + n\omega_c)/|k_z|v_{th}^2$. Here k_z represents the z-component of the wave vector. The dielectric tensor is expressed as:

$$\bar{\bar{\epsilon}}_r = \bar{\bar{I}} + \frac{\omega_{pe}^2}{\omega^2} \sum_{n=-\infty}^{n=\infty} \begin{bmatrix} \epsilon_{xx} & \epsilon_{xy} & \epsilon_{xz} \\ \epsilon_{yx} & \epsilon_{yy} & \epsilon_{yz} \\ \epsilon_{zx} & \epsilon_{zy} & \epsilon_{zz} \end{bmatrix} \tag{23}$$

where $\epsilon_{xx} = \frac{n^2}{ŕ}\overline{I_n}Z_n \epsilon_{xy} = -\epsilon_{yx} = in\overline{I'_n}Z_n$; $\epsilon_{xz} = \epsilon_{zx} = -n\sqrt{\frac{2}{ŕ}}\overline{I_n}(1 + \xi_n Z_n)$; $\epsilon_{yy} = \left(\frac{n^2}{ŕ}\overline{I_n} - 2\,ŕ\overline{I'_n}\right)Z_n$; $\epsilon_{yz} = \epsilon_{zy} = -i\sqrt{2ŕ\overline{I_n}}(1 + \xi_n Z_n)$; $\epsilon_{zz} = 2\xi_n\overline{I_n}(1 + \xi_n Z_n)$.

Here $Z_n = Z(\xi_n)$ and $\overline{I_n} = e^{-ŕ}I_n(ŕ)$ where I_n is the modified Bessel's function of nth order and Z is a function that is written as [15] $Z(\xi_j) = \frac{1}{\sqrt{\pi}}\int_{-\infty}^{\infty}\frac{e^{-s^2}}{s - \xi_j}$.

This dielectric tensor coming from warm plasma approximation has few new features that affect the wave propagation unlike the cold plasma approximation case. It can be seen from Eq. (23) that the dielectric tensor is not only a function of ω_{pe} and ω but also a function of the plasma temperature and the k-vector components. So, the dispersion relation yields a new kind of solution that is called electrostatic wave.

2.4 Electrostatic wave

Usually, the electric field of an electrostatic wave does not change with time. This fact is known from the derivation of the electrostatic field from a scalar potential ($\tilde{E} = -\tilde{\nabla} V$), which makes $\tilde{\nabla} \times \tilde{E}$ to be zero. However, the time-dependent electric field is sometimes denoted as electrostatic wave as it obeys $\tilde{E} = -\tilde{\nabla} V$ relation. So a concrete example for calling an electric field to be electrostatic is the k-vector direction of electrostatic wave with respect to its electric field. If the k-vector is parallel to the electric field, it is called as an electrostatic wave. It means $\tilde{\nabla} \times \tilde{E} = \tilde{k} \times \tilde{E}$ that makes the $\frac{\partial \tilde{B}}{\partial t}$ component to be zero. Therefore, the electrostatic wave does not have any magnetic field component.

A commonly occurred electrostatic wave in a warm plasma condition is named as the Langmuir wave [16]. The Langmuir wave is the main constituent of un-magnetized plasma that appears together with the ion-acoustic wave (IAW). In case of magnetized plasma, electrostatic waves are also present. In this case, if the electrons are displaced by some force, an electric field builds up to restore the electrons back to their initial position to maintain the plasma quasi-neutrality condition. Due to the very low inertia, the electrons will show an overshoot and oscillate around an equilibrium position. The frequency of oscillations is equivalent to the electron plasma frequency of the plasma. The dispersion relation of the Langmuir wave is written as [16]:

$$\omega_L^2 = \omega_{pe}^2 + \frac{3}{2} k^2 v_{th}^2 \tag{24}$$

As the electron plasma oscillates very fast compared to the massive ions present in plasma, the massive ion motion is considered to be fixed in the GHz frequency scale (Langmuir frequency range). Although the frequency of the massive ion motion is very low compared to the Langmuir wave, the massive ions part will take part in the oscillations due to the electric field build-up. This low-frequency oscillations fall usually in the range of ion-acoustic wave frequency. The ion wave dispersion is obtained from the fluid equation as,

$$\frac{\omega_{\text{IAW}}}{k_{\text{ion}}} = \{k_B(T_e + \gamma_i T_i)/M_i\}^{1/2} \tag{25}$$

Usually, the plasma oscillations in the ion-acoustic frequency range lie in between few kHz to tens of MHz.

There exists another kind of electrostatic wave in magnetized plasma, which is known as electron Bernstein waves (EBW). EBW exist in warm plasma conditions when the electron temperature has finite value. It is known that the superposition of the static magnetic field with the oscillating electric field of the plasma waves can make the electron orbit to be elliptical [17, 18]. Now, if the magnetic field is increased further, the electron orbit will become a circular one as the Lorentz force dominates the electrostatic component [17, 18]. The presence of EBW makes the electron gyrophase to organize in such a manner that the space charge distribution in plasma obtains a minima and maxima in the direction perpendicular to the externally applied magnetic field. It was shown [19] that the space charge accumulation is periodic. The charge accumulation propagates with a wavelength that is four times the electron larmor radius [19]. As the wavelength of the EBW is much lower than the length of a typical Langmuir probe tip, used for the plasma characterization, the Langmuir probe is unable to detect the EBW wave directly [20]. The dispersion of electron Bernstein wave (EBW) can be written as:

$$\frac{k_\perp^2}{2\omega_{pe}^2/v_{th}^2} = 2\omega_c^2 \sum_{n=1}^{\infty} n^2 \frac{e^{-b} I_n(b)}{\omega_{MW}^2 - n^2\omega_c^2} \tag{26}$$

Here $k_{\perp}$, v_{th}, and I_n represent k-vector perpendicular component, electron thermal velocity and the Bessel's function, respectively. From Eq. (26), it can be seen that the k-vector becomes infinite at the cyclotron harmonic, i.e., $\omega_{MW}^2 = n^2\omega_c^2$. Therefore, the EBW waves can have resonance at the harmonics of the cyclotron resonance frequencies. In microwave ion source, the resonance absorption of the EBW is possible at the harmonics of the cyclotron frequencies of the ECR magnetic field (B_{ECR}) value.

2.5 Mode conversion theory

In microwave-generated magnetized plasma, the presence of plasma density gradient and the variation of the magnetic field make the wave propagation and its energy absorption unpredictable. It is difficult to estimate the wave trajectory from the simple linear uniform plasma theory [21]. It is natural that the wave would cross the boundaries shown in the CMA diagram by travelling up or down depending on the magnetic field variation and plasma density distribution. Inhomogeneous and anisotropic plasma can exhibit a wide variety of possibilities for the cut-off, resonance, cut-off-resonance and/or the back-to-back cut-off pairs. In inhomogeneous plasma, two or more waves can coexist that propagates in the plasma having density gradient. Although their polarization and propagation vector are different from one another, they can exhibit identical characteristics at some particular plasma regions having particular plasma loading conditions. At those particular scenarios, the waves can remain no longer distinguishable and therefore can convert into another. The mode conversion theory deals with establishing resonance characteristic in inhomogeneous plasma considering two different waves present in the plasma by taking into account the wave reflection, cut-off, resonance and absorption conditions. As the microwaves that are present in the microwave ion source plasma is dominated by the ordinary- and extraordinary-type microwave, the mode conversion theory is mainly focused upon considering the cut-off-resonance pair condition in plasma. The *X* mode microwave is unable to propagate the dense plasma because it is reflected at R cut-off. On the other hand, O-mode microwave is able to propagate in the dense plasma, where it converts into the X mode microwave under certain condition that is obtained from Eq. (18). As per CMA diagram, the *X* mode can have resonance at the UHR region. Therefore, before entering the UHR region, the launched microwave can convert into the *X* mode based on the following mechanisms:

- If the *X* mode microwave is launched from the high magnetic field side, the *X* mode will not see the R cut-off (see **Figure 1**). It will propagate toward the UHR zone where it can convert into an EBW and an ion wave as per the literature. This method is known as *X-B* conversion process.

- Another method in generating the EBW and ion wave is the *O-X-B* mode conversion process. In this process, the *O* mode microwave launched from the vacuum side crosses the R cut-off and converts into a slow *X* mode after tunneling through the evanescent layer. The generated slow *X* mode is then converted into an EBW and an ion wave near the UHR layer.

In microwave ion source plasma under mirror magnetic field configuration, there can coexist two types of components (*O* and *X* modes) of the launched microwave. The ion source cavity acts as a resonator having comparable cavity dimension with respect to the launch microwave wavelength. Therefore, different types of propagating modes can coexist in the presence of plasma. This makes it difficult to identify which mechanism is actually allowing the X mode to reach the UHR region. However, it is possible to estimate this mode conversion efficiency. For the case of X mode microwave, the refractive index is written as:

$$\frac{c^2k^2}{\omega^2} = 1 - \frac{\omega_{pe}^2}{\omega^2}\frac{\omega^2 - \omega_{pe}^2}{\omega^2 - \omega_{UHR}^2} \tag{27}$$

Let us suppose the X mode wave sees cut-off at the electron plasma frequency, ω_{pe1}. The wave cut-off condition yields

$$\frac{\omega_{pe1}^2}{\omega^2} = 1 - \frac{\omega_c}{\omega} \tag{28}$$

After reflection at the cut-off, the wave propagates in the inhomogeneous and anisotropic plasma in the location where it will find a resonance (refractive index = ∞). The resonance is called upper hybrid resonance where electron plasma frequency is ω_{pe2}. The corresponding Eq. (27) becomes then

$$\frac{\omega_{pe1}^2}{\omega^2} = 1 - \frac{\omega_c^2}{\omega^2} \tag{29}$$

By dividing Eq. (28) by Eq. (29), one gets

$$\frac{\omega_{pe2}^2}{\omega_{pe1}^2} = 1 + \frac{\omega_c}{\omega} = \frac{n_{e2}}{n_{e1}} \approx 1 + \frac{\Delta x}{L} \tag{30}$$

where Δx and L denote the distance between the cut-off and the upper hybrid resonance layer and the length parameter, respectively. From Eq. (30), one obtains

$$\Delta x = \frac{\omega_c}{\omega} L \tag{31}$$

Now, for the wavelength $\lambda \ll L$, the Δx will be many times the wavelength. So the wave will not be able to reach the upper hybrid resonance point. The mode conversion efficiency is determined by the penetration depth of the wave into the plasma. The effective mode conversion can be achieved when the parameter L is comparable to the wavelength, λ. In another way, from Eq. (31), it is also seen that mode conversion can be improved if the magnetic field is reduced.

If a strong electromagnetic field is present in the ion source cavity, the plasma particle follows the relation $v_E \leq v_{the}$; $v_E = eE/m\omega$, where v_E, E, e, m and v_{the} are wave phase velocity, electric field intensity, electron charge, electron mass and electron thermal velocity, respectively. This condition leads to plasma parameters to vary with time. In effect, the non-linear effect such as parametric instability comes into play. Off-course, the parametric decay happens above certain threshold value of the electric field and the microwave energy is shared among the plasma waves and the microwave through the non-linear interaction phenomenon.

2.6 Parametric decay

As the plasma parameters vary with time under the conditions of intense electric field of the microwave, the corresponding velocity becomes close to the electron thermal velocity. It is known that the X mode electric field accumulates near the

UHR region with time due to its propagation from the cut-off region to the UHR region. The accumulated electric field will be able to share some part of its energy through non-linear interactions to the oscillating modes present in the plasma. This non-linear energy coupling between the X mode microwave and the oscillation modes is known as parametric instability. It is demonstrated that the X mode through this type of instability couples energy to the Bernstein wave and ion wave near the UHR region. The X mode normally contains the longitudinal component of the electric field (TM type). Considering the x-component of the X mode propagation from Eq. (17), the dispersion of X mode wave can be written as follows:

$$\left(\omega^2 - \omega_{UHR}^2\right)E_x + i\frac{\omega_{pe}^2\omega_c}{\omega}E_y = 0 \tag{32}$$

From Eq. (11), as the X mode approaches the UHR, the term $\omega^2 - \omega_{UHR}^2 \rightarrow 0$ and so E_y would be zero. Therefore, the component E_x will remain non-zero at the UHR, which is directed in the direction of propagation of X mode. This means the X mode at the UHR becomes electrostatic. The X mode microwave that becomes an electrostatic wave at the UHR matches with the Bernstein wave and the ion wave. With the two oscillation modes that are coupled with the X mode electric field (E_0), the motion of one of the modes (Bernstein wave) can be expressed in terms of the simple harmonic oscillator.

$$\frac{d^2x_1}{dt^2} + \omega_1^2x_1 = C_1x_2E_0\cos\omega_0t \tag{33}$$

Here the motion of the first oscillator (amplitude x_1 and the resonant frequency ω_1) is driven by the time-dependent electric field of the X mode and the amplitude (x_2) of the second oscillator. The equation of the motion for the second oscillator x_2 is:

$$\frac{d^2x_2}{dt^2} + \omega_2^2x_2 = C_2x_1E_0\cos\omega_0t \tag{34}$$

Let us assume, $X_1 = \overline{x_1}\cos\omega' t$ and $X_2 = \overline{x_2}\cos\omega'' t$. In the absence of any non-linear interaction, it can be expressed as $\omega' = \omega_1, \omega'' = \omega_2$ and $\omega_0 = \omega_1$. But in the presence of non-linear interaction, this is incorrect. For the case of non-linear interaction, as the driving terms cause frequency shifting, the frequency ω_2 is approximately equal to ω''. However, in the absence of non-linear interaction, the solution from the system of coupled Eqs. (33) and (34) is obtained as:

$$\omega'' = \omega_0 \pm \omega' \tag{35}$$

Now if ω' is small such that both values of ω'' lie within the bandwidth of oscillator, x_2, there exist two oscillators: $x_2(\omega_0 + \omega')$ and $x_2(\omega_0 - \omega')$. Under this assumption, solving the coupled Eqs. (33) and (34), the following frequency matching condition is arrived:

$$\omega_0 \cong \omega_2 \pm \omega_1 \tag{36}$$

As the oscillators are waves, the '?t' term is replaced by '(?t-$\tilde{k}.\tilde{r}$). Following this, a new matching condition of k-vector is reached, $\tilde{k_0} = \tilde{k_2} \pm \tilde{k_1}$.

The frequency and k-vector matching conditions correspond to the energy conservation and momentum conservation following the quantum mechanics theory. It is proved that [15] the X mode generates two waves through parametric decay: a high-frequency Bernstein wave and a low-frequency ion wave following the frequency matching condition. This leads to the generation of secondary peaks around the launched microwave frequency, which is called as sideband. Therefore, the simultaneous presence of the sideband peaks around the launched frequency signifies the parametric decay near the mode conversion region.

The parametric decay occurs above a certain threshold value, which actually depends on the damping rate of the oscillator. If the damping rates Γ_1 and Γ_2 for two oscillators x_1 and x_2, respectively, are introduced in Eqs. (33) and (34), one can conclude that the lowest threshold occurs at the exact frequency matching condition. The corresponding threshold electric field is obtained as [19]:

$$C_1 C_2 E_{0threshold}^2 = 4\omega_1\omega_2\Gamma_1\Gamma_2 \tag{37}$$

2.7 Damping of electrostatic and electromagnetic energy

Electrostatic waves generated in the plasma through the parametric decay instability can damp their energy to the plasma particles and thus increase the plasma density. When the phase velocity of the electrostatic wave becomes comparable to the thermal velocity of the plasma particles, the energy is transferred from the wave to the plasma particles and is known as Landau damping mechanism. In microwave ion source plasma, density can be increased 2–3 times more than the ECR heating mechanisms through the off-resonance heating mechanism. For this reason, the off-resonance condition is used to create favorable conditions of the upper hybrid resonance heating. Under certain plasma temperature, the electrostatic wave can transfer energy resonantly to the plasma particles if the wave phase velocity matches the plasma particle velocity. In some cases, the plasma particle velocity can be higher than the wave phase velocity. Under this condition, the plasma particle can transfer energy to the wave. The Landau damping mechanism follows the equation written below:

$$I_m\left(\frac{\omega}{\omega_{pe}}\right) = -0.2\sqrt{\pi}\left(\frac{\omega_{pe}}{kv_{th}}\right)^3 e^{-12k^2\lambda_D^2} \tag{38}$$

The exponential term on the right hand side represents that the Landau damping will be small for small value of $k\lambda_D$. The Landau damping phenomenon is applicable for both case of electrons and ions. The electrostatic wave whose frequency falls in the range of electrons can heat the electrons only in the parallel direction with respect to the magnetic field. In case of electrostatic ion wave, such restriction does not apply. This is because the electrons are strongly magnetized and so will not be able to move across the magnetic field.

In case of a compact microwave plasma device where the microwave wavelength becomes comparable to the device dimension, the cavity resonant mode can also play crucial role in damping the electromagnetic energy to the plasma particles. The presence of multiple cavity modes in the plasma can produce modulated wave due to the interaction between each pairs of the cavity resonant modes. The generated modulated wave propagates in the plasma and damps its energy to the plasma particles where the frequency of the modulated wave matches the local plasma frequency of the plasma particles.

3. Simulation modeling of MW interaction in plasma

For the MW interaction modeling during the plasma evolution, a schematic of the computational domain is shown in **Figure 2**. The computational domain consists

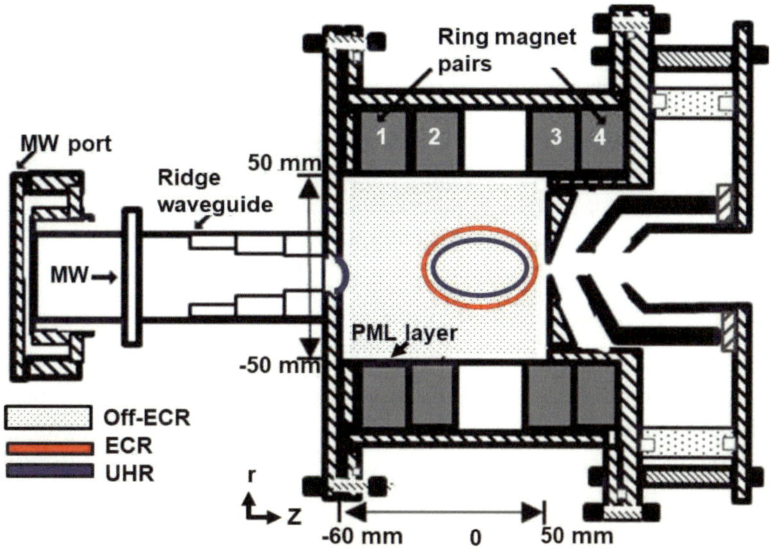

Figure 2.
Simulation domain of the MW ion source.

of a microwave coupled reactor, which is a cylindrical plasma cavity. A MW of frequency 2.45 GHz is injected into the microwave coupled reactor through a ridge waveguide port (on the left side of **Figure 2**) to ionize the gaseous particles, thus forming the plasma that is confined under the mirror magnetic field configuration. The reactor has dimension of 107-mm and 88-mm diameter. The microwave power is fed into a cavity resonator through a tapered waveguide. The waveguide is tapered by embedding four ridge sections having different ridge length, ridge gap and ridge width on the inner sidewalls of the waveguide. The ridge dimensions are optimized from the analytical calculation as well as from the electromagnetic simulation. The mirror magnetic field is created by using two pairs of ring magnets that surround the microwave coupled reactor [22]. On the right side of the microwave coupled reactor (**Figure 2**), the ion beam extraction system is attached through a 5-mm hole on the wall of the reactor. The similar computational domain is used in the experimental set-up (Section 5) to validate the simulated data. Here, the finite element method (FEM)-based COMSOL model is used [22].

The MW propagation and the plasma evolution are assumed to be decoupled to each other during the simulation modeling in the temporal scale [10]. The MW electric fields ($\vec{E}$) are averaged for some MW periods before putting their value in the plasma model and the resultant field is given as input to the plasma model. The electron's momentum equation is time integrated along with Maxwell's equations for some MW periods until the MW model of the FEM gets a periodic solution of the equation to transfer an average power to the particles over such a period. The FEM model continues this process until it gets a steady state solution for at least $\sim 10^3$ ($\omega/\nu_m \sim 10^3$) MW periods. Here, ω is MW frequency and ν_m is electron's momentum transfer frequency. Since the electrons stay in the ECR zone for a very short time (transit time) duration, it causes non-local kinetic effects. This results in the de-phasing between the velocity and field oscillations that becomes very difficult to describe using the fluid model. This problem is resolved by introducing the effective collision frequency (ν_{eff}) in the simulation to converge the solution. In the low pressure condition, for the collision-less heating, ν_{eff} has to be in the order of inverse transit time for the electrons [10].

In case of MW model, the equations for the electric field are solved in frequency domain while keeping the other parameters in time domain. In the beginning, the FEM model started from the Maxwell's equations in order to justify the modeling approach. The MW electric fields are changing with time at a frequency of $\omega/_{2\pi}$. During the plasma evolution, the total electric field (E) value in the μs time zone is regarded as a resultant quantity (or total) that comes from the superposition between the MW and ambipolar-type electric field of the plasma. The modification of the resultant electric field follows the equation that is given below:

$$\nabla \times \mu^{-1} (\nabla \times E) - k_0^2 \left(\varepsilon_r - \frac{i\sigma}{\omega\varepsilon_0} \right) E = 0 \tag{39}$$

The notations used in the above equations $k_0, \varepsilon_r, \sigma \; and \; E$ are the vacuum wavenumber, relative permittivity, the total electric field and the plasma conductivity in full tensor form, respectively. This plasma conductivity is a function of plasma density, collision frequency and B-field. So, all types of total electric field components are possible to estimate as the FEM model can determine the above parameters for some particular plasma loading conditions during its evolution. The model also considers that the $\tilde{E}$-field evolution in the sub-nanosecond region is well separated from the total E-field quantity. This means the total E-field changing during the μs region is composed of the MW electric field and the ambipolar electric field.

For representing the MW propagation in an infinite space, the perfectly matched layers (PML) are introduced in the computational domain as shown in **Figure 1**. The present FEM model considers the electron transport properties to follow the Boltzmann distribution function. The distribution is an integro-differential equation in phase space (r, u) that cannot be solved efficiently. For this reason, the FEM model assumes the plasma as a fluid following the drift diffusion approximation. The assumption is adapted from the Boltzmann equation that is multiplied by some weighing function and then integrated the resulted function over the velocity space. This exercise yields completely three-dimensional and time dependent equations [10, 23]. During the modeling, FEM assumes that ion motion is negligible with respect to the electron motion in the timescale (ns) of MW. Additionally, the electron density is constant spatially within the ECR surface. The Debye length is also assumed to be much smaller than the interaction length of the MW. The average value of the electron velocity on the microwave timescale is obtained from the assumption of Maxwellian distribution function and from the first derivative of the Boltzmann equation. The following drift-diffusion equations are shown that are used to compute the electron density and electron energy density:t

$$\frac{\partial n_e}{\partial t} + \nabla.[-n_e(\mu_e.E) - D_e.\nabla n_e] = R_e \tag{40}$$

$$\frac{\partial n_\varepsilon}{\partial t} + \nabla.[-n_\varepsilon(\mu_\varepsilon.E) - D_\varepsilon.\nabla n_\varepsilon] + E.\Gamma_e = R_\varepsilon \tag{41}$$

The term in Eq. (42) is $E.\Gamma_e = en_e v_e \cdot E_{ambipolar} - \Pi$. This is a heating term $E.\Gamma_e$ that comprises two components. The first component means the electrons are getting energy through the ambipolar field during the plasma evolution. The second component (Π) signifies the MW absorption power ($n_e \langle v_e \cdot \tilde{E} \rangle$) that is carried by the electrons in the plasma. The other terms $n_e, v_e, R_e, \Gamma_e, \mu_e, \mu_\varepsilon, E, D_e, n_\varepsilon, R_\varepsilon, \Gamma_\varepsilon, n_\varepsilon$ and D_ε are the electron density, average electron velocity, the source term of the electrons, electron flux, the mobility of the electrons ($4 \times 10^4 \; m^2/(Vs)$), electron energy mobility, total electric field, electron diffusivity, electron energy density, energy loss/gain from inelastic collisions, electron energy flux, electron energy mobility and electron energy diffusivity, respectively. Some of the terms described above, the electron's diffusivity, mobility and energy diffusivity, are estimated following the relations that are given below:

$$D_e = \mu_e T_e; \mu_\varepsilon = (5/3)\mu_e \text{ and } D_\varepsilon = \mu_\varepsilon T_e \tag{42}$$

The above written electron transport properties represent full tensor parameters in which the tensor term electron mobility is influenced by the magnetic field. The electron mobility without the magnetic field $(\equiv 1 \times 10^{25}/N_n)$ and with the magnetic field is the known parameters. Here, N_n represents the neutral density of the gaseous particles. The electron source term in Eq. (40) is obtained from the relevant plasma chemistry that is expressed below with the help of the rate coefficients $R_e = \sum_{j=1}^{M} x_j k_j N_n n_e$. The presented symbols M, x_j and k_j represent the number of reactions, the mole fraction of a specific species and the rate co-efficient for the reaction j, respectively [10]. The energy loss terms as shown in Eq. (41) is written for all the reactions as $R_\varepsilon = \sum_{j=1}^{M} x_j k_j N_n n_e \Delta\varepsilon_j$. The notation $\Delta\varepsilon_j$ denotes the energy loss that occurs from the reaction j. The energy source and the loss terms are calculated inherently in the FEM model. The rate coefficients mentioned in the above expressions are taken from the cross-section data as per the relation $k_k = \gamma\int_0^\infty \varepsilon\sigma_k(\varepsilon)f(\varepsilon)d\varepsilon$, where γ is $(2q/m_e)^{1/2}$. In this expression, the symbols m_e, ε, σ_k and f are represented as the mass of electron, energy of electron, cross-section for the reaction and the electron energy distribution function (Maxwellian) for the electrons, respectively.

The description for the boundary conditions taken during the plasma simulation is as follows. The plasma chamber wall is kept at ground potential. The reflections, secondary emission and also the thermal emission from the electrons are assumed to be negligible at the wall boundaries. In effect of that the electron flux and electron energy flux at wall boundary can be written as $n.\Gamma_e = (\frac{1}{2} v_{e,th} n_e)$ and $n.\Gamma_\varepsilon = (\frac{5}{6} v_{e,th} n_e)$, respectively. In addition, the heavy plasma particle losses at the boundary walls and their migration are considered to be originating from only the surface reactions and ambipolar electric field, respectively. A fixed power is maintained that is absorbed by the plasma ($P_{absorbed}$). This is done by re-adjusting the normalization factor (α) during the moment of the plasma evolution (ns to µs). The normalization of the plasma absorbed power follows the relation $P_{absorbed} = \alpha \iiint n_e P_{set}\, dV$. Here, $P_{set} \equiv -e\langle \tilde{v}_e.\tilde{E}\rangle_t$ is represented as the average set power applied to the cylindrical plasma cavity. This normalization helps in convergence of the solution and avoids any disproportionate absorption of the power by the plasma. Here, a fixed plasma absorbed power of 70 W is chosen to benchmark its results with the experimental findings that are reported taking the boundary conditions [23, 24] similar to the computation system and the operating conditions.

In the MW-plasma simulation model, the instant of MW launch is taken as reference (t = 0 s) when the MW is launched into the cavity. The MW is launched in right (R) hand mode (R mode is extraordinary type) using the four step ridge waveguide. This makes the $\tilde{E}$-field intensity to be maximum in the center of the cavity that is propagating in parallel to the externally applied magnetic field [25, 26]. As soon as the MW is launched, it continues to interact with the gas particles on their propagation timescales (ns). To understand and visualize the profile modification of the electrical field from the start of MW launch to the steady state plasma generation and also their impact on the power coupling to the plasma with time, total breakdown time is split into some discrete values such as 10 ns, ~67 ns, ~158 ns, 452 ns, 630 ns, ~2 μs, 2.5 μs, 3 μs, ~5 μs, ~8 μs, ~20 μs, ~40 μs, 250 μs and 300 μs. The simulation is exercised with an argon gas. The initial conditions for the calculation of the plasma parameter evolution are plasma density ($1 \times 10^{12}\ m^{-3}$) and plasma temperature (~4 eV). These parameters are used initially to estimate the plasma conductivity σ. Then the electric field is estimated by putting the above initial conditions in Eq. (39). The estimated field values are later utilized to find out the temporal variation of different plasma parameters and so electric field components' self-consistency.

The present FEM model uses different solvers sequentially to compute the magnetic field distribution first throughout the computational domain. Then the solvers related to the frequency-transient analysis are used to calculate the MW-plasma parameters. A complete computational flowchart for the magnetic field as well as the MW-plasma is depicted in **Figure 3**. The typical number of degrees of freedom that is used for solving the magnetic field is approximately 54,525. The FEM magnetostatic model uses the equation-based mesh adaptation technique to generate the extremely fine mesh size on the ECR surfaces. The mesh element has

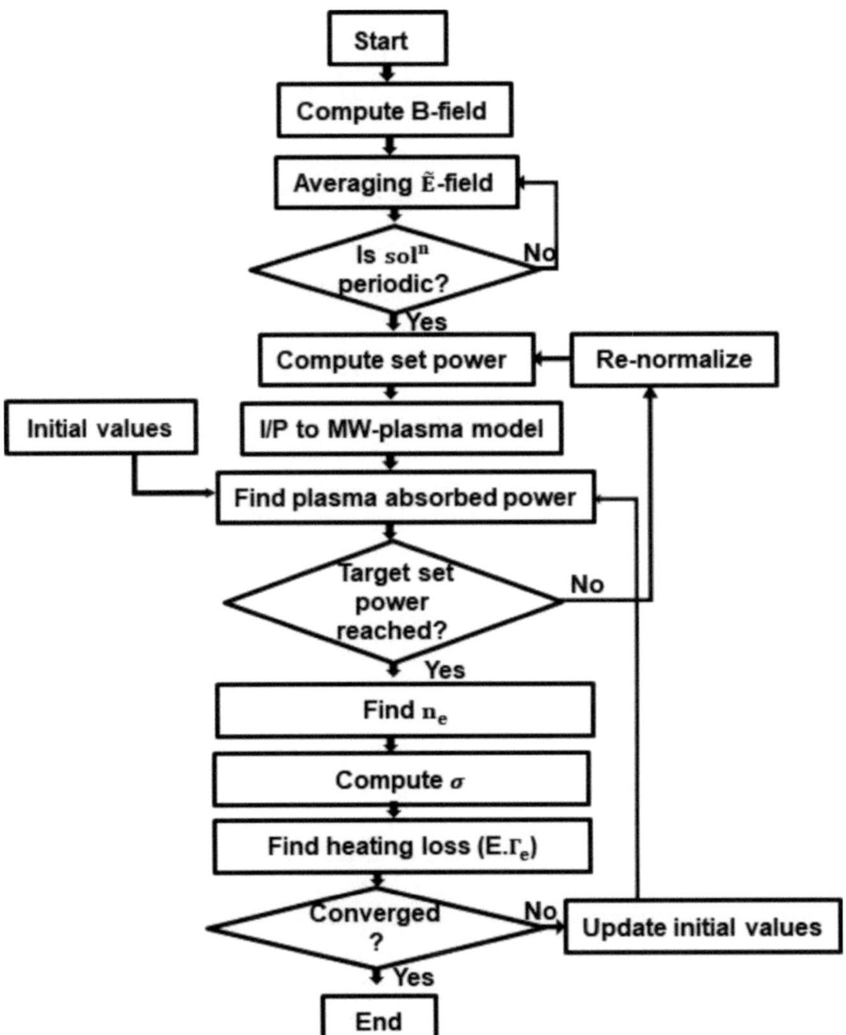

Figure 3.
Simulation flowchart of MW plasma interaction in COMSOL Multiphysics.

minimum and maximum size for the magnetic field and microwave-plasma calculations are 0.2 mm and 2.4 mm, respectively. The effect of the different edges that are involved in the computational domain is taken into account by keeping the maximum and minimum mesh element size at 0.5 mm and 0.0055 mm, respectively. The magnetic field estimated is used in the MW-plasma model to estimate the tensor plasma parameters. For the case of MW-plasma simulation, the number of degrees of freedom used for the solution is approximately 47,005.

Due to the dephasing as discussed above, the current FEM model uses the concept of effective collision frequency (ν_{eff}) to describe the sudden phase decoherence. In the dephasing situation, the phase relationship between velocity and the electric field oscillation is destroyed in the temporal scale due to which the electrons experience a large field variation in the ECR surface. The accelerating electric field transfers energy to the electrons that are residing on the ECR surface only for a small time duration in the time range of resonant cyclotron motion of the electrons. The electrons also experience spatial density variations while oscillating across the ECR surface. A large density variation across the ECR surface also generates the radial ambipolar electric field on the resonance surface and so there is a possibility of de-phasing that can happen at the resonance zone.

As the phase de-coherence may happen between electron gyro motion and the MW oscillatory˜E-field, the electrons get accelerated and deccelerated assymetrically. This temporal asymmetry of acceleration and decceleration in opposite cycle are responsible for transferring an effective energy to the electrons. The effective collision frequency ν_{eff} is estimated using the relation $\nu_{eff} = \sqrt{v_e\omega/\delta_B}$, where v_e is the electron thermal speed and δ_B is the gradient of magnetic field. The value of ν_{eff} is few orders more than that of actual collisional frequency that an electron encounters with the gaseous particles. However, it is to be noted that the effect of ν_{eff} is insignificant on the power absorption profile. Instead it helps in overcoming the numerical instability. For obtaining a steady-state solution, the number of MW periods has to be in the order of ω/ν_{eff}. This makes computation less rigorous than the case of actual collisions, if considered.

The evolution of the radial (E_r) and axial (E_z) components of the electric fields during the plasma evolution plays an important role in the MW coupling to the plasma and so the plasma parameters as discussed above. To support these facts, the

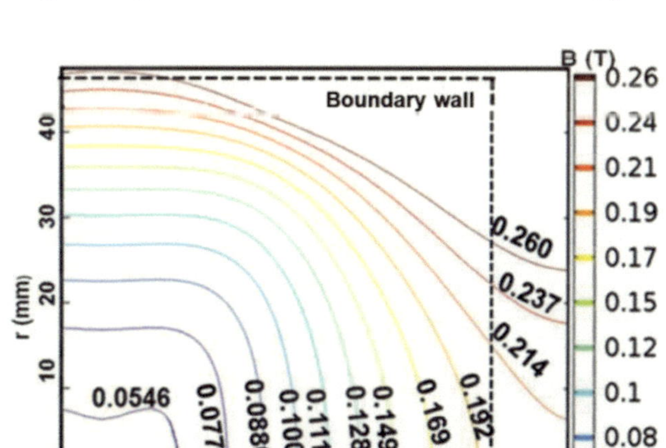

Figure 4.
Magnetic (B) field contour inside the quadrant section of the plasma chamber, which has cylindrical axis symmetry. The B-field is also simulated using COMSOL software. The narrow contour area near 0.088 T line is the ECR zone (corresponding to 0.0875 T). The span of ECR zone is around z = ±24 mm and r = ±28 mm.

evolution of the different components of the electric fields and correspondingly the plasma parameters is shown in the results and discussion section given below to study their effects on the plasma parameters. To understand the resonance zone in the plasma chamber, the required magnetic field contours are shown in **Figure 4**.

4. Results and discussion

This section of the chapter shows the temporal behavior of the electric fields, power deposition and the corresponding variation of plasma density and hot electron temperature from the start of MW launch to the steady state condition.

4.1 Behaviors of total electric field ($|E|$) during plasma evolution

Figure 5(a) and **(b)** shows the evolution of the radial profiles of the total electric field taken on the different axial locations for two time instances, 3 μs and 20 μs, after the MW launch. The radial profiles show the total electric field intensity to be more near the MW launching port (z = −40 mm). As one moves toward the extraction (z = −60 mm), its values is decreased because of the plasma shielding effect.

A sharp change in the E-field is shown in **Figure 5(a)** at the time, t = ~3 μs near the 2.45 GHz ECR surface (r ≈ 23 mm) for the planes, z = ±20 mm. The strong inhomogeneity in the E-field implies the absorption of the MW power at the same locations [27]. The power absorption location is also dependent on the magnetic field profile as well within the cylindrical cavity. Another figure, **Figure 5(b)** shows the radial E-field pattern across the different planes for time, t = 20 μs. The inhomogeneous part of the E-field profile looks similar to **Figure 5(a)** for the case of z = −20 mm plane. But the intensity of the E-field is being reduced with time due to the plasma shielding. One can observe that the inhomogeneous part of the E-field is shifted toward the off-ECR regime (r ≈ 28 mm) from the ECR surface with the increase of time. Therefore, the power absorption region is being shifted from the ECR zone to the off-ECR zone. This effect is visible in the power deposition location throughout the cylindrical cavity in **Figure 6**. **Figure 5(a–d)** shows the corresponding shifting of the power deposition location from the ECR zone to the off-ECR zone or UHR zone with time. As the plasma reaches steady state during the time, t = 20 μs, the evolution of the plasma density and temperature is shown in **Figure 7**.

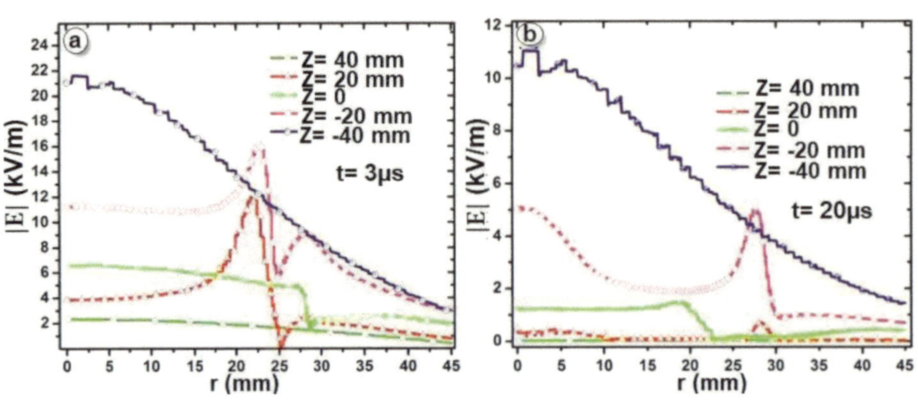

Figure 5.
(a) Total electric field (|E|) magnitude during gas ignition time of ~3 μs, (b) total electric field (|E|) magnitude during gas ignition time of ~20 μs.

4.2 Time evolution of plasma with power deposition

Figure 6a shows that the MW power is being deposited exactly on the ECR (~0.0875 T) surface corresponding to the launch MW frequency of 2.45 GHz when the plasma density is low. But as time passes (see **Figure 6b–d**), the power deposition location gets shifted in the off-ECR or upper hybrid resonance (UHR) regime. The UHR zone is a region where the two conditions $n_e < n_{crit}$ and $B < B_{ECR}$ are satisfied [26–28]. The term n_{crit} represents the critical density for the MW frequency, 2.45 GHz that is 7.4×10^{16} m^{-3}. If one can compare **Figures 6(c)** and **7(a)**, one can visualize that the plasma density (~1.3×10^{17} m^{-3}) crosses the critical density from ~2.5 μs onwards and the plasma density that is above the critical density is denoted as overdense plasma. So as the overdense plasma is achieved, the electrons get accelerated by less amount of MW energy on the ECR surface. Correspondingly, the plasma bulk temperature increases from the start of MW launch (t = 0 s) to the instant of ~630 ns. It is evident from **Figure 7a** that the plasma bulk electron temperature increases and becomes steady near a value of ~80 eV. Then the plasma bulk temperature decreases in a faster way with further increase of time. Hence, one can conclude that heating through ECR process is being ceased to occur with the further increase in time. The causes to increase the absorbed power density is expected to be highest (~5×10^7 W/m^3) on the ECR surface during the time, t = 630 ns.

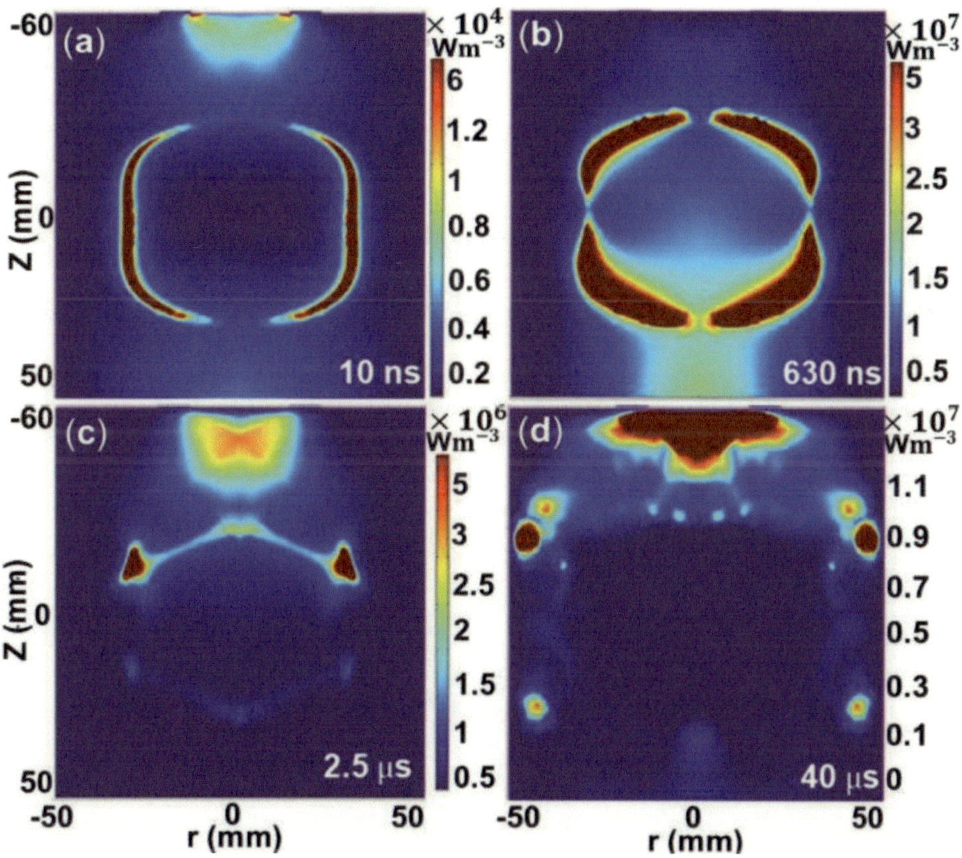

Figure 6.
Power deposition density at different time steps for 70 W of absorbed power. (a) t = 10 ns, peak power density is 6.6×10^4 W/m^3, (b) t = 630 ns, peak power density is 1.43×10^7 W/m^3, (c) t = 2.5 μs, peak power density is 5.12×10^6 W/m^3 and (d) t = ~40 μs, peak power density is 1.67×10^7 W/m^3.

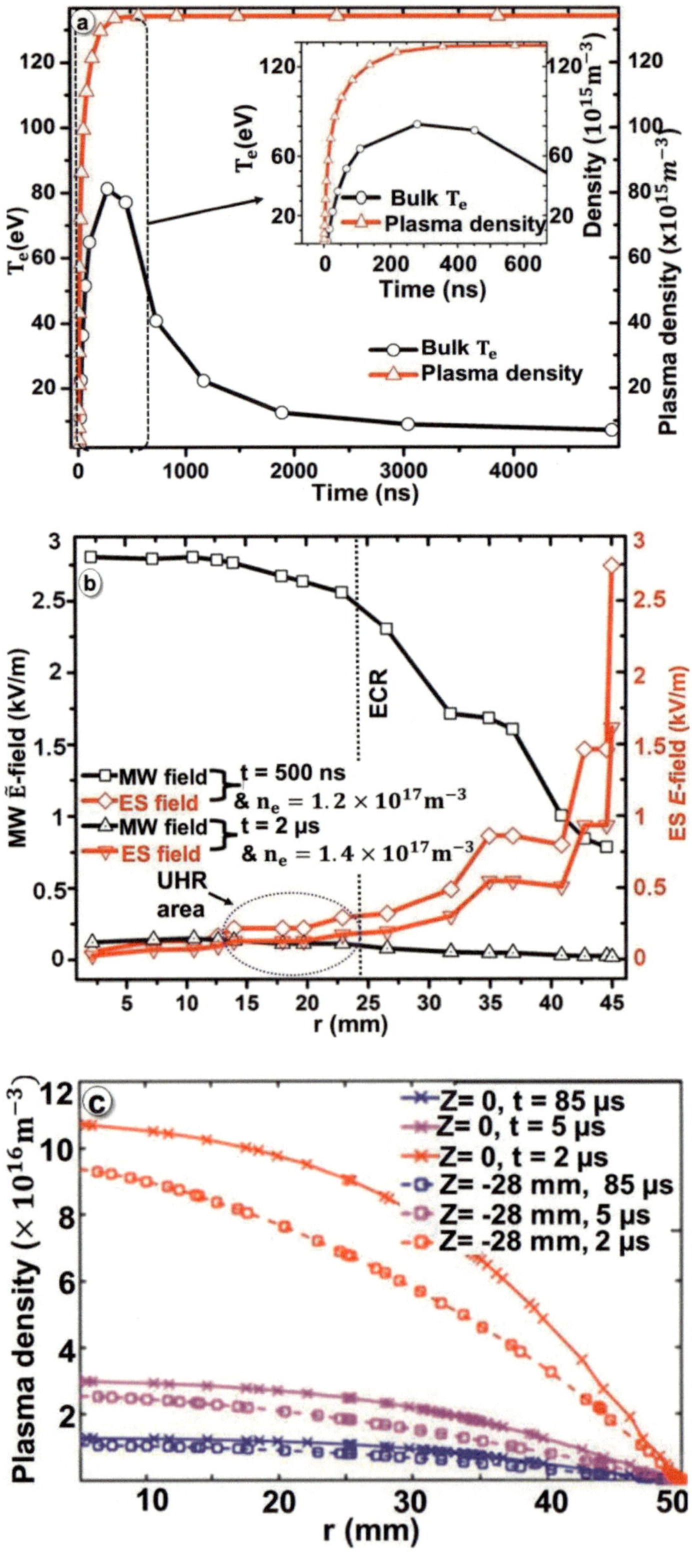

Figure 7.
(a) Temporal evolution of electron density and temperature during plasma formation time at point (r = 0, z = −28 mm) with gas pressure, 2 × 10^{-3} mbar. (b) Temporal variation of MW electric field and plasma generated ambipolar electric field, the two main constituents of total electric field. (c) Radial profile of plasma density at three different time instants (t = ∼2 μs, ∼5 μs and 85 μs), at two axial locations during evolution of the plasma.

Correspondingly, the plasma density also reaches above the critical density. The decrease in the plasma bulk temperature and the corresponding increase of the plasma density even above the critical density are attributed to be occurring from the off-ECR or ES surface wave heating mechanisms [29, 30]. In other words, although the plasma density is approaching steady state (**Figure 7a**), the temperature is not yet stabilized during that time.

To confirm the off-resonance or electrostatic heating methods as discussed above, the evolution of the MW electric field and the electrostatic electric field is shown simultaneously in **Figure 7(b)**. The radial distribution of the two types of the electric field is given in that figure for two discrete time instants and the corresponding plasma densities. One can observe that the MW electric field is significant throughout the cavity and is more than the electrostatic electric field for the time, t = 500 ns. For another instance, t = 2 μs, the electrostatic electric field becomes more than the MW electric field throughout the cavity. The electrostatic field is even more in the upper hybrid resonance locations than the MW field as shown in **Figure 7(b)**. This evidence confirms that the electrostatic heating is being taking place at the UHR region where the magnetic field and plasma density satisfy the above-mentioned conditions [28]. To visualize the plasma density pattern due to these electric field behaviors, the radial distribution of the density is shown in **Figure 7(c)** for different time instances (i.e., 2, 5 and 85 μs) and axial planes on the cylindrical cavity during the plasma evolution. The plane z = −28 mm on the cavity is situated near the MW launching port. **Figure 7(c)** shows the plasma density to be more at the central location (z = 0 mm) than the location z = −28 mm that is located toward the MW launch side.

From the above-mentioned results (**Figures 6** and 7), it can be commented that the power is absorbed by the ECR mechanism especially in the plasma condition where the density is below (underdense plasma) the critical density and slightly above the critical density (overdense plasma). If one notices the plasma parameters for the underdense conditions, one can observe that the density remains below the critical density from the time, t = t = 45 ns to t = 110 ns. In this case, the plasma electrons are magnetized and hence are following the magnetic field lines. The field free zones that are located near (r, z) = (0, 0) are being filled by the plasma particles because of the diffusion processes. Due to the ECR heating, the electron temperature is being increased in the field free zones in the underdense plasma situation, t < 110 ns. As the magnetic field lines are stronger (B ~ 2300–2600 G) near the radial locations of the cavity, i.e., in the gaps of the two pairs of ring magnets, the plasma bulk electron temperature exhibits a sharp gradient in those regions. The maximum plasma bulk electron temperature achievable is ~85 eV that occurs during the time, t = 280 ns. The high energy part of the plasma bulk electrons is being concentrated completely in the same gap as mentioned before during this time.

It is observed that the plasma bulk electron temperature increases in the radial direction at the regions mainly in between the two pairs of the ring magnets with the increase in time from t = 280 ns to 730 ns [22]. Therefore, it can be summarized that with the increase in plasma density (or time, t = ~45 ns to t = ~280 ns) from underdense to overdense state, the plasma bulk temperature is increased by an amount of ~80 eV mainly in the radial direction near the region, 24 mm < r < 40 mm, −25 mm < z < 25 mm. This is because the ECR surfaces lie in those regions. The continuous heating through ECR in this location causes the high energy part of the plasma electron temperature to be concentrated on the same location even in the overdense plasma state during the time t = 730 ns.

The anisotropic behavior of the plasma bulk electron temperature even in the overdense plasma signifies the ECR heating [31, 32]. With further increase of time after the MW launch, i.e., near t = 2000 ns, the plasma bulk temperature (T_e) is decreased to ~10–12 eV. This range lies close to the thermal electron temperature. Therefore, the reduction of the high energy component of the plasma bulk temperature can be argued as the minimal occurrence of ECR heating process and correspondingly the initiation of the electrostatic heating mechanisms in the same time frame [29–31, 33]. The plasma reaches steady state after 2000 ns (2 μs) (**Figure 7a**). At the saturation, the plasma density is approximately 2 times more (~1.3×10^{17} m^{-3}) than the critical density. During the same instant of time, the plasma electron temperature and its gradient reduce in the location where the magnetic field is relatively lower.

In order to investigate the different coupling mechanisms involved during the plasma evolution process and their impacts on the plasma parameters, the behavioral pattern of the different components of the electric field (E_r and E_z) is shown in **Figures 8–10**. The power coupling mechanisms are ECR, UHR and polarity reversal related to ES heating. The particular type of coupling mechanism is understood from the electric field evolution pattern throughout the cavity. The electric field characters can be useful to characterize the plasma parameters based on the theoretical and experimental proofs. The proofs say the higher the plasma density, the more the power absorbed by the plasma.

4.3 E_r-field evolution with time

In **Figure 8(a–c)**, the radial electric field (E_r) variations are shown for three different z-planes of the cylindrical cavity. The figures also show the evolution of the E_r-field profile at different time instances after the MW launching. **Figure 8(a)** gives the E_r profile close to the ECR surface (z = −28 mm), which is also near to the MW launch point (z = −60 mm). Another figure (**Figure 8**(b)) shows the E_r-field profile along the central plane (z = 0) of the cavity, whereas **Figure 8(c)** gives the E_r-field evolution along the plane z = 28 mm that is close to the ion beam extraction point.

In all the figures, a strong inhomogeneity in the E_r-field profile signifies the power absorption locations. Taking a time instant t = 67 ns in all the figures, the presence of the E_r-field throughout the cavity can be visualized since the plasma density at this time has not reached above the critical density. As the plasma density crosses the critical density with advancement of time (>500 ns), the E_r-field profile is being modified following the density evolution pattern. The inhomogeneity in the E_r-field profile is more pronounced in **Figure 8**(b) than in the case of other figures. The difference in the radial variation of the E_r-field comes from the variation of resonance magnetic field contour as shown in **Figure 3**. The strong inhomogeneous part of E_r-field initially very near the ECR surface indicates that the electrons are accelerated by the ECR phenomenon. Then after >2 μs, this inhomogeneous part gets shifted toward the UHR $\left(f_{UHR} = \sqrt{\left({f_{pe}}^2 + {f_{ce}}^2\right)}\right)$ regions where the MW frequency matches the f^{UHR}. The E_r-field value after ~2 μs at r = 0 mm and z = −28 mm location is negligible and so the plasma density reached overdense state, whereas for some regions (UHR regions), as shown in **Figure 8(a)**, the E_r-field component is showing significant intensity. The high value of E_r-field, near the location ($B < B_{ECR}$), represents that the plasma is still underdense in this region during time t = ~3 μs.

Therefore, the dual conditions ($B < B_{ECR}$ and $n_e < n_{crit}$) are satisfied for the UHR heating process. This phenomenon is also being reflected in **Figure 7(a)** and **(b)** as discussed earlier. **Figure 7(b)** shows the electrostatic electric field to grow with the plasma density and correspondingly the MW electric field decreases after

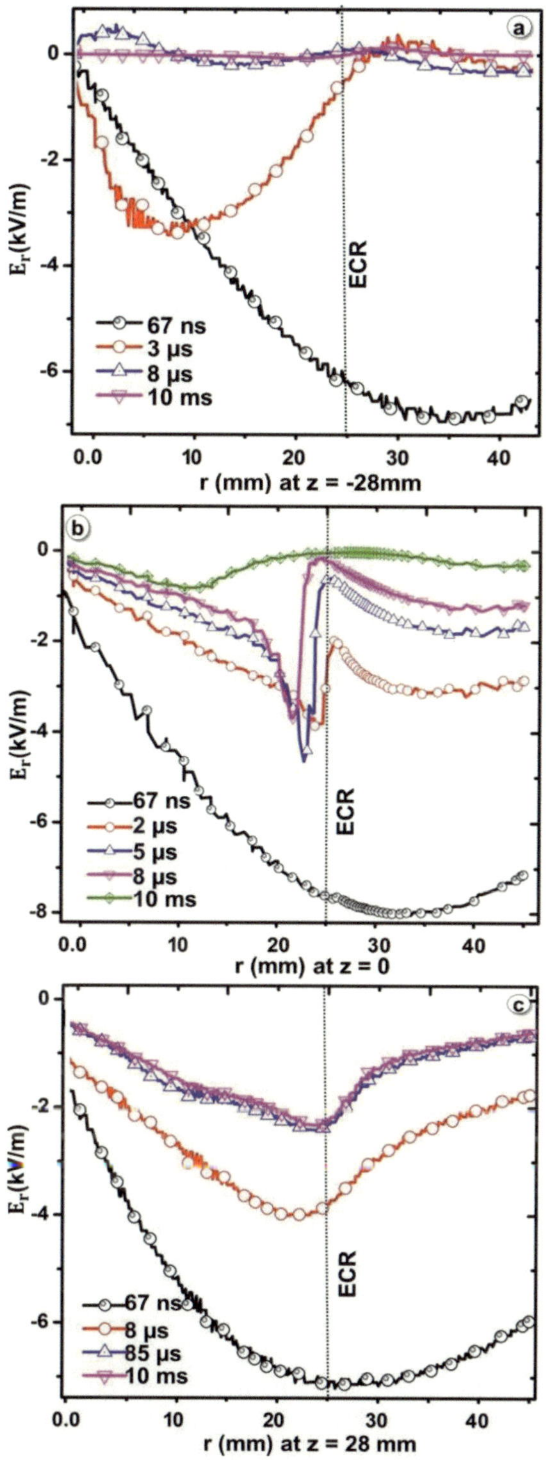

Figure 8.
E_r-field evolution and its pattern during plasma generation (a) r = −28 mm, (b) r = 0 mm and (c) r = 28 mm.

500 ns. The electrostatic electric field, the component of the total electric field, is getting increased due to the increase of the plasma ambipolar potential. So this gives a clear evidence of power transferring from MW coupling to the electrostatic wave coupling to the plasma. Since the electrostatic wave does not face any density

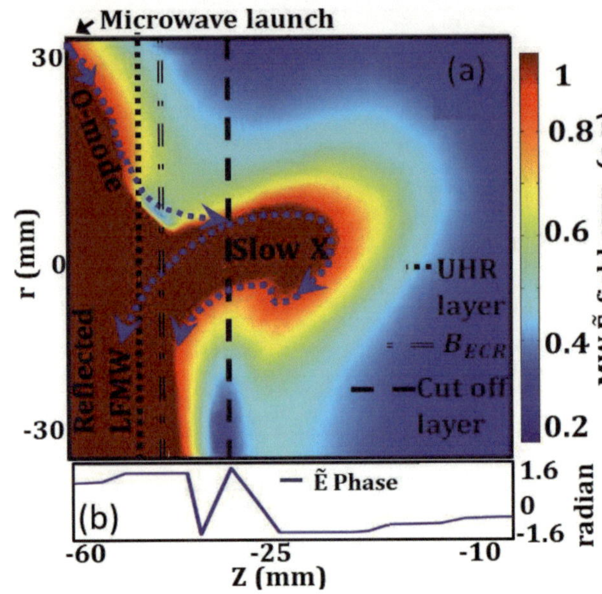

Figure 9.
(a) Propagation and mode conversion of MW $\tilde{E}$-field corresponding to an ordinary (O)-type resonant mode are shown and (b) Phase angle ($\approx \tan^{-1}(Im\ (\tilde{E})/Re\ (\tilde{E}))$) of MW $\tilde{E}$-field.

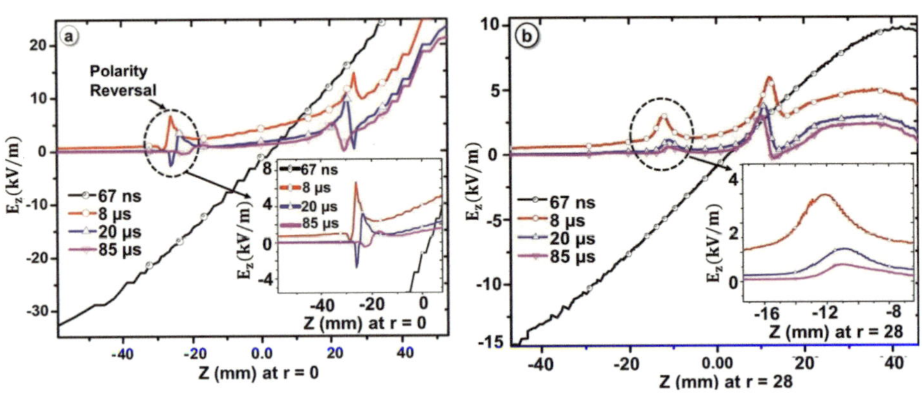

Figure 10.
Spatio-temporal evolution of E_z-field profile obtained along the axis of the cylindrical cavity. A sharp change in the two places for both the figures implies the power absorption locations that can be checked with ***Figure** 5* *described above. (a) r = 0 mm, (b) r = 28 mm.*

barrier, it is capable of penetrating the core dense plasma and transferring the energy through damping to the plasma particles. As demonstrated by other groups [23, 28], this change in the power coupling mode from the electromagnetic to electrostatic case is a signature of the UHR heating. Due to the heating at UHR, the plasma electron temperature and also the density are possible to enhance further.

It can be observed in **Figure 8** that the magnitude of the E_r-field at time t = 67 ns (**Figure 8**) is greater than 7 kV/m along different z-planes within the cavity near the ECR locations. This intensity is sufficient for creating ionization in the Argon gaseous particles. In the later instant (~2 μs) of time, the magnitude of E_r-field changes in between 3.75 kV/m and 0.15 kV/m on the ECR regions. This type of nonlinearity in the E_r-field has also been demonstrated by Hopwood et al. [4] before. The power coupling phenomena involved during the plasma evolution can also be transmitted into the extracted ion beam performance in terms of spike or instabilities. It was experimentally demonstrated by Ropponen et al. [1] that a sharp transient in the ion current density during the preglow mode can cause a sharp fall in the high energy component of the plasma electron temperature.

Since the inhomogeneous part of E_r-field is shifting toward the off-resonance locations as shown in **Figure 8(b)**, it can also produce high-energy electrons there if certain conditions are satisfied as per Gammino et al. [34]. The conditions say the MW electric field has to be above a certain transition value for being capable to energize the electrons that also depend on the magnetic field gradient present in the same location.

The generated high-energy electrons can interact with the slow extraordinary-type microwave and produce the cyclotron range instability in the plasma [1]. It was also proved experimentally by Mansfeld et al. [35] that the extracted ion beam current from the microwave ion source can gain oscillation due to the presence of cyclotron-type instability of plasma during the afterglow operation mode. The slow extraordinary mode microwave is produced from the mode conversion of ordinary-type microwave near the UHR region. As the mode conversion layer is present in the present plasma cavity, the ordinary mode microwave crosses the evanescent layer and some part of its energy is converted into the slow X mode. Since the plasma is confined in the cavity under mirror magnetic field configuration, the injected MW will have two components, extraordinary mode and ordinary mode. **Figure 9(a)** shows the MW ordinary mode is propagating toward the overdense plasma region from the underdense launching point. At some point, it will encounter a cut-off corresponding to the ordinary-type MW. At the cut-off, some part of the ordinary mode MW energy is evanescently transformed into a slow extraordinary mode following the CMA diagram concepts. For that reason, a bend in the MW propagation in the slow extraordinary (X) mode is seen in the electric field simulation (**Figure 9b**). This slow X mode then propagates toward the UHR region and hence the electric field is being accumulated there, as shown in **Figure 9(a)**. The accumulation of the electric field at this location increases its intensity at some plasma condition and can cross the corresponding parametric decay threshold condition. The parametric decay of the slow X mode near the UHR region can generate ion- and electron-type electrostatic waves as per the literature [36].

As a supportive evidence of the generation of electrostatic ion wave, **Figure 8(b)** shows that the inhomogeneous part of the E_r-field is shifting toward the off-resonance region with a velocity of 1250–1500 m/s. This range of velocities falls in the range of ion acoustic speed. Based on the published reports [37], the electric field propagating perpendicularly with respect to the external magnetic field as in the present case (**Figure 9a**) encounters the UHR layer in the overdense plasma state. In the present computation, the ordinary mode (O mode) electric field after converting into the slow extraordinary mode (X mode) shows a bending in the perpendicular direction and reaches the UHR region. This is known as 'O – slow X' conversion process that is responsible for generating ion waves and makes the inhomogeneous electric field to shift at the same ion acoustic speed [38].

4.4 E_z-field evolution in plasma with time

The spatio-temporal evolution of the axial component (E_z) of the electric field is shown in **Figure 10** throughout the cylindrical cavity along the planes r = 0 mm and 28 mm, respectively. It is seen that only a portion of single wavelength of the wave electric field is present that has significant intensity throughout the plasma cavity after the immediate instant (67 ns) of MW launching. The magnitude of the E_z field becomes almost zero throughout the cavity as soon as the plasma starts creating after the MW launching. Therefore, the E_z field does not play any significant role in accelerating the electrons, and if this would happen, the electrons losses will occur in the axial direction. **Figure 10(a–b)** evidences that the E_z field becomes almost zero everywhere except near the ECR surface after ~8 μs. The E_z-field magnitude shows minimum values where the plasma density is of maximum values.

It can be observed in **Figure 10(a)** that the polarity in the inhomogeneous part of the electric field is getting opposite for two different time instances, 20 and 85 μs. The reversal in the polarity of electric field occurs near the ECR surface. The polarity reversal is caused by the ambipolar field produced from the plasma density gradient. The plasma density gradient is computed from the electron momentum equation using drift-diffusion approach in the present FEM model [38]. The shifting of the inhomogeneous part of the electric field is in the inward direction. The speed of displacement of the inhomogeneous part of the electric field is estimated in the range of $\sim 10^3$ m/s that lies in the range of ion sound speed. Similar shifting at the same velocity is also observed before in **Figure 8**(b) corresponding to the E_r-plots. These observations indicate that the plasma density gradient near the ECR surface is accompanied by the generation of ion acoustic waves that are electrostatic in nature [39, 40]. Hence, the electric field polarity reversal associated with electrostatic ion wave heating is being initiated during this period of the plasma evolution after the microwave launch.

5. Validation with experiment

Experiments are performed to cross-check the above-mentioned plasma parameters obtained during the gas ignition moment. The present section of the chapter provides the details of the experimental methods, analysis of the experimental results and also a comparative study of the experimental data with the simulation.

5.1 Methods of experiment

Experiments are carried out in a microwave ion source system that has similar system configuration, magnetic field distribution, MW conditions and also the operating conditions. The simulated temporal plasma parameters, such as the plasma density and hot electron temperature, are validated with the experiment [23]. In the present experiment, MW-plasma reactor of the experimental set-up is a cylindrical cavity (**Figure 11**) of 107-mm length and 88-mm diameter. The plasma in the reactor is generated by coupling microwave through the electron cyclotron resonance (ECR) heating as well as off-ECR heating methods, as discussed before. The complete experimental set-up consists of a cylindrical cavity, microwave system, ion beam extraction system and two pairs of ring magnets (each magnet has pole strength ~1.38 T) assembly [23]. The plasma cavity/reactor is surrounded by two pairs of ring magnets to generate a mirror-type magnetic field to confine the plasma inside the cavity.

To generate plasma, the MW is produced by a magnetron (power: 0–2 kW, make: Richardson Electronics, Model no. NL10250-7), which is operated either in continuous or in pulsed mode. MW power is fed to the reactor through a combination of a four-step ridged waveguide, a HV break and vacuum window assembly, an impedance tuner unit, directional coupler and an isolator with water dummy load (**Figure 11**). The plasma impedance is matched by a 3-stub tuner to get maximum $\tilde{E}$-field at the center of the MW-plasma reactor. MW is coupled to the cylindrical plasma reactor by a four-step ridged waveguide (WR 284) [23]. The

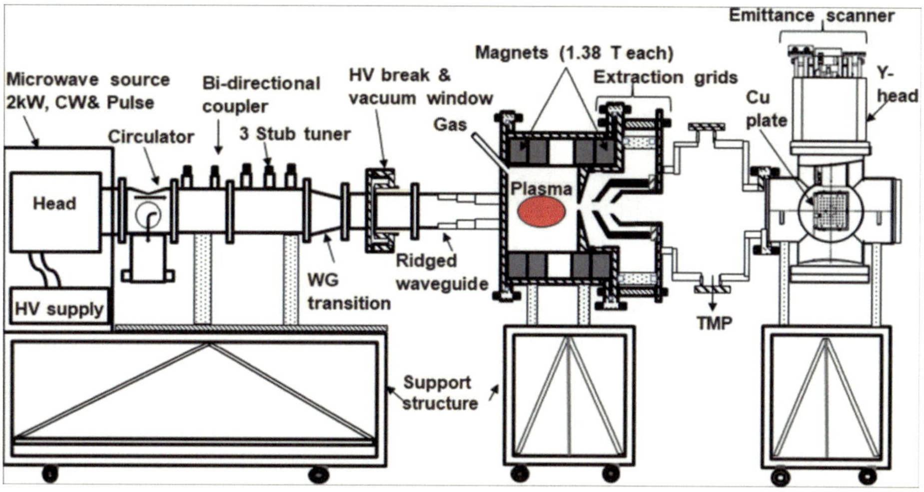

Figure 11.
Schematic view of the experimental system.

impedance of WR284 waveguide (four-step ridge waveguide) having 220-mm length is made comparable to the typical plasma impedance by embedding four ridge sections on both the inner side walls of it. The ridge sections of different lengths but same width (48 mm) are placed consecutively at the broader walls of the waveguide. As shown in **Figure 11**, the ridge waveguide is mechanically connected to one side of a high voltage (HV) break cum vacuum window, made of Teflon having a dimension of 35 mm diameter and 6 mm thickness, whose other side is connected to the ion source cavity (**Figure 11**). The other side of the cavity is connected to the conventional pierce geometry-based 3-grid extraction system, housed in a drift duct vacuum chamber that consists of a plasma grid, an extraction grid and a grounded grid (**Figure 11**). The ion source is evacuated by a combination of Turbo Molecular Pump, which is backed by a dry-scroll rotary pump, connected to the drift duct vacuum chamber in the downstream side of the extraction system. The gas feed system comprises a needle valve, mass flow controller and the required gas cylinder. In the present experiment, nitrogen gas is used because of availability. Other gases, like argon or hydrogen, can also be used. The experiment is carried out for the MW power in the range of 50–700 W and gas pressure in the range of 1×10^{-3} mbar to 1×10^{-4} mbar. The Langmuir probe (LP) diagnostics is used to characterize the plasma parameters within the pressure range varying from 2×10^{-4} to 1×10^{-3} mbar [23].

All the MW power ranges mentioned throughout the chapter are considered as set power. The difference of forward and reflected power is considered to be plasma-absorbed power. MW reflection varies from 5 to 10% within the above set power range. These ranges of plasma reflection with similar experimental set-ups and operating environments are reported in [10]. The accuracy at low set power levels of magnetron is tested by repeated measurements of its output power (forward) before the experiment is performed. An extra component, named isolator with water-cooled dummy load (make: National Electronics, Model: 2722-162-10311, isolation: 26 dB, reflection rating: 6.5 kW) is placed in the experimental set-up (not shown in **Figure 11**) before the HV break. The power and carrier frequency signals are measured by microwave spectrum analyzer (model: FSH8, make: ROHDE & SCHWARZ, band: 100 Hz–8 GHz) at the directional coupler port. Signals are attenuated by 60 dB ($\equiv 10^6$, power ratio) at the directional coupler ports before coming to microwave spectrum analyzer through the high frequency (0–40 GHz), low loss and low VSWR cable of length ~1 m.

To generate plasma in the low power range, magnetron's low power testing is required. In the present experiment, low power testing ensures the variation of full width at half maximum of 2.45 GHz frequency is within 5–12 MHz (within the specified ± 25 MHz bandwidth as per Sairem data). Also, the set power fluctuations are within 2–5% as shown in **Figure 12(a)**, which is considered to be stable operating conditions. After performing the inverse Fourier transform of the MW spectrum analyzer data, a graph of detected MW power vs. time is shown in **Figure 12 (b)** to verify the power levels at the first few milliseconds for the comparison of plasma parameters with the simulated data. The rise time of the pulse is obtained as ~2.2 μs.

The magnetron's output (set power) in the low power range is checked in time scale prior to the Langmuir probe diagnostic [23]. The response of the magnetron set power at 2.45 GHz frequency is recorded at the directional coupler port by a microwave spectrum analyzer (SA) circuitry. The circuitry consists of a high frequency cable, a band pass filter, and spectrum analyzer and FSH4 View software.

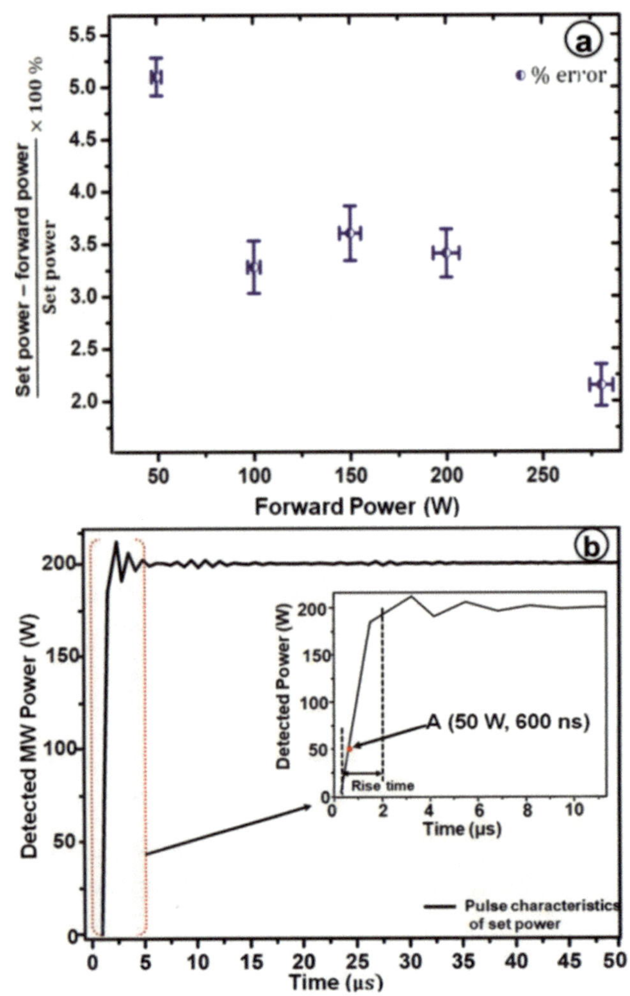

Figure 12.
(a) Percentage error of the magnetron set power fluctuations. Data for 280 W have been benchmarked with the Sairem company data and (b) variation of the detected MW power vs. time for the set power of 200 W.

The magnetron's pulse response at a fixed set power, 200 W, is obtained by taking the inverse Fourier transform of the microwave spectrum analyzer data. The magnetron's rise time is ~2.2 μs. This exercise of measuring the pulse response of a particular set power can help to pick up the temporal values at different set power levels following the pulse response of magnetron [23].

To compare the simulated hot electron temperature and density, their parameters are noted down at different instant of time during the plasma evolution. Taking the time instant to be same as that of the simulation, the set power values are noted down in an experimental datasheet. Then, a single Langmuir probe measures the plasma floating potential at the noted set power values as mentioned above [23]. The Langmuir probe measurement is performed in steady state plasma condition. Since the real-time measurement of the plasma parameter requires in the ns-time scale very sophisticated and expensive hardware that has faster time responses (ns range), steady state Langmuir probe measurements are performed to avoid those expensive diagnostics.

5.2 Comparison of experimental results with simulation

The line plots are shown in **Figure 13(a)** for the simulated hot T_e. The simulated data point lies in a region that is very close to the experimental data points. The simulated hot T_e is approximately 78 eV at 70 W of plasma absorbed power during time t = 600 ns. One can also observe that the set power for the magnetron reaches 50 W at the same time instant. The measurement of the hot T_e by a Langmuir probe diagnostic at the same set power level shows it to be ~ 72 eV. Similarly, the hot T_e was measured for two other power levels, 70 and 130 W. The hot T_e at these two power levels are ~ 36 and 28 eV, respectively. Hence, the experimental results fit well with the simulated values described above [23]. On the other hand, the experimental data for the plasma density are also shown in **Figure 13(b)**. The plasma densities for the above set power levels are observed to be ~1.8–2 times less than the simulated data. The deviation of the measured plasma density from the simulated values is caused due to the difference in the absorbed power that is absorbed by the plasma in both the cases (experiment and simulation). The experimental results for the plasma density are shown in **Figure 13(b)** for three different set powers. The measured plasma density ($\sim 1.1 \times 10^{17} m^{-3}$) at set power of 200 W agrees well with the simulated plasma density ($\sim 1.3 \times 10^{17} m^{-3}$).

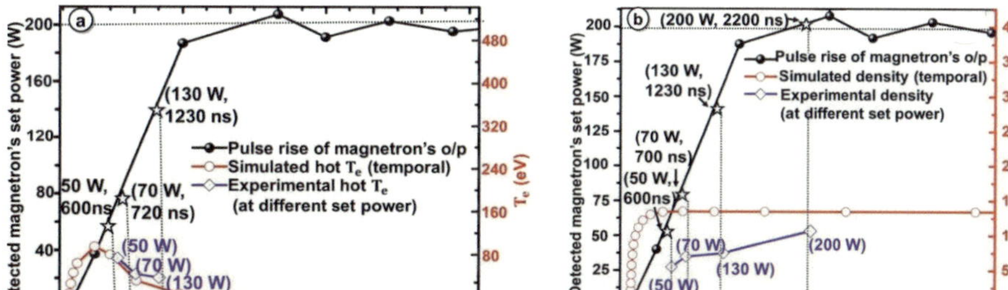

Figure 13.
(a) Simulated temporal hot T_e variation along with experimental data, taken at discrete set power values. Plasma-absorbed power during simulated hot T_e evolution is fixed at 70 W. Experimental set powers 50, 70 and 130 W correspond to plasma-absorbed powers at 40, 50 and 70 W, respectively [14]. Pulse rise of set power is also shown to verify the power levels during the ns to few μs periods. (b) Simulated temporal plasma density variation along with experimental results, obtained at the same set power values.

6. Conclusion

The evolution of the spatial and temporal plasma parameters are presented in the current chapter during the gas ignition process made by injecting the 2.45 GHz microwave to an ion source cavity. The simulated results supported by the experiment confirm that the plasma parameters are influenced significantly by the electric fields during the plasma ignition period. Due to the shifting of the microwave coupling mechanisms to the plasma, the plasma density and the hot fraction of the electron temperature are also getting affected during the plasma generation period. The initial rise of the hot fraction of the plasma electron temperature from the start of microwave interaction into the plasma is argued to be caused by the electron cyclotron resonance heating phenomenon. After certain instant of microwave launch, the slight increase in the plasma density and the decrease in the plasma electron temperature are proved to be happening from the electrostatic heating mechanism. The electrostatic heating near the upper hybrid resonance region causes to shift the inhomogeneous part of the electric field at the velocity of the ion acoustic speed when the plasma density reaches above the critical density corresponding to the launch microwave, 2.45 GHz.

The experimental plasma parameters are obtained in an experimental set-up that has similar system configuration and the operating environment as that of the simulation. The comparison shows a reasonable agreement with the simulated results. The plasma density especially in the overdense plasma condition is found to be agreeing more in the overdense plasma than in the underdense plasma condition. During plasma evolution after the microwave launch, the microwave coupling mechanisms are modified following the corresponding electric field (electrostatic and electromagnetic) distribution pattern throughout the ion source cavity. Initially, the electron cyclotron resonance heating comes into play to ionize the gaseous particle and generate plasma that contains the maximum fraction of the high-energy electrons in the ion source plasma. Then, as the density reaches in the overdense condition, the coupling mechanisms are the electrostatic wave heating in the ion acoustic frequency range. As the electrostatic wave does not suffer any density cut-off, the density is increased further above the critical density. Simultaneously, the polarity of the axial electric field is reversed near the electron cyclotron resonance region signifying the creation of the plasma density gradient due to the generation of strong ambipolar electric field near the resonance region. In future, it is intended to study the different power coupling mechanisms in the overdense plasma state that are caused due to the heating at the cyclotron harmonics by the generated electrostatic waves in the present experimental device.

Author details

Chinmoy Mallick[1,2*], Mainak Bandyopadhyay[1,2] and Rajesh Kumar[1,2]

1 Institute for Plasma Research (IPR), Gandhinagar, Gujarat, India

2 Homi Bhabha National Institute (HBNI), Mumbai, Maharashtra, India

*Address all correspondence to: chinmoyju1990@gmail.com

IntechOpen

© 2020 The Author(s). Licensee IntechOpen. This chapter is distributed under the terms of the Creative Commons Attribution License (http://creativecommons.org/licenses/by/3.0), which permits unrestricted use, distribution, and reproduction in any medium, provided the original work is properly cited.

References

[1] Ropponen T, Tarvainen O, Izotov I, Noland J, Toivanen V, Machicoan G, et al. Studies of plasma breakdown and electron heating on a 14 GHz ECR ion source through measurement of plasma bremsstrahlung. Plasma Sources Science and Technology. 2011;**20**:055007. DOI: 10.1088/0963-0252/20/5/055007

[2] Cortázar OD, Megía-Macías A, Vizcaíno-de-Julián A. Experimental study of breakdown time in a pulsed 2.45-GHz ECR hydrogen plasma reactor. IEEE Transactions on Plasma Science. 2012;**40**:3409. DOI: 10.1063/1.4819875

[3] Cortazar OD, Megía-Macías A, Tarvainen O, Vizcaíno-de-Julián A, Koivisto H. Plasma distributions observed in a 2.45 GHz hydrogen discharge. Plasma Sources Science and Technology. 2014;**23**:065028. DOI: 10.1088/0963-0252/23/6/065028

[4] Hopwood J, Wagner R, Reinhard DK, Asmussen J. Electric fields in a microwave-cavity electron-cyclotron-resonant plasma source. Journal of Vacuum Science and Technology A. 1990;**8**:2904. DOI: 10.1116/1.576647

[5] Stenzel RL. A new probe for measuring small electric fields in plasmas. The Review of Scientific Instruments. 1991;**62**:130. DOI: 10.1063/1.1142514

[6] Celona L, Gammino S, Ciavola G, Maimone F, Mascali D. Microwave to plasma coupling in electron cyclotron resonance and microwave ion sources. Review of Scientific Instruments. 2010; **81**:02A333. DOI: 10.1063/1.3265366

[7] Consoli F, Celona L, Ciavola G, Gammino S, Maimone F, Barbarino S, et al. Influence of microwave driver coupling design on plasma density at Test-bench for Ion sources Plasma Studies, a 2.45 GHz electron cyclotron resonance plasma reactor. Review of Scientific Instruments. 2008;**79**:02A308. DOI: 10.1063/1.2805665

[8] Takahashi K, Kaneko T, Hatakeyama R. Polarization-reversal-induced damping of left-hand polarized wave on electron cyclotron resonance. Physical Review Letters. 2005;**94**: 215001. DOI: 10.1103/PhysRevLett.94.215001

[9] Becker MM, Kählert H, Sun A, Bonitz M, Loffhagen D. Advanced fluid modeling and PIC/MCC simulations of low-pressure ccrf discharges. Plasma Sources Science and Technology. 2017; **26**:044001. DOI: 10.1088/1361-6595/aa5cce

[10] Hagelaar GJM, Makasheva K, Garrigues L, Boeuf JP. Modelling of a dipolar microwave plasma sustained by electron cyclotron resonance. Journal of Physics D: Applied Physics. 2009;**42**: 194019. DOI: 10.1088/0022-3727/42/19/194019

[11] COMSOL Multiphysics Reference Manual, Version 5.2. COMSOL, Inc. Available from: www.comsol.com

[12] Eric J. Vanzura et al, The NIST 60 Millimeter Diameters Cylindrical Cavity Resonator: Performance Evaluation for Permittivity Measurements', NIST Technical Note 1354. 1993

[13] Celona L et al. Observations of resonant modes formation in microwave generated magnetized plasmas. European Physical Journal D: Atomic, Molecular, Optical and Plasma Physics. 2011;**61**:107-115

[14] McDonald KT. An Electrostatic Wave. arXiv:physics/0312025

[15] Chen FF. Introduction to the Plasma Physics and Controlled Fusion: Plasma

Physics. UK: London Press, London; 1986. Addison-Wesley; 1994

[16] Muller A et al. On the charge state distribution of multiply charged ions extracted from electron beam ion sources. Nuclear Instruments and Methods. 1977;**140**:181

[17] Stix TA. Waves in Plasmas. New York: Springer-Verlag; 1992

[18] Mascali D. A new approach to the study of the ECR heating and particle dynamics in the plasma of electron cyclotron ion sources [PhD thesis]. UniversitA degli studi di Catania; 2008

[19] Castro G et al. Study of innovative plasma heating methods and applications to high current ion sources [PhD thesis]. Universita degli Studi di Catania;

[20] Castro G et al. Experimental investigation of non-linear wave to plasma interaction in a quasi-flat magnetostatic field. Review of Scientific Instruments. 2016;**87**:02A507

[21] Stix TH, Swanson DG. Propagation and mode conversion for waves in non-uniform plasmas. In: Galeev A, Sudan RN, editors. Plasma Physics Lab. Rep. PPPL-1903 (June 1982); Handbook of Plasma Physics. Vol. 1. North Holland, Amsterdam: Princeton University; (to be published)

[22] Mallick C, Bandyopadhyay M, Kumar R. Spatio-temporal evolution of electric field inside a microwave discharge plasma during initial phase of ignition and its effect on power coupling. Physics of Plasmas. 2019;**26**: 043507. DOI: 10.1063/1.5085204

[23] Mallick C, Bandyopadhyay M, Kumar R. Plasma characterization of a microwave discharge ion source with mirror magnetic field configuration. The Review of Scientific Instruments. 2018;**89**:125112. DOI: 10.1063/1.5048292

[24] Kinder RL, Kushner MJ. Wave propagation and power deposition in magnetically enhanced inductively coupled and helicon plasma sources. Journal of Vacuum Science and Technology A. 1999;**17**:2421. DOI: 10.1116/1.581978

[25] Mascali D, Celona L, Gammino S, Miracoli R, Castro G, Gambino N, et al. Electrostatic wave heating and possible formation of self-generated high electric fields in a magnetized plasma. Nuclear Instruments and Methods in Physics Research A. 2011;**653**:11-16. DOI: 10.1016/j.nima.2011.02.04

[26] You HJ, Meyer FW, Chung KS. The cold and hot electron populations, temperatures and their transports in the edge plasma of the ORNL CAPRICE ECR ion source. Plasma Sources Science and Technology. 2009;**18**: 015004. DOI: 10.1088/0963-0252/18/1/015004

[27] Chen L, Zhao Y, Wu K, Wang Q, Meng Y, Ren Z. On the generation of magnetic field enhanced microwave plasma line. Physics of Plasmas. 2016;**23**: 123509. DOI: 10.1063/1.4971446

[28] Mascali D, Celona L, Gammino S, Castro G, Miracoli R, Romano FP, et al. An investigation on the formation of suprathermal electrons in a B-min ECR machine and a novel method for their damping. Plasma Sources Science and Technology. 2013;**22**:065006. DOI: 10.1088/0963-0252/22/6/065006

[29] Sugai H, Ghanashev I, Mizuno K. Transition of electron heating mode in a planar microwave discharge at low pressures. Applied Physics Letters. 2000;**77**:3523. DOI: 10.1063/1.1329322

[30] Shalashov AG, Viktorov ME, Mansfeld DA, Golubev SV. Kinetic instabilities in a mirror-confined plasma sustained by high-power microwave radiation. Physics of Plasmas. 2017;**24**: 032111. DOI: 10.1063/1.4978565

[31] Bodendorfer M, Altwegg K, Wurz P, Shea H. Field structure and electron life times in the MEFISTO electron cyclotron resonance ion source. Nuclear Instruments and Methods in Physics Research B. 2008;**266**:4788. DOI: 10.1016/j.nimb.2008.07.019

[32] Izotov I, Tarvainen O, Mansfeld D, Skalyga V, Koivisto H, Kalvas T, et al. Microwave emission related to cyclotron instabilities in a minimum-B electron cyclotron resonance ion source plasma. Plasma Sources Science and Technology. 2015;**24**:045017. DOI: 10.1088/0963-0252/24/4/045017

[33] Chelvam PP, Raja LL. Modeling of gas breakdown and early transients of plasma evolution in cylindrical all-dielectric resonators. Journal of Physics D: Applied Physics. 2017;**50**:474003. DOI: 10.1088/1361-6463/aa925f

[34] Gammino S, Mascali D, Celona L, Maimone F, Ciavola G. Considerations on the role of the magnetic field gradient in ECR ion sources and build-up of hot electron component. Plasma Sources Science and Technology. 2009; **18**:045016. DOI: 10.1088/0963-0252/18/4/045016

[35] Mansfeld D, Izotov I, Skalyga V, Tarvainen O, Kalvas T, Koivisto H, et al. Dynamic regimes of cyclotron instability in the afterglow mode of minimum-B electron cyclotron resonance ion source plasma. Plasma Physics and Controlled Fusion. 2016;**58**: 045019. DOI: 10.1088/0741-3335/58/4/045019

[36] Castro G, Mascali D, Gammino S, Torrisi G, Romano FP, Celona L, et al. Overdense plasma generation in a compact ion source. Plasma Sources Science and Technology. 2017;**26**: 055019. DOI: 10.1088/1361-6595/aa61c4

[37] Van Niekerk EG, Krumm PH, Alport MJ. Electrostatic ion cyclotron waves driven by a radial electric field. Plasma Physics and Controlled Fusion. 1991;**33**:375-388

[38] Schulze J, Donkó Z, Lafleur T, Wilczek S, Brinkmann RP. Spatio-temporal analysis of the electron power absorption in electropositive capacitive RF plasmas based on moments of the Boltzmann equation. Plasma Sources Science and Technology. 2018;**27**: 055010

[39] Dieckmann ME, Sarri G, Doria D, Ynnerman A, Borghesi M. Particle-in-cell simulation study of a lower-hybrid shock. Physics of Plasmas. 2016;**23**: 062111

[40] Fischer P, Gauthereau C, Godiot J, Matthieussent G. Ion acoustic wave generation by a standing electromagnetic field in a sub-critical plasma. Journal de Physique. 1987;**48**: 233-238

Section 4

Physics of Plasma Antennas

Chapter 4

Plasma Antennas

Theodore Anderson

Abstract

We have demonstrated that one or two plasma tubes can be used to focus, spread, and steer antenna beams. We have also shown that we can simulate convex and concave plasma lenses by using cylindrical plasma tubes. Focusing by a plasma is useful because it can be used to increase the gain of an antenna, and to quickly reconfigure the beamwidth as needed without physically moving the antenna. With this technology, there is no need for phased arrays to steer and focus an antenna beam. Beam steering with a plasma allows tuning to different frequencies which is a difficult task for standard antennas. Our experimental results with 44 GHz showed a dramatic improvement in beam steering and focusing characteristics compared to beam focusing and steering at 24 GHz. The shorter wavelength compared to the spatial variation in plasma density over the radius of the plasma tube, the easier it is to steer and focus antenna beams. These results have been incorporated in a new smart plasma antenna design.

Keywords: plasma antenna, antenna beam focusing, antenna beam steering, physics of refraction through a plasma

1. Introduction

Plasma antennas use partially or fully ionized gas as the conducting medium instead of metal to create an antenna. The advantages of plasma antennas are that they are highly reconfigurable and can be turned on and off. Hence research to reduce the power required to ionize the gas at various plasma densities is important and this has been achieved by various techniques including pulsing techniques. The power requirements for plasma antenna operation continue to decrease.

The same geometric resonances apply to plasma antennas as metal antennas. Plasma antennas of the same shape, length, and frequency of corresponding metal antennas will have the same radiation patterns. Plasma antennas have the advantage of reconfigurability.

High frequency antennas can transmit and receive through lower frequency plasma antennas eliminating or reducing co-site interference. Because of this principle, higher frequency plasma antennas can be nested inside lower frequency plasma antennas and the higher frequency plasma antennas can transmit and receive through the lower frequency plasma antennas. Higher frequency plasma antenna arrays can transmit and receive through lower frequency plasma antenna arrays. Co-site interference occurs when larger frequency antennas block or partially block the radiation patterns of smaller higher frequency antennas. With plasma antennas, co-site interference can be reduced or eliminated. The interference among plasma antennas can be reduced or eliminated by turning all the plasma antennas off (extinguishing the plasma) except the plasma antennas that are transmitting and/or receiving. This is not possible with metal antennas. A general rule is that when an incident electromagnetic wave upon a plasma antenna is such that the frequency of the incident electromagnetic wave is greater than the plasma frequency of the plasma, the incident electromagnetic wave passes through the plasma without attenuation. If the incident electromagnetic wave has a frequency much less than the plasma frequency, the plasma behaves similar to a metal. The frequency at which plasma behaves like a metal or a dielectric is reconfigurable. The plasma frequency is a natural frequency of the plasma and it is a measure of the amount of ionization in the plasma. It is defined and used throughout this book. Both plasma antennas and metal antennas increase in size as the frequencies they operate goes down to maintain geometric resonance and high efficiency. However as the frequency of operation of the plasma antenna decreases, the density of the plasma needed to operate the plasma antenna also goes down. A rule of thumb is that the plasma frequency should be about twice the operating frequency of the plasma antenna. Hence the plasma frequency goes down as the frequency of the plasma antenna goes down. As the plasma frequency decreases, the plasma antenna becomes transparent to a greater bandwidth of electromagnetic waves. In short as the plasma antenna increases in size, the RCS of the plasma antenna goes down whereas for the corresponding metal antenna, the RCS goes up as the metal antenna increases in size. This gives the plasma antenna some great advantages at low frequencies over the corresponding metal antenna. In addition plasma antennas do not receive electromagnetic noise greater than the plasma frequency since these frequencies pass through the plasma antenna.

Related to plasma antennas, plasma frequency selective surfaces, plasma wave-guides, and plasma co-axial cables have been developed. Unlike metal frequency selective surfaces, plasma frequency selective surfaces have the properties of reconfigurable filtering of electromagnetic waves. This could have tremendous advantages to radome design. Plasma frequency selective surfaces can be reconfigured by varying the plasma density, varying the shape of the elements, or tuning any number of the plasma FSS elements on or off. Plasma wave guides and plasma co-axial cables can be stealth like plasma antennas, and they can operate at low frequencies, and be invisible at high frequencies. Plasma waveguides and co-axial cables can be feeds for plasma antennas. Plasma feeds as well as the plasma antennas have reconfigurable impedances. If the impedance of the plasma antenna is changed, the impedance of the plasma antenna feeds can be changed to maintain impedance matching.

Thermal noise in a plasma antenna is less than the thermal noise in a metal antenna at the higher frequencies. Higher frequencies mean that there is a point in the RF spectrum in which the thermal noise of plasma antennas is equal to the thermal noise of metal antennas. At higher frequencies than this point, the plasma antenna thermal noise decreases drastically compared to a metal antenna. Below this point the thermal noise of the plasma antenna is greater than a metal antenna. For a fluorescent tube which has been built as a plasma antenna, the point where the thermal noise of the plasma antenna is equal to the metal antenna is about 1 GHz. This point can be decreased in frequency by decreasing the plasma density and/or gas pressure. The plasma in the plasma antennas are inert gases that operate at energies and frequencies in which Ramsauer Townsend Effects apply.

Ramsauer Townsend Effects mean that the electrons in the plasma diffract around the ions and neutral atoms in the plasma. This means that the collision rate of the unbound electrons in the plasma with ions and neutral atoms is small and much smaller than in a metal. This phenomenon contributes to the lower thermal noise plasma antennas have over corresponding metal antennas.

Satellite plasma antennas benefit from the lower thermal noise at the frequencies they operate. Ground based satellite antennas point at space where the thermal noise is about 5° K. A low thermal noise, high data rate satellite plasma antenna system is possible with low noise plasma feeds and a low noise receiver. Satellite plasma antennas can operate in the reflective or refractive mode. Satellite plasma antennas need not be parabolic but can be flat or conformal and effectively parabolic. The effective plasma parabolic dish antenna is part of the scope of future work. Electromagnetic waves reflecting off of a bank of plasma tubes get phase shifted as a function of the plasma density in the tube. This becomes an effective phase array except that the phase shifts are determined by the plasma density. If the plasma density in the tubes is computer controlled, the reflected beam can be steered or focused even when the bank of tubes is flat or conformal. In the refractive mode, the refraction of electromagnetic waves depends upon the density of the plasma. In the refractive mode, steering and focusing can be computer controlled even when the bank of tubes is flat. This eliminates the problem of the blind spot and feed losses caused by the feed horn and receiver in front of a metal satellite antenna.

Pulsing techniques instead of applying continuous energy were developed to increase the plasma density and decrease the amount of energy to maintain the plasma.

In the history of antennas, it has been difficult to develop low frequency directional and electronically steerable antennas that fit on land vehicles and aircraft. Low frequency means the wavelength is on the order or larger than the vehicle. With plasma antennas this is possible with multipole expansions of clusters of plasma antennas that are all within a wavelength of each other. This depends on the ability of turning plasma antennas on or off (extinguishing the plasma) to create reconfigurable multipoles of plasma antennas that can be rotated in time creating directional and steerable antenna beams. This is not possible with metal antennas because they cannot be turned on and off.

Several groups have done work in using numerical techniques to plot plasma antenna radiation patterns. Zhou et al. [1] used FDTD Method techniques. Bogachev et al. [2] predicted radiation patterns for plasma asymmetrical dipole antenna. Zhivko Kiss'ovski [3] calculated the radiation pattern of miniaturized plasma antennas. Golazari et al. [4] did measurements and simulations of a loop plasma antenna in UHF band Barro et al. [5] did simulations to get the radiation patterns of cylindrical plasma antennas. Kumar et al. [6] have done simulations of a plasma antenna array. Melazzi et al. [7] have developed a plasma antenna numerical code called ADAMANT. An overview of experimental and numerical research is Anderson et al. [8]. Mansutti et al. [9] have done numerical work on metal-plasma L band antenna.

The phase speed of electromagnetic waves in a plasma is given by:

$$v_p = \frac{c}{\sqrt{1 - \omega_p^2/\omega^2}},$$

Where the plasma frequency is given by:

$$\omega_p = \sqrt{\frac{n_e\, e^2}{\epsilon_0\, m_e}}$$

In this paper, we are experimenting in the region where the antenna frequency is greater than the plasma frequency:

$$\omega > \omega_p$$

In this region refraction and not reflection takes place.

The phase speed of electromagnetic waves in a plasma is greater than in free space. The greater the density of the plasma the greater the phase speed. Since the plasma density can be reconfigured, the steering and focusing of antenna beams by the physics of refraction through a plasma is reconfigurable [10, 11]. The amount of refraction through a plasma depends on the path length through a plasma and the change in plasma density over that path length [12]. This physical process can also be considered as a plasma lens [13].

Refraction in a plasma depends on:

1. Plasma density

2. Path length

3. Gradient of plasma density

We have very good results at 24 GHz and above using COTS plasma tubes. We made custom plasma tubes with larger diameter and refraction that worked well at 10 GHz.

At 24 GHz, two plasma tubes were used to get antenna beam focusing and one plasma tube was used to get antenna beam spreading. Antenna beam steering was achieved with one and two plasma tubes at 24 GHz. In another case our beam steering experiments from using the physics of refraction through a plasma were done at 44 GHz.

2. Focusing antenna beams with the physics of refraction through plasma

In the following sections we show our work on antenna beam focusing, beam spreading, and beam steering using refraction of RF waves in a plasma. This is our first iteration of the plasma lens work and it can only improve. We found it was easier to show the lensing effects of plasma at 24 GHz since the size and shape of COTS plasma tubes are amenable to a 24 GHz. These effects all scale according to wavelength, but cylindrical annular rings of plasma are the best way to control the plasma density variations of plasma to optimize the engineering effects of plasma refraction to control beam focusing, beam spreading, and beam steering.

We have demonstrated the ability to use a plasma for manipulation of a microwave signals by focusing a wide beam into a more narrow beam and also by steering the beam.

Figure 1 shows the experimental set-up for beam steering and lensing. A narrow-beam 24 GHz signal is directed into the side of two 1.5 inch diameter plasma tubes which focuses the antenna beam into higher directivity, gain, and range.

This change in velocity of the signal inside the plasma results in a lensing effect if the beam passes through varying lengths of plasma similar to light passing through glass of varying thickness to make a lens. But there is an important and interesting difference between an ordinary lens made of glass or plastic and a plasma lens: The glass lens slows-down the signal while a plasma lens speeds-up the signal. Therefore a convex glass lens focuses a signal to a point while a convex plasma diverges the signal similar to a concave glass lens that diverges the signal while a concave plasma lens focuses to a point.

We have built converging and diverging plasma lenses using plasma tubes as shown in **Figures 1** and **2**. A single plasma tube with the beam passing through its diameter acts as a diverging plasma lens (**Figures 2** and **3**) and two plasma tubes side-by-side form a converging (focusing) lens (**Figures 1** and **4**).

We have built the "concave" set-up for beam shown in **Figure 1**, and **Figure 4** is a 24 GHz, 5 mW Gunn diode is used as the microwave source with the signal

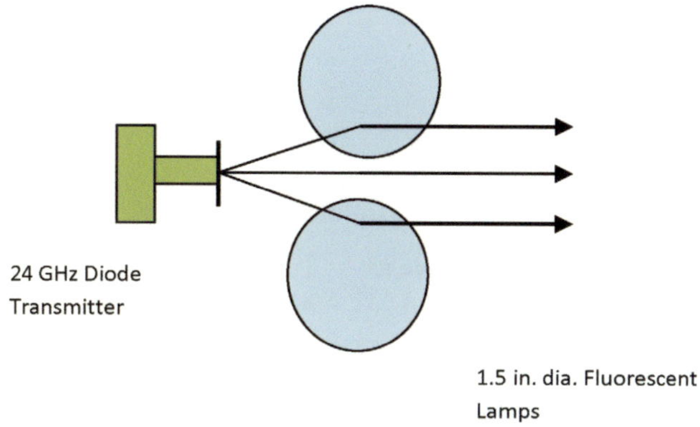

Figure 1.
Schematic for experimental setup of antenna beam focusing with tubes with plasma using two COTS (commercial off the shelf) tubes.

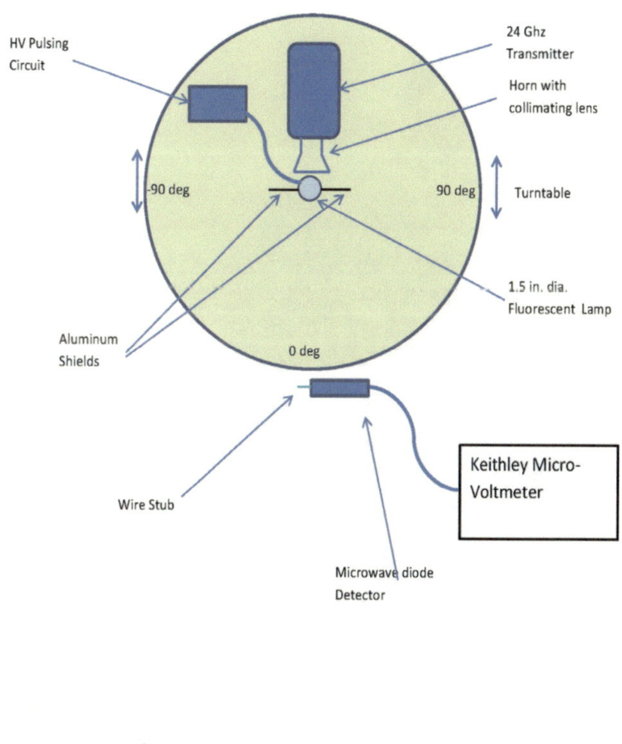

Figure 2.
Schematic for experimental setup for antenna beam spreading with tubes with plasma in one COTS (commercial off the shelf) tubes.

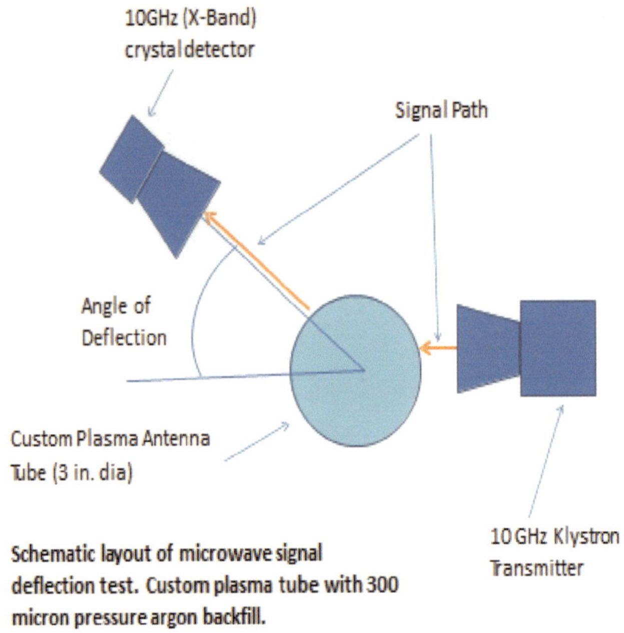

Figure 3.
Schematic showing antenna beam spreading with one 3-inch diameter custom made plasma tube at 10 GHz.

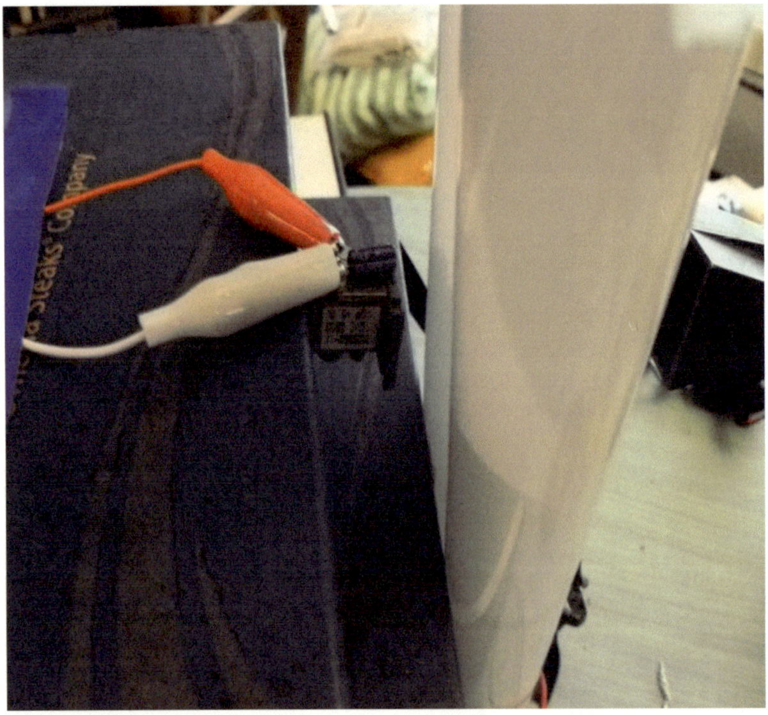

Figure 4.
Plasma focusing experimental setup from a different angle. Gunn diode 24 GHz transmitter with fluorescent tubes used for plasma beam focus.

radiating from the open waveguide, which gives a directional but unfocused microwave output. This setup allows us to focus the beam in the forward direction resulting in a gain of 2 (3 dB). This can be seen in **Figure 5**.

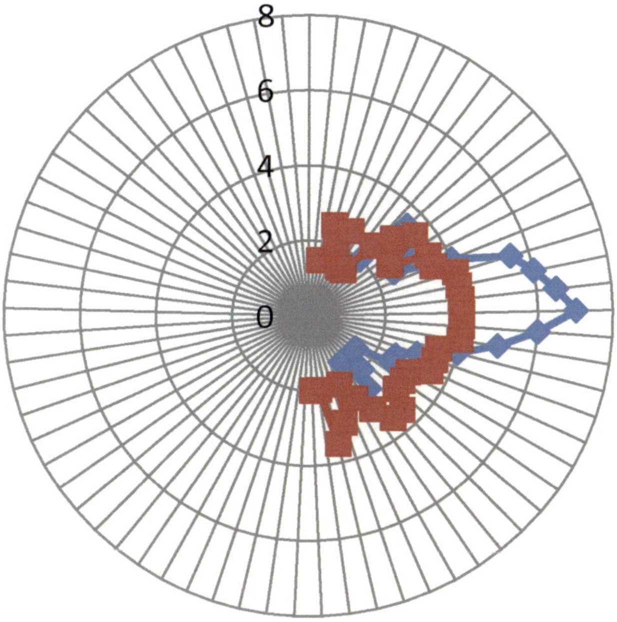

Figure 5.
24 GHz beam focus with plasma. Red line is with no plasma. Blue line is with plasma. Note that the plasma focus increases beam amplitude by a factor of two compared to no plasma (~3 dB gain). A crystal waveguide detector was used as a receiver. Amplitude numbers are relative voltage readings from the crystal detector.

We are the only group utilizing a pulsed high voltage power supply to give a much higher average plasma density and much lower average power to ionize the gas into plasma. This ability to tune the focusing of a RF beam is very useful because the same lensing structure can be used with different frequencies and to vary focal length as needed.

3. Steering antenna beams with the physics of refraction through plasma

We have done microwave beam steering using a cylindrical plasma tube. Those tests were done using fluorescents lamps with a diameter of 1.5 in (3.8 cm); and with a 24 GHz microwave beam. A shortcoming of the set-up was that the 24 GHz signal has a wavelength of 1.25 cm, which is 1/3 the diameter of the tube. For a properly working lens, the wavelength should be small compared to the physical dimensions of the lens, or in our case the plasma tube.

This work involved using a much higher frequency (44 GHz, 0.68 cm) and therefore shorter wavelength. This wavelength is a factor of 5.6 smaller than the 3.8 cm tube diameter which should result in a more ideal lensing action. Our testing has confirmed this and has shown significantly narrower beam steering and less signal loss.

Of course all the dimensions scales with wavelengths used so going down to lower frequencies requires larger diameter tubes or other shaped plasma containers. This is why it is important to make custom made plasma tubes or other geometries.

Figure 6 is a schematic for the experimental setup, and **Figures** 7 and 8 are photos of the setup. A 33–50 GHz HP Microwave Signal Generator is used to generate the incident microwave beam. The plasma tube (fluorescent lamp) is placed ~1 inch in front of the open-ended waveguide, and an aluminum shield is placed against the side the tube to ensure that most of the microwave signal goes

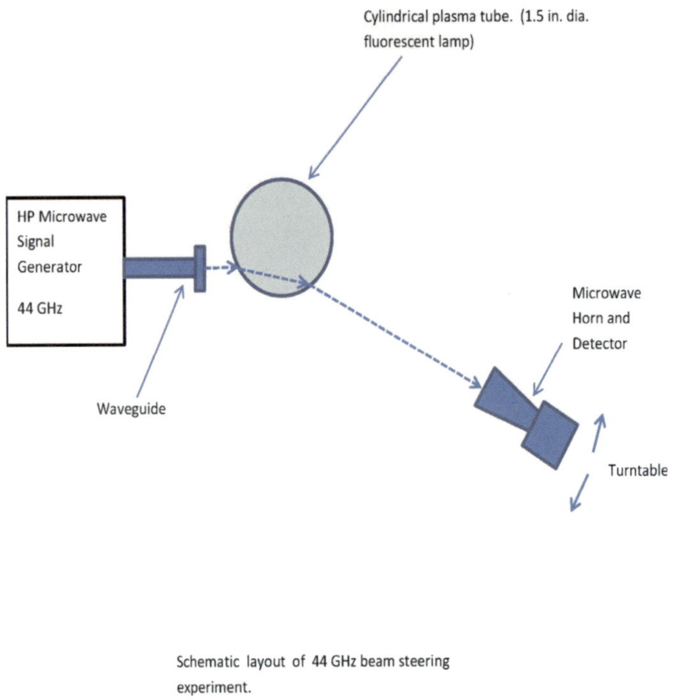

Figure 6.
Schematic for antenna beam steering using one COTS tube with plasma.

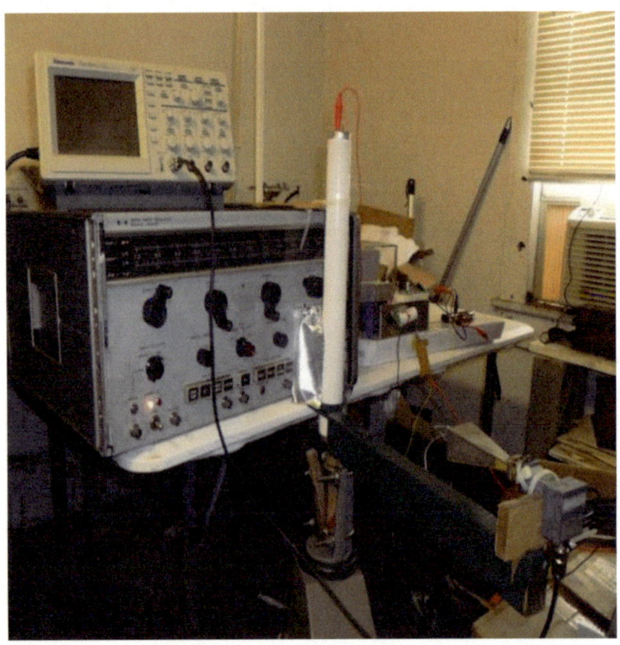

Figure 7.
Plasma beam steering experiment. Antenna rotator with holder (green) and receiver horn/detector is in the foreground. The oscilloscope used to monitor the signal waveform is on top of the HP signal generator. The solid state pulsing circuit is to the right of the signal generator.

through the plasma and is not bypassed to the side of the tube. A 33–50 GHz microwave horn with HP crystal detector is placed on a rotating arm that is scanned in an arc around the plasma tube by an antenna rotator. The arm places the horn receiver about 16 inches from the plasma tube. This corresponds to 60 wavelengths between the steering plasma tube and the detector and clearly qualifies as far field.

In order to generate a plasma with a density high enough to interact with RF signals in microwave frequencies, we use short current pulses (~1 μs) that quickly ionize the plasma; then rely on the plasma ions rather slowly migrating to the wall of the plasma tube. Using this technique (developed by Dr. Theodore Anderson and the late Professor Igor Alexeff) we generate a much higher average density plasma with a low average current and power.

Our high current pulses have so far had a period of 1 ms. With this time separation between pulses, the plasma density decays by about a factor of two

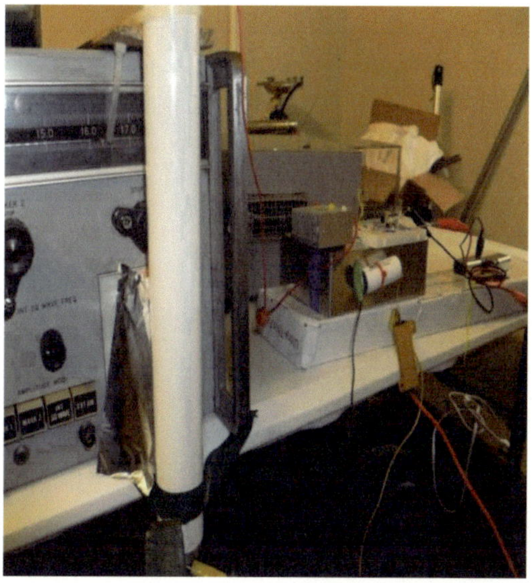

Figure 8.
Fluorescent plasma tube is located in front of the output waveguide on the HP signal generator. Aluminum shield on left of tube prevents stray RF from bypassing the plasma. High voltage pulser is on the right.

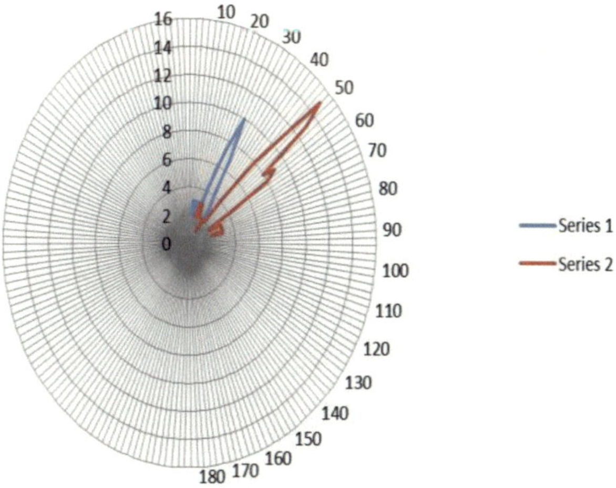

Figure 9.
Beam steering (44 GHz) for two different plasma ionizing currents. Blue line: 5 A peak. Red line: 8 A peak. A crystal waveguide detector is used as a receiver. Amplitude numbers are relative voltage readings from the crystal detector.

before the next pulse comes to refresh the plasma back to peak density. This has not caused a problem with the smart antenna because we are using the plasma to totally block the RF signal, like a shutter. But lensing and beam steering require a plasma of a specific density to get a consistent beam deflection angle.

We are avoiding this problem by recording the received microwave signal immediately after the current pulse; therefore measurements are taken at a constant plasma density, which is also the maximum density before significant decay of density occurs.

Our recent experimental results with 44 GHz show dramatic improvement in beam deflection characteristics compared to previous testing at 24 GHz. The shorter wavelength compared to tube dimensions has clearly resulted in cleaner and more consistent beam steering.

Figure 9 shows beam deflection at two peak ionization currents. The blue curve shows about 26 deflection with a current of 5 A peak. The red curve shows about 50 deflection with a current of 8 A peak.

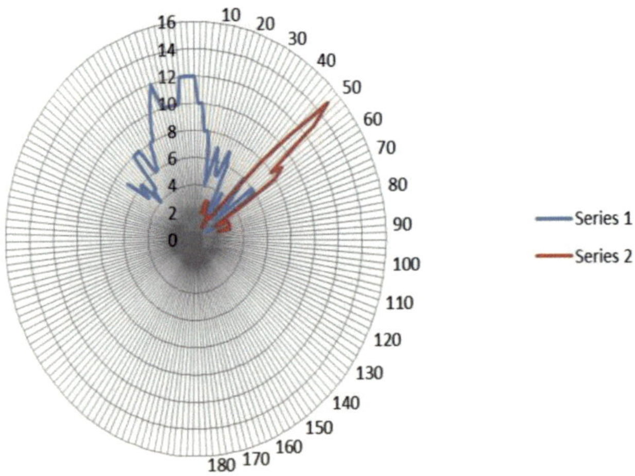

Figure 10.
Plasma beam steering. Beam is steered ~45° clockwise. Blue line: No plasma. Red line: 8 A peak ionizing current. A crystal waveguide detector is used as a receiver. Amplitude numbers are relative voltage readings from the crystal detector.

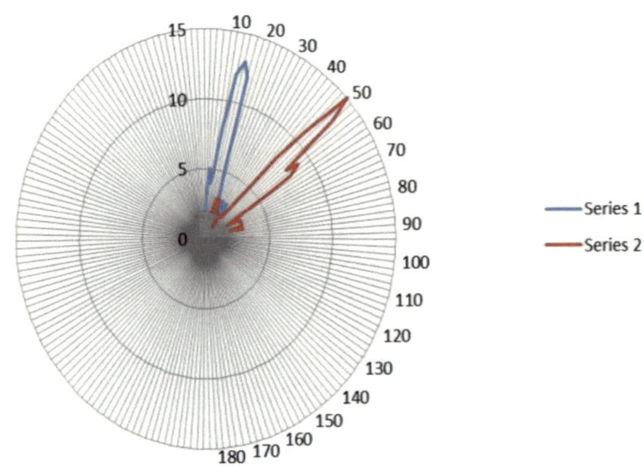

Figure 11.
Beam steering for two different plasma ionizing currents. Blue line: 3 A peak. Red line: 8 A peak. A crystal waveguide detector is used as a receiver. Amplitude numbers are relative voltage readings from the crystal detector.

Figure 9 demonstrates that we can vary the angle of deflection by changing plasma ionizing current; but the most striking result shown in **Figure 9** is the very narrow beam-width of the deflected signals. This is quite surprising. **Figure 10** shows the 50 deflected beam (red line) along-side the un-deflected beam with no plasma. The incident un-deflected beam has a much wider beam-width compared to the deflected beam with plasma turned on.

Figure 11 shows the beam deflection (blue line) with a lower peak plasma current of 3 A with a beam deflection of 15°. For comparison the 50° beam deflection with 8 A peak is shown in red.

4. Pulsed plasma antenna circuitry

A plasma antenna operating in the microwave frequency range requires higher operating currents (Greater than ~1A) and consequently can overheat when used continuously. Our plasma antennas are able to work CW at high frequencies (>1 GHz) because of a concept invented by Igor Alexeff and Theodore Anderson that uses fast high-current pulses; instead of DC current.

The plasma initiates quickly in less than a microsecond, but when plasma ionizing current is turned off, the ions take about a millisecond to recombine with electrons. Therefore plasma density stays high for almost a millisecond even though the ionizing current is no longer on. We use an even short pulse width (~1 μs), and therefore less power is required to run the antenna.

We developed a pulsed voltage doubler circuit, allowing us to use a lower voltage DC power supply for the input power to the pulsing circuit using a modified Marx Generator. A Marx Generator is a pulsed voltage multiplier. A series of capacitors was charged in parallel and then discharged in series through spark gaps.

Figure 12 shows a two stage voltage doubler circuit. We have built and tested a modified Marx Generator that replaces the spark gaps with an IGBT electronic switch.

Keeping the second spark gap in the circuit results in a faster rise time than in our previous pulsing circuits. The voltage doubler allows the use of a lower voltage DC power supply than would otherwise be required.

A simple non-voltage multiplying IGBT pulsing driver circuit is shown in **Figure 13**. A CMOS timer IC is used to generate short 1 μs pulses with a repetition time of 750 μs. An IXYS brand 2500 V IGBT is the high voltage switch. This pulsing arrangement allows us to use a factor of 750 less DC current and power from the DC power supply.

We have been driving our plasma antenna tube with the fast, high-current pulses described here to allow operation at higher frequencies than is possible with CW current. Our pulsing circuit had required that the both electrodes of the plasma tube operate at high voltage; with the positive electrode at constant maximum DC voltage (2–3 kV). We have modified the circuit to allow the negative electrode to remain grounded while a positive 1 μs pulse is applied to the anode electrode. This change offers several advantages:

Improved electrical safety because one electrode remains grounded and the other electrode is supplied with short pulses. Less capacitive loss in the current leads allowing the use of faster pulses; since one electrode is grounded and can be attached to a ground plane.

Eliminating the need for a negative lead wire means less high voltage wiring. Lower EMI caused by high current pulses in wiring.

Grounding one side of the plasma tube requires that the high voltage switcher (IGBT) be able to float up and down in voltage with the tube's anode. To do this we

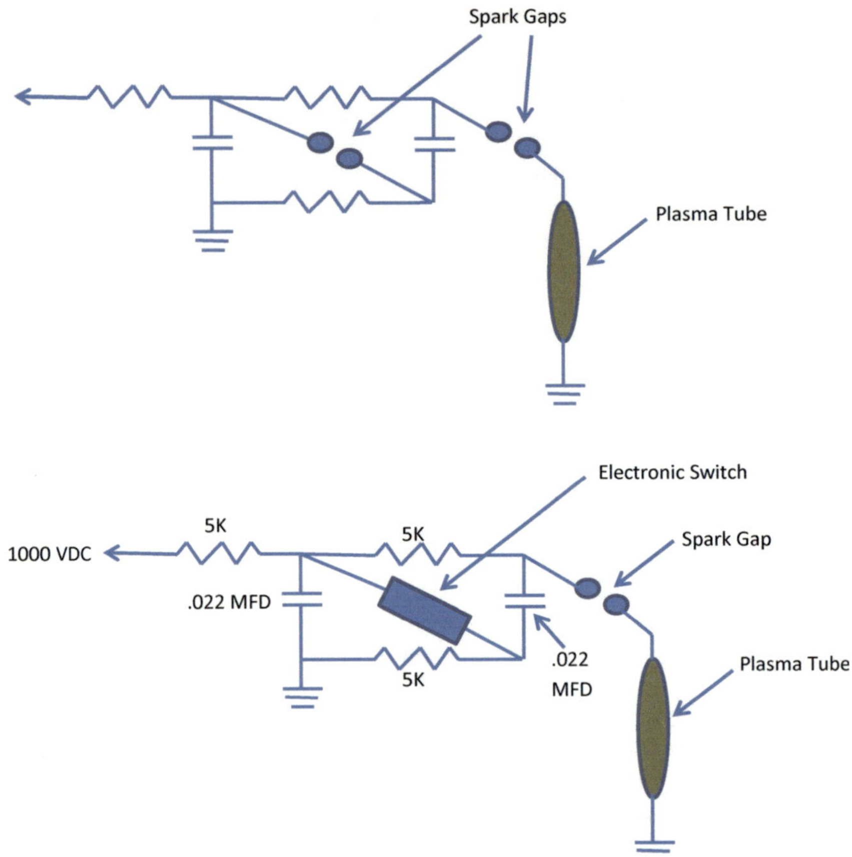

Figure 12.
1000 V DC to 2000 V pulse circuit using modified Marx generator with spark gaps in the top photo and with an electronic switch and spark gap in the bottom figure. Two electronic switches was also built but not shown.

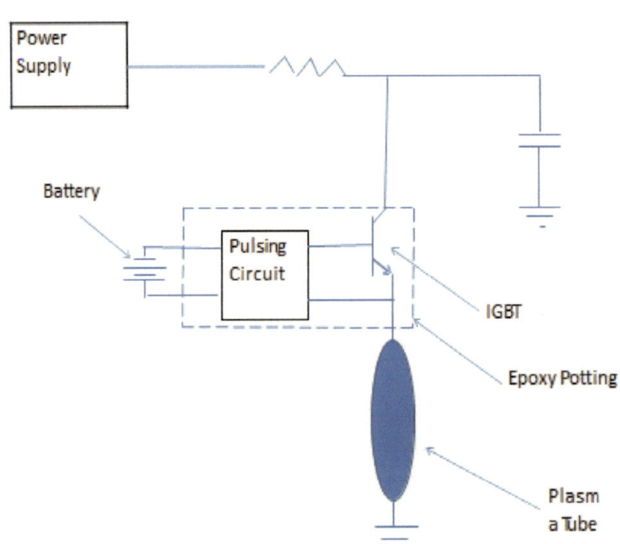

Figure 13.
Basic pulsing circuit. A DC power supply charges the capacitor, the IGBT pulsing circuit switching delivers short ~1 μs pulses to the plasma tube.

Figure 14.
Prototype pulsing circuit potted in epoxy. High voltage leads are at top and bottom, and the battery connection is on the right.

have electrically isolated the switcher by using a battery to power the electronics and by potting the IGBT and electronics in epoxy.

Figure 14 shows our first prototype with steel-filled epoxy; not the best choice of epoxy but one that is working quite well so far. Potting the circuit is advantageous for airplane and aerospace applications, providing mechanical ruggedness as well as electrical isolation. Epoxy has much higher thermal conductivity than air, but not as good as a metal heat sink. The steel filled epoxy has about a factor of 10 higher thermal conductivity than air. Highly thermally conductive epoxies can have a factor of 100 higher conductivity than air. Potting in epoxy in our case allowed operation without an additional metal heat sink, saving space and eliminating the need for electrical isolation between the IGBT and a metal heat sink. We tested the ruggedness of our epoxy-potted circuit by dropping it on a concrete floor from a height of 6 ft. without damage to the circuit.

We ran the circuit shown in **Figure 13** (with no additional heat sink) with peak current of 20 A and a pulse period of 1000 μs.

After ½ hour of operation, the epoxy and IGBT were warm to the touch but with no indication of over-heating.

5. Power, current, and voltage requirements in pulsing excitations

The plasma antenna requires a relatively high voltage, low current power supply.

Short pulses are applied to the terminals of the plasma tube. Peak current is about 1 A with 5 μs pulse width and a time between pulses of about 1 ms.

This duty cycle of 1/200 results in an average power of about 5 W.

Overall average power drain from the battery driving the plasma antenna will be much less than 5 W. Two standard 9 V batteries and/or one 6 V.75 AH SLA battery can operate a plasma antenna. The smart plasma antenna can operate on a 12 V car battery which is enough voltage to ionize the plasma in 12 tubes and run the computer.

The power losses of the supply voltage connected to the plasma antenna are because the VSWR numbers in many cases indicate a very good match between the feeds and antennas.

Anyone trying to build a plasma antenna according should consult a licensed electrical safety expert before proceeding. After consulting a licensed electrical safety expert, proceed as follows. Use a three-wire grounded power cord and securely attach the green ground wire to the metal enclosure. Install an appropriately sized fuse or circuit breaker to protect from short circuits or overloads. Always unplug the unit before modifying or working inside.

6. Conclusions

We have demonstrated that one or two plasma tubes can be used to focus, spread, and steer antenna beams. We have also shown that we can simulate convex and concave plasma lenses by using cylindrical plasma tubes. Focusing by a plasma is useful because it can be used to increase the gain of an antenna, and to quickly reconfigure the beamwidth as needed without physically moving the antenna. With this technology, there is no need for phased arrays. Beam steering with a plasma allows tuning to different frequencies which is a difficult task for standard antennas. Our experimental results with 44 GHz showed a dramatic improvement in beam steering and focusing characteristics compared to beam focusing and steering at 24 GHz. The shorter wavelength compared to the spatial variation in plasma density over the radius of the plasma tube, the easier it is to steer and focus antenna beams. These results have been incorporated in a new smart plasma antenna design which appears in another paper.

Driving the plasma with short high-current pulses allows CW operation at higher frequencies with a minimum amount of ionization power and higher plasma densities. Circuits for pulse forming and voltage multiplication are presented. The maximum frequency that a plasma antenna can operate CW has previously been limited by the high DC current needed to ionize the plasma. We minimize the average ionization power and increase the plasma density by using fast current pulsing with a short duty cycle. The average current is much lower but the average plasma density remains high than in the DC mode.

Author details

Theodore Anderson
Haleakala R&D, Inc., Brookfield, Massachusetts, USA

*Address all correspondence to: tedanderson@haleakala-research.com

IntechOpen

© 2020 The Author(s). Licensee IntechOpen. This chapter is distributed under the terms of the Creative Commons Attribution License (http://creativecommons.org/licenses/by/3.0), which permits unrestricted use, distribution, and reproduction in any medium, provided the original work is properly cited.

References

[1] Zhou J, Fang J, Lu Q, Liu F. Research on radiation characteristic of plasma antenna through FDTD method. Scientific World Journal. 2014;**2014**: 290148. DOI: 10.1155/2014/290148

[2] Bogachev NN, Gusein-Zade NG, Nefedov VI. Radiation pattern and radiation Spectrum of the plasma asymmetrical dipole antenna. Plasma Physics Reports. 2019;**45**(4):372-375

[3] Kiss'ovski Z, Vachkov V. Model of a miniature plasma antenna. International Journal of Engineering and Advanced Technology (IJEAT). 2015;**4**(6): 2249-8958

[4] Golazari SS, Amiri N, Kashani FH. Design, simulation andmeasurement of loop plasma antenna in UHF band. In: 24th Telecommunications Forum (TELFOR). Belgrad; 2016

[5] Barro OA. Mohammed Himdi, and Olivier Lafond, reconfigurable cylindrical plasma antenna. Progress In Electromagnetics Research (PIER) M. 2018;**66**:65-72

[6] Kumar R, Kumar P. Study of array plasma antenna parameters. AIP Advances. 2018;**8**:045306. DOI: 10.1063/1.5018660

[7] Melazzi D, Lancellotti V, De Carlo P, Manente M, Pavarin D, Anderson T. Numerical investigation into the performance of two reconfigurable gaseous plasma antennas. EuCAP 2014 Symposium Journal; 2014

[8] Anderson T, Melazzi D, Lancellotti V. An overview of experimental and numerical results on plasma antenna arrays. EuCAP 2015 Symposium Journal; 2015

[9] Giulia M, Paolo R, Mohammad Abdul H, Federico B, Capobianco AD, De Carlo P, et al. Design of a reconfigurable metal-plasma L-band transmit-array antenna. IEEE International Symposium on Antennas and Propagation and USNC-URSI Radio Science Meeting - AP-S/URSI 2019 Symposium Journal; 2019

[10] Anderson T. Plasma Antennas. Artech House. 2011:187-193

[11] Anderson T. Plasma Antennas. Artech House. 2011:177-185

[12] Anderson T. Plasma devices for steering and focusing antenna beams. U.S. Patent Issue Number: 8,384,602

[13] Linardakis P, Borg G, Martin N. Plasma-based lens for microwave beam steering. Electronics Letters. 2006; **42**(8):444-446

Section 5

Plasma Diagnostic Methods

Chapter 5

Plasma Diagnostic Methods: Test Charge Response in Lorentzian Dusty Plasmas

Shahid Ali and Yas Al-Hadeethi

Abstract

Different plasma diagnostic methods are briefly discussed, and the framework of a test charge technique is effectively used as diagnostic tool for investigating interaction potentials in Lorentzian plasma, whose constituents are the superthermal electrons and ions with negatively charged dust grains. Applying the space-time Fourier transformations to the linearized coupled Vlasov-Poisson equations, a test charge potential is derived with a modified response function due to energetic ions and electrons. For a test charge moving much slower than the dust-thermal speed, there appears a short-range Debye-Hückel (DH) potential decaying exponentially with distance and a long-range far-field (FF) potential as the inverse cube of the distance from test charge. The FF potentials exhibit more localized shielding curves for low-Kappas, and smaller effective shielding length is observed in dusty plasma compared to electron-ion plasma. However, a wakefield (WF) potential is formed behind the test charge when it resonates with dust-acoustic oscillations, whereas a fast moving test charge leads to the Coulomb potential having no shielding around. It is revealed that superthermality and plasma parameters significantly alter the DH, FF, and WF potentials in space plasmas of Saturn's E-ring, where power-law distributions can be used for energetic electrons and ions in contrast to Maxwellian dust grains.

Keywords: kinetic model, test charge technique, DA waves, superthermal tails, dynamical shielding

1. Plasma diagnostic methods

To understand a plasma state and its characteristics, numerous experimental techniques, mechanisms, devices, theoretical models, and computational packages have been developed as diagnostic tools for measuring the plasma parameters such as the plasma electron density and temperature [1, 2] as well as their spatial profiles and dynamics. These diagnostic techniques are used to adequately describe both low-temperature and high energy density plasmas. In some situations, the measurements by these techniques cause perturbations in plasmas and are termed as active diagnostic techniques, while passive ones do not perturb plasmas. Based on the degree of ionization, the plasmas can broadly be classified into cold and hot plasma states, which accordingly demand for various types of diagnostics to

IntechOpen

precisely estimate the plasma parameters for optimal understanding of the physics of plasmas. This includes both theoretical and experimental findings. The most common techniques for cold and hot plasmas ($T_e \geq$ few keV) are the Langmuir probes in the form of planar, cylindrical, or spherical electrodes in plasmas with a goal to monitor the plasma parameters. The probes can be of several types, namely, single and/or double probes, which are used for density, temperature, and floating potential measurements. Emissive and magnetic probes work more efficiently for plasma potential measurement and wave field amplitude and phase diagnostics, respectively, while the Rogowski for antenna current measurements. There are many complications, for example, the plasma potential and density can fluctuate or drift during the time of probe measurements. In some complications, since the probe draws a large amount of currents from a plasma and perturbs the initial state of the plasma, it may lead to the erroneous measurements; even then, the Langmuir probe diagnostics are widely used and this is because of the fact that they are relatively simple to use, cheap, and give reliable values of important plasma parameters.

On the other hand, in certain plasma sources like tokamak plasmas, strong currents are generated, which give rise to various kinds of magnetohydrodynamic (MHD) instabilities. For this, a magnetic probe is used, which is beneficial especially for measuring either local magnetic fields or its fluctuations not only in tokamaks but also in laser-produced plasmas (LPP). Furthermore, the amplitude of current flowing into the plasma can be estimated by integrating the induced magnetic field around the plasma column by utilizing the so-called Rogowski coil. However, in some plasmas (especially high temperature), it is not feasible to utilize material probes for determining the plasma parameter like plasma electron density. Therefore, a nonperturbing approach is needed to diagnose the plasma. In such a scenario, the electromagnetic spectrum is utilized. But the electromagnetic wave intensity must be low enough to the level that it will not result in plasma perturbation. For probing the high-density plasmas, a lower wavelength is required as a probe. This justifies the utilization of infrared radiation in tokamak and ultraviolet radiation for measuring the plasma electron density in LPP. The variation of the polarizing angle involving the beam probe in the presence of magnetic field can also be used for diagnostics of tokamak plasmas.

Interestingly, the evaluation of appropriate plasma parameters may be carried out by spectroscopy of emitted radiations as used generally from the beginning of plasma physics. This technique for emission measurements has been particularly making significant contributions over the past five decades for the fact that plasmas produced for nuclear fusion research exhibit intense emission in the X-ray region. Astrophysical applications further justify the wide interest in X-ray emission from plasmas. The phase soft X-ray (so defined due to their low penetrating power) indicates electromagnetic radiations with a wavelength in range $1Å \leq \lambda \leq 300Å$ (or, in terms of photon energy $h\nu$, $300\ eV \leq h\nu \leq 10\ keV$). Hard X-rays below 1 Å are occasionally produced in plasmas for highly accelerated electrons, like runaway electrons in tokamak plasmas and suprathermal electrons in LPP [3]. The characteristics of soft X-ray spectra like line intensities, line profiles, and continuum intensities can be investigated to determine the electron densities by Stark broadening, while the ion densities from the absolute radiation intensities and ion temperatures using the Doppler broadening of spectral lines [2].

The particle measurement method is another scheme for investigating the characteristics of plasmas by using the beam of fast particles. It has received much attention in the studies of inertial confinement fusion and energy deposition in a medium driven by cluster-ion and fast heavy-ion beams, as well as in plasma accelerators and low-temperature laboratory plasmas.

2. Dusty plasma and test charge technique

The most common ingredient of astrospace plasmas is the dust component in addition to electrons and ions, found everywhere in earth atmospheres, in comets, in planetary rings, in interstellar clouds, in interplanetary space, in interstellar medium, etc. Dust grains may exist in the form of ice particles, metallic and dielectric materials, and are highly charged species due to different charging processes. For instance, the absorption of ambient electrons and ions on dust grain surface may lead to the negatively charged dust grains, while thermionic and secondary electron emissions as well as ultraviolet photoionization give rise to positively charged dust grains. Thus, a multispecies dusty plasma can be assumed as more complex plasma than conventional electron-ion plasma, for dust size, mass, and charge variations. Being an abundant component of the space and industrial plasmas [4, 5], dusty plasma has always attracted lots of interests for studying new distinct features of plasma modes [6, 7] with a static and dynamic background of dust grains both analytically and experimentally [8–10]. Numerous linear and nonlinear [viz., solitons, shocks, vortices, etc.] dusty modes and associated instabilities are investigated using the frameworks of perturbative and nonperturbative schemes.

The behavior of charged particles in plasmas can be described by the well-known fluid and kinetic theories [4, 11], essentially helpful for studying the basic properties of plasma waves and instabilities, depending strongly on the observed phenomena. Laboratory plasmas have effectively been modeled by fluid description, where charged fluids of plasma species are assumed in temporal and spatial configurations. But, it has been observed that fluid theory does not account for velocity space coordinate and is insufficient to study the wave phenomena in non-equilibrium plasmas, where particle distributions show significant deviations from the equilibrium states. Hence, fluid theory is unable to explain the wave-particle interactions that could lead to collisionless Landau damping phenomenon and many other interesting features of collective modes and instabilities. Conversely, kinetic theory adequately describes the physical phenomena in real time and phase space configurations, providing all information about plasma waves, instabilities, plasma equilibrium, Landau damping rate, etc.

Test charge techniques [12, 13] can be utilized to study the shielding of test charges in collisional [14] and turbulent [15] plasmas, the electric field [16], and far-field potential of a test charge in a nonuniform magnetoplasma [17], the wake-field excitations in charge fluctuating dusty plasmas [18], the two-body correlations [19], the energy loss of test charges [20], etc. If a test particle is projected into the plasma with a constant speed, its charge density is coupled with the plasma charge density by the space charge effects. Consequently, the test charge is screened by a cloud of opposite sign charges leading to the short-range Debye-Hückel (DH) potential. Of course, the speed of test charge significantly matters in plasmas when it is considered with respect to thermal speeds of plasma species. The interaction potentials and energy loss of charged particles have been recognized in many research areas, for example, in ion-cluster interaction with condensed matter [21, 22], in inertial confinement fusion [23–25], in particle acceleration [26], in low-temperature laboratory plasmas [27, 28], and in dense plasmas for heavy-ion energy deposition [29].

Montgomery et al. [13] employed the test charge technique to obtain far-field potential distribution around a test charge, which decays as the inverse cube of the distance from test charge in electron-ion Maxwellian plasmas. Subsequent investigations of shielded potentials have been phenomenally influenced by the ionic motion [30], electron-electron collisions [14, 31], and plasma turbulence [15]. The electrostatic potential [32] due to small and large test charge velocities has been investigated to display the excitation of long-range wakefields in Maxwellian plasmas. Shivamoggi and Mulser [33] examined the effects of magnetic field, collisions, and plasma inhomogeneity on the potential due to slowly and rapidly moving test charges [17, 34], and Lakshmi et al. [35] discussed the Debye shielding phenomenon in a dusty plasma by considering the Boltzmannian electrons and ions with cold negative dust grains. It was revealed that plasma parameters significantly alter the characteristics of small and large amplitude potentials. Later, Shukla [36] reported the FF potential for a slowly moving test charge in a Maxwellian dusty plasma and showed the impact of dust-charge variation on the dipole-like FF potential. Moreover, oscillatory wake-field can be excited behind the test charge [37] in a collisionless unmagnetized plasma with Maxwellian electrons and ions. Nambu et al. [38] extended this work to dusty plasmas and explained the resonant phenomenon of DA waves with a test charge, resulting in the long-range WF potential. Later, Shukla and Rao [39] analyzed the WF, DH, and FF potentials of test charge in a colloidal Maxwellian plasma accounting for the streaming ions and dust grains. It was found that external magnetic field and ion-streaming effects [40, 41] strongly affect the positive/negative potential regions in plasmas. To explore the effects of two-body correlations, dust-charge perturbations and dust-neutral collisions, various geometries have been designed for propagating test charges [18–20] in an unmagnetized Maxwellian dusty plasma. In all above investigations, the plasma particles are described by the Maxwellian distribution function.

The shielding phenomenon is one of the main objectives of this chapter to unfold many intrinsic properties of the Lorentzian space dusty plasma, which discerns it from the standard Maxwellian plasmas. It plays a key role in setting up the basic criteria for Lorentzian dusty plasmas. Any plasma medium can physically be polarized by the test charge to give rise to perfect screening if thermal agitations are absent in the plasma system. Conversely, an imperfect shielding occurs if the plasma particles get enough thermal energy to escape from the edge of screening cloud. The interaction potentials caused by the test charge are not only strongly influenced by different test charge speeds in comparison with the thermal speeds but also lead to the possibility of dust crystallization and dust coagulation in space Lorentzian dusty plasmas.

3. Power-law Lorentzian distribution function (df)

In some circumstances, the behavior of plasma particles cannot be described by the usual Maxwellian distribution function (df) but often modeled by the power-law df. When all or some of the plasma particles move faster than their thermal speeds, the plasma particles are known as superthermal/suprathermal species, showing high energy and velocity tails in the distribution. They are mostly accelerated by wave-particle interactions, modulational instabilities and Langmuir turbulence [42], beam-plasma interactions [43], solar wind where type III solar radio emissions occurs [44], intense microwave-plasma interactions [45], ionospheric heating experiments [46], etc. Recognizing the role of superthermal energetic particles in plasmas, the wave dynamics and instabilities need to be re-investigated with a power-law df that gives a better fit to empirical data from space plasmas. A 3D isotropic Kappa-df [47] for superthermal particles can be expressed as

$$f_{s0}^{K}(\mathbf{v}_s) = A_{\kappa_s}\left(1+\frac{v_s^2}{\kappa_s\theta_{Ts}^2}\right)^{-\kappa_s-1}. \tag{1}$$

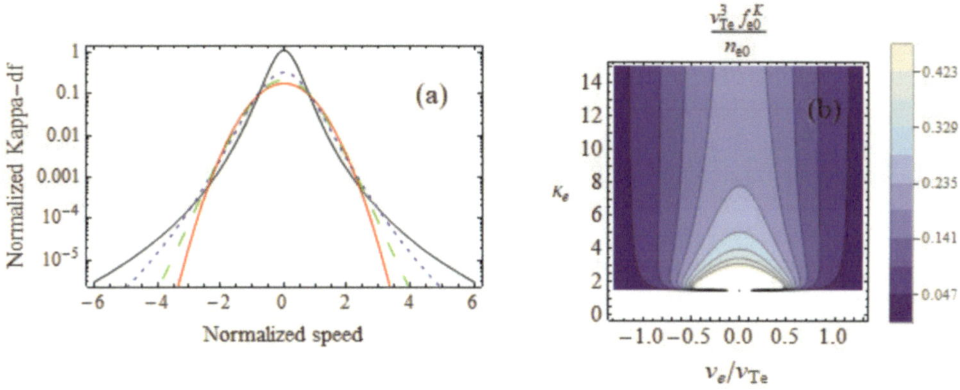

Figure 1.
Kappa-df against the electron speed for different values of kappa index both in 2D and in contour plots. In (a), the solid-black, dotted-blue, dashed-green, and solid-red curves correspond to $\kappa_e = 2, 4, 10$, and $\kappa_e \to \infty$, respectively. However, in (b), the light colors indicate more number of electrons and vice versa for dense colors.

The normalization constant and effective thermal speed are denoted by $A_{\kappa_s} = n_{s0}\pi^{-3/2}\kappa_s^{-3/2}\theta_{Ts}^{-3}\,\Gamma(\kappa_s+1)/\Gamma(\kappa_s-1/2)$ and $\theta_{Ts} = \{2(\kappa_s - 3/2)/\kappa_s\}^{1/2}(T_s/m_s)^{1/2}$, respectively. The symbol Γ indicates the Gamma function, and $\mathbf{v}_s$ and n_{s0} are the velocity and equilibrium number density, whereas T_s (m_s) stands for the temperature (mass) of the *sth* species ($s = e$ for electrons and i for positive ions). The effective thermal speed θ_{Ts} is always realistic in the limit $\kappa_s > 3/2$, where κ_s is the spectral index showing the deviation from the Maxwellian df. **Figure 1** displays the normalized Kappa-df [as given by Eq. (1)] for the electrons as function of normalized electron speed with varying κ_e-index both in 2D and contour plots. See that superthermal electrons exhibit high energy tails at $\kappa_e = 2$, 4, and 10 in distribution curves, which tend to the Maxwell-Boltzmann distribution curve for $\kappa_e \to \infty$. Therefore superthermality effects are only significant for low values of Kappa, and for its infinite values, the Kappa-df exactly converges to the Maxwell-df $\left[viz., f_{s0}^K(\mathbf{v}_s) \to f_{s0}^M(\mathbf{v}_s)\right]$. It may be noted from contours (see **Figure 1(b)**) that light-colored regions correspond to more electrons at low speeds, and while moving toward the dense-colored regions, the number of electrons decreases but comparatively has high speeds. In 1968, for the first time, Vasyliunas [47] pointed out the implications of Kappa-df by fitting empirical data from solar wind and showed the significance of low values of electron spectral index, that is, $\kappa_{s=e} \sim 2-4$. The effects of high energy tails have significantly modified the dispersive properties of waves and instabilities [48, 49] in Lorentzian plasmas. Recently, Ali and Eliasson [50] investigated the impact of suprathermal hot electrons on the electrostatic potential of slowly moving test charge in a two-temperature electron plasma and extended the model for Lorentzian dusty plasmas [51].

4. Kinetic model for Lorentzian dusty plasmas

To compute the potential distributions around a test charge, we consider a collisionless Lorentzian dusty plasma, containing the suprathermal electrons and ions with negatively charged dust grains following the Maxwell-df. The plasma is also assumed to be field-free in the sense that there is no external electric or magnetic field (viz., $E_0 = 0 = B_0$), so that the equilibrium electrostatic potential $\phi_0 = 0$. The quasi-neutrality condition at equilibrium demands $n_{e0} = n_{i0} - Z_{d0}n_{d0}$, where Z_{d0}

being the equilibrium dust-charge state and n_{j0} denote the particle number densities of the jth species [j equals $s(=e,i)$ for Kappa-distributed electrons and ions while $j=d$ for negatively charged dust grains]. In this model, all the dust grains are assumed to be spherical in shape with constant size of radius r_d and mass m_d.

The Lorentzian dusty plasma in the presence of a test charge can be described by the following linearized coupled set of Vlasov-Poisson equations:

$$(\partial_t + \mathbf{v}_d \cdot \nabla) f_{d1} + \frac{q_{d0}}{m_d} \mathbf{E}_1 \cdot \nabla_{\mathbf{v}_d} f_{d0}^M = 0, \tag{2}$$

$$(\partial_t + \mathbf{v}_s \cdot \nabla) f_{s1} + \frac{q_s}{m_s} \mathbf{E}_1 \cdot \nabla_{\mathbf{v}_s} f_{s0}^K = 0, \tag{3}$$

and

$$\nabla^2 \phi_1 + 4\pi \left(\sum_{s=e,i} \rho_s + \rho_d + \rho_T \right) = 0, \tag{4}$$

where $\mathbf{E}_1 (= -\nabla\phi_1)$ is the induced electric field with perturbed potential ϕ_1, $q_{d0}(= -Z_{d0}e)$ is the charge of the negative dust grains, and $q_s(= -e, e)$ being the charge of electrons and positive ions. $\rho_d = q_{d0} \int f_{d1} d\mathbf{v}_d$, $\rho_s = q_s \int f_{s1} d\mathbf{v}_s$, and $\rho_T = q_T \delta(\mathbf{r} - \mathbf{v}_T t)$ identify the dust-charge density, electron-ion charge densities, and test charge density, respectively. The symbol δ stands for a 3D Dirac's delta function, and $f_{d0}^M(\mathbf{v}_d)$ and $f_{s0}^K(\mathbf{v}_s)$ are the Maxwell-df and Kappa-df with their perturbed parts $f_{d1}(\mathbf{r}, \mathbf{v}_d, t)$ and $f_{s1}(\mathbf{r}, \mathbf{v}_s, t)$, such that $|f_{d1}| \ll f_{d0}^M$ and $|f_{s1}| \ll f_{s0}^K$. Also note that test particle has a charge q_T which moves with a constant velocity $\mathbf{v}_T$ along the z-axis in a Lorentzian dusty plasma.

Taking space–time Fourier analysis of Eqs. (2), (3), and (4), we obtain the Fourier transformed potential in this form

$$k^2 D(\mathbf{k}, \omega) \phi_1(\mathbf{k}, \omega) = 8\pi^2 q_T \delta(\omega - \mathbf{k} \cdot \mathbf{v}_T). \tag{5}$$

The modified longitudinal dielectric constant can be defined by

$$D(\mathbf{k}, \omega) = 1 + \sum_s \frac{\omega_{ps}^2}{k^2} \int \frac{\mathbf{k} \cdot \nabla_{\mathbf{v}_s} f_{s0}^K(\mathbf{v}_s)}{(\omega - \mathbf{k} \cdot \mathbf{v}_s)} d\mathbf{v}_s + \frac{\omega_{pd}^2}{k^2} \int \frac{\mathbf{k} \cdot \nabla_{\mathbf{v}_d} f_{d0}^M(\mathbf{v}_d)}{(\omega - \mathbf{k} \cdot \mathbf{v}_d)} d\mathbf{v}_d, \tag{6}$$

where $\omega(k)$ being the angular frequency (wave number) and $\omega_{pj} = \left(4\pi q_j^2 n_{j0}/m_j\right)^{1/2}$ is the plasma oscillation frequency. It is important to mention that if Lorentzian dusty plasma does not contain any test charge, viz., $q_T = 0$, then Eq. (5) simply implies that $D(\mathbf{k}, \omega) = 0$, showing a modified linear dispersion relation of electrostatic waves to account for superthermal electrons and ions. However, the inverse Fourier analysis of Eq. (5) leads to the standard form of electrostatic potential due to a test charge in a dusty plasma [11, 32].

$$\phi_1(\mathbf{r}, t) = \frac{q_T}{2\pi^2} \int \frac{d\mathbf{k}}{k^2} \frac{\exp\left[i\mathbf{k} \cdot (\mathbf{r} - \mathbf{v}_T t)\right]}{D(k, \mathbf{k} \cdot \mathbf{v}_T)}. \tag{7}$$

The dielectric constant in terms of dielectric susceptibilities $\left(\text{viz., } D = 1 + \sum_{s=e,i} \chi_s + \chi_d\right)$ can be expressed as

$$D(k, \mathbf{k} \cdot \mathbf{v}_T) = 1 + \sum_{s=e,i} \frac{1}{k^2 \lambda_{Ds}^2} \left\{ 1 - \frac{1}{2\kappa_s} + C_s Z_{\kappa_s}(C_s) \right\} + \frac{1}{k^2 \lambda_{Dd}^2} \{ 1 + C_d Z_M(C_d) \}, \quad (8)$$

where $\lambda_{Ds} (= \theta_{Ts}/\sqrt{2}\omega_{ps})$ and $\lambda_{Dd} (= v_{Td}/\omega_{pd})$ are the Debye shielding lengths associated with the electron-ion effective speed θ_{Ts} and dust thermal speed $v_{Td} = (T_d/m_d)^{1/2}$. The standard plasma dispersion functions for Kappa-distributed electrons and ions [48] and for Maxwellian dust grains [52], respectively, can be given by

$$Z_{\kappa_s}(C_s) = \frac{1}{\sqrt{\pi}} \frac{\Gamma(\kappa_s + 1)}{\kappa_s^{3/2} \Gamma(\kappa_s - 1/2)} \int_{-\infty}^{\infty} d\beta_s \frac{\left(1 + \beta_s^2/\kappa_s\right)^{-\kappa_s - 1}}{(\beta_s - C_s)}, \quad \text{Im}(C_s) > 0,$$

and

$$Z_M(C_d) = \frac{1}{\sqrt{\pi}} \int_{-\infty}^{\infty} d\beta \frac{\exp\left(-\beta^2\right)}{(\beta - C_d)}, \quad \text{Im}(C_d) > 0,$$

with their corresponding arguments $C_s = \mathbf{k} \cdot \mathbf{v}_T/\sqrt{2}|k|\theta_{Ts}$ and $C_d = \mathbf{k} \cdot \mathbf{v}_T/\sqrt{2}|k|v_{Td}$.

To proceed further, we shall consider two limiting cases of Eq. (8) by imposing certain limitations on the test charge speed in comparison with the thermal and acoustic speeds and simplify the interaction potentials [as given by Eq. (7)] in Lorentzian dusty plasmas.

4.1 Slow moving test charge response

For a slow test charge propagation in a Lorentzian dusty plasma, we assume that test charge speed (v_T) is much slower than the dust thermal speed (v_{Td}). As a result, the test charge is shielded by all the plasma species, for example, electrons, ions, and dust grains. Since mass of the dust grains is larger than the mass of electrons and ions, therefore the dust thermal speed is much smaller than the electron and ion thermal speeds. Thus, imposing the inequalities $v_T \ll v_{Td} \lll \theta_{Ts}$, we consider only small argument expansions, that is, $|C_s| \ll 1$ and $|C_d| \ll 1$, in the plasma dispersion functions to obtain a simplified expression for dielectric constant. The inverse of the latter eventually yields the following result

$$D^{-1} \simeq \frac{k^2 \lambda_{D\kappa}^2}{k^2 \lambda_{D\kappa}^2 + 1} - i \frac{\mu v_T}{v_{Td}} \left(\frac{\pi}{2}\right)^{1/2} \frac{k^2 \lambda_{D\kappa}^4}{\lambda_{Dd}^2 \left(k^2 \lambda_{D\kappa}^2 + 1\right)^2}. \quad (9)$$

The modified effective Debye length $\lambda_{D\kappa}$ can be simplified in this form

$$\lambda_{D\kappa} = \lambda_0 \left(c_{\kappa e} \frac{n_{e0}}{n_0} + c_{\kappa i} \frac{T_e}{T_i} + Z_{d0}^2 \frac{n_{d0}}{n_0} \frac{T_e}{T_d} \right)^{-1/2}, \quad (10)$$

with superthermality parameters attributed to electrons and ions as

$$c_{\kappa e} = \frac{2\kappa_e - 1}{2\kappa_e - 3} \quad \text{and} \quad c_{\kappa i} = \frac{2\kappa_i - 1}{2\kappa_i - 3}.$$

The usual Debye length in electron-ion plasma is denoted by $\lambda_0 = (T_e/4\pi n_0 e^2)^{1/2}$ with $n_0 = n_{i0} \equiv n_{e0} + Z_{d0} n_{d0}$ and $\mu (= \cos\theta_k)$ representing the angle between the vectors $\mathbf{k}$ and $\mathbf{v}_T$. For Maxwellian plasmas, the superthermality parameters $c_{\kappa e,i} \to 1$ as long as $\kappa_{e,i} \to \infty$, implying that the effective shielding length exactly coincides with the earlier result [35] for static cold dust grains. It may be noticed from Eq. (9) that the inverse of dielectric constant is significantly influenced by the dust Landau damping rate [13]. The importance of the latter was first pointed out [13] in 1968, and it was suggested that if a test charge moves slowly in comparison to the dust thermal speed, the dust Landau damping term appearing in Eq. (9) cannot be ignored. For a static test charge, we set $v_T = 0$ and insert Eq. (9) into Eq. (7) to derive the short-range DH potential [53, 54] in the form $\phi_{DH} = (q_T/|r|)\exp(-|r|/\lambda_{D\kappa})$, where $|r| = (\rho^2 + \xi^2)^{1/2}$ is the distance from the test charge to an observer in terms of radial and axial distances ρ and $(\xi = z)$, respectively.

For numerical analyses, we can choose the data from the dusty plasma near Saturn's E ring, cited in Refs. [55–58] and many references therein. The data essentially corresponds to the Radio and Plasma Wave Science (RPWS) instruments onboard the Cassini spacecraft, containing the plasma parameters, such as $n_{d0} = 0.1\text{cm}^{-3}$, $n_{e0} = 70\text{cm}^{-3}$, $Z_{d0} = 300$, $T_d = T_e/10$, $T_i = T_e/2$, and $T_e = 4.642 \times 10^5\text{K}$. The computation further helps us in finding the magnitude of the effective shielding length $\lambda_{D\kappa}(= 15.649\text{cm})$ at the near-Maxwellian electrons and ions with $\kappa_{i,e} = 100$. The impact of superthermal tails in the electron and ion distributions only appear at lower values of the Kappa that may result into the reduction of the shielding length $\lambda_{D\kappa} \approx 15.4787\text{cm}$ for fixed $\kappa_i = 1.6$ and $\kappa_e = 100$, as well as $\lambda_{D\kappa} \approx 15.5888\text{cm}$ with $\kappa_e = 1.6$ and $\kappa_i = 100$. Thus, we notice from **Table 1** that for infinite values of the spectral indices, that is, $\kappa_{e,i} \to \infty$, the superthermality parameters $(c_{\kappa e,i})$ tend to unity, implying that the effective shielding length exactly coincides with the previous results [35] for static cold dust grains. However, the variation due to superthermal electrons and ions is shown in three different combinations (see **Table 1**) to affect the normalized values of the effective Debye length and DH potential almost 4 digits beyond the decimal point. At lower Kappa-values, the impact is relatively enhanced and in turn, suprathermal ions more efficiently modify the effective shielding length and DH potential as compared to suprathermal electrons because the ions may take

S #	Spectral-index for electrons κ_e	Spectral-index for ions κ_i	Electron-superthermality parameter c_{κ_e}	Ion-superthermality parameter c_{κ_i}	Effective Debye length $\frac{\lambda_{D\kappa}}{\lambda_0}$	DH Potential $\frac{\phi_{DH}}{(q_T/\lambda_0)}$
1	100	100	1.01015	1.01015	0.0332829	0.0122817
2	6	100	1.22222	1.01015	0.0332802	0.0122756
3	4	100	1.4	1.01015	0.0332779	0.0122705
4	3	100	1.66667	1.01015	0.0332745	0.0122629
5	2	100	3	1.01015	0.0332573	0.0122249
6	1.6	100	11	1.01015	0.0331548	0.0119997
7	100	100	1.01015	1.01015	0.0332829	0.0122817
8	100	6	1.01015	1.22222	0.03322751	0.0122644
9	100	4	1.01015	1.4	0.0332686	0.0122499
10	100	3	1.01015	1.66667	0.0332588	0.0122281
11	100	2	1.01015	3	0.0332098	0.0121203
12	100	1.6	1.01015	11	0.0329206	0.0114957
13	100	100	1.01015	1.01015	0.0332829	0.0122817
14	6	6	1.22222	1.22222	0.0332724	0.0122583
15	4	4	1.4	1.4	0.0332636	0.0122387
16	3	3	1.66667	1.66667	0.0332503	0.0122095
17	2	2	3	3	0.0331843	0.0120643
18	1.6	1.6	11	11	0.0327966	0.0112346

Table 1.
The electron and ion Kappa-indices affect the values of the effective Debye length and DH potential at fixed values of $r = 0.2$, $\eta = 10^{-3}$, *and* $Z_{do} = 300$.

more time to attain the Maxwellian equilibrium due to their larger mass compared to electrons.

The effect of dust concentration [denoted through the parameter $\eta(= n_{d0}/n_0)$] is shown on the contours of normalized effective shielding length [as given by Eq. (10)] in **Figure 2** for (a) $\eta = 0$ and (b) $\eta = 10^{-3}$ as function of electron-to-ion temperature ratio in the range $1 \leq T_e/T_i \leq 6$ and electron spectral index $1.6 \leq \kappa_e \leq 5$ at near-Maxwellian ions $\kappa_i = 100$. Since the speed of the test charge is much lower than the dust thermal speed (viz., $v_T \ll v_{Td}$), the test charge is therefore screened by all the plasma species, viz., the electrons, ions, and dust grains, hence effectively modifying the shielding length. The effective shielding length is shown to be decreased in the dusty plasma in comparison with traditional electron-ion plasma and is strongly influenced by the suprathermal tails of electrons. The impact of the superthermal electrons at lower Kappa values is more effective in the case of electron-ion plasma than dusty plasma. In the plots, one can easily observe that light-colored regions correspond to higher values of the effective shielding length while dense-colored regions determine the lower values of the shielding length. However, in **Figure 2(c)** and **(d)**, a reduction in the effective shielding length is revealed by the influence of suprathermal ions at near-Maxwellian electrons

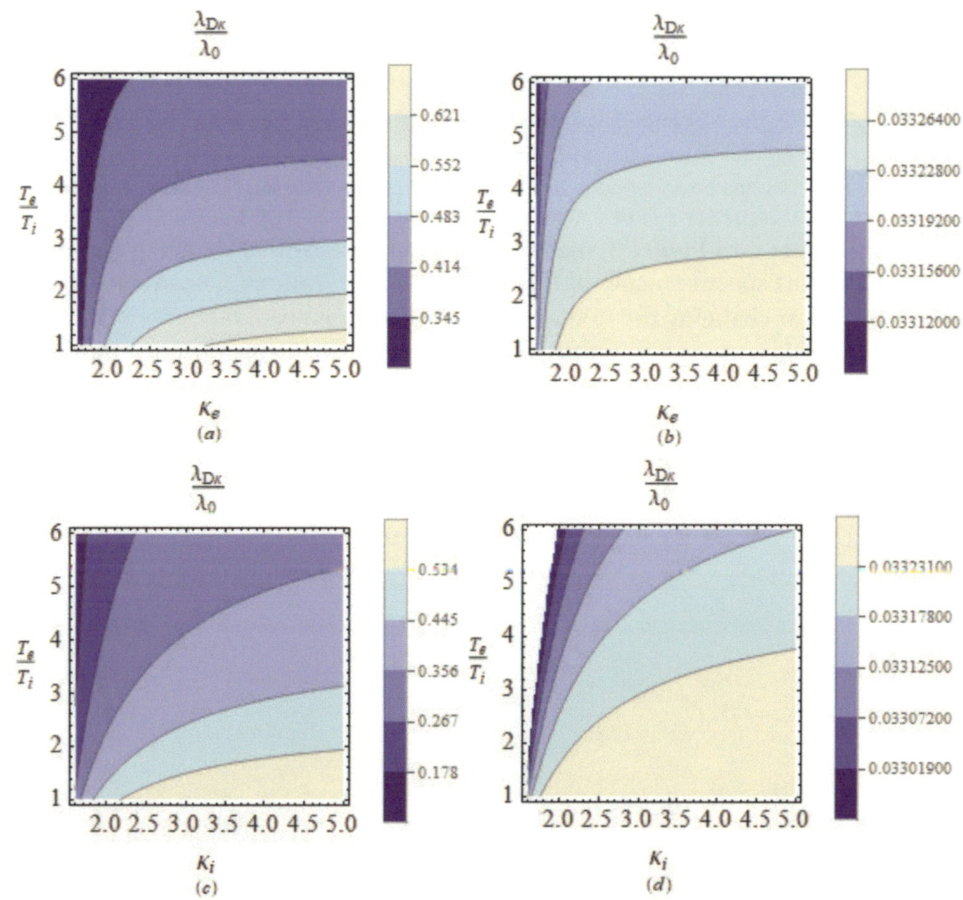

Figure 2.
Contours represent the effective shielding length $[\tilde{\lambda}_{D\kappa} = \lambda_{D\kappa}/\lambda_0]$ *against the temperature ratio* (T_e/T_i) *and electron spectral index* (κ_e) *for (a)* $\eta(= n_{do}/n_o) \equiv 0$ *(electron-ion plasma) and (b)* $\eta = 10^{-3}$ *(dusty plasma) with* $\kappa_i = 100$ *and* $Z_{do} = 300$*. contours in (c) and (d) vary for* $\tilde{\lambda}_{D\kappa}$ *against the temperature ratio* (T_e/T_i) *and ion spectral index* (κ_i) *for (c)* $\eta = 0$ *(electron-ion plasma) and (d)* $\eta = 10^{-3}$ *(dusty plasma) with* $\kappa_e = 100$ *and* $Z_{do} = 300$*. Other parameters are mentioned in Section 1.4.1.*

$\kappa_e = 100$. It may be noted that due to the variation of electron-to-ion temperature ratio against the ion spectral index, the magnitudes of effective shielding length become smaller at lower values of κ_i than the case of κ_e. However, the effective shielding length approaches to the maximum value in the limit $T_e \simeq T_i$ and decreases by enhancing the electron-to-ion temperature ratios.

4.2 Short-range DH and long-range FF potentials

To study short and long-range shielded potentials of a slowly moving test charge along the z-axis in a Lorentzian dusty plasma, we use Eq. (9) into Eq. (7) and spherical polar coordinates, as $\mathbf{v}_T = (0, 0, v_T)$, $\mathbf{k} = (k \sin\theta_k \cos\varphi_k,\ k \sin\theta_k \sin\varphi_k,\ k\cos\theta_k)$, and $\mathbf{r} = (r \sin\theta_r \cos\varphi_r,\ r \sin\theta_r \sin\varphi_r,\ r\cos\theta_r)$ to finally obtain the total potential $[\phi_1(\mathbf{r}, t) = \phi_{DH} + \phi_{FF}]$ as

$$\phi_1(\mathbf{r}, t) = \frac{q_T}{r} \exp\left(-\frac{r}{\lambda_{D\kappa}}\right) + \frac{2\sqrt{2} q_T}{\sqrt{\pi} r} \frac{v_T}{v_{Td}} \frac{\xi \lambda_{D\kappa}}{\lambda_{Dd}^2} \frac{\lambda_{D\kappa}^3}{r^3} \tag{11}$$

The first part of Eq. (11) corresponds to the short-range Debye-Hückel (DH) potential, which accounts for the short distances between the test charge and observer, whereas the second part represents the long-range far-field potential in the limit $r \gg \lambda_{D\kappa}$ decaying as the inverse cube of the distance to the test charge. Here $r = \left(\rho^2 + \xi^2\right)^{1/2}$ is the distance from the test charge to observer with radial and axial positions ρ and $\xi(= z - v_T t)$, respectively. When $\kappa_{e,i} \to \infty$, the effective Debye length $\lambda_{D\kappa}$ approaches to λ_D with $c_{\kappa e,i} = 1$, and consequently Eq. (11) exactly coincides with the earlier result [36] in the limit $\cos(\gamma) = \xi/r$, having null dust-charge fluctuations. In **Figure 3**, the magnitude of the effective shielding length varies against the specific ranges of the electron concentration (μ_e) and dust concentration (η) for changing the (a) electron-to-ion temperature ratios $\frac{T_e}{T_i} (= 1, 2, 4, 8)$ at near-Maxwellian electrons and ions $\kappa_{e,i} = 100$ and (b) electron-to-dust temperature ratios $\frac{T_e}{T_d} (= 10, 15, 20, 25)$ with fixed $\frac{T_e}{T_i} = 2$ for Kappa-distributed electrons and ions (i.e., $\kappa_{e,i} = 1.6$), respectively. Note that for an electron-ion plasma, the effective shielding length is approached to unity [35] as the maximum value for isothermal case $\frac{T_e}{T_i} = 1$ (black dotted curve) (see **Figure 3(a)**), which then decreases with respect to the electron concentration. For non-isothermal values, that is, $\frac{T_e}{T_i} = 2$

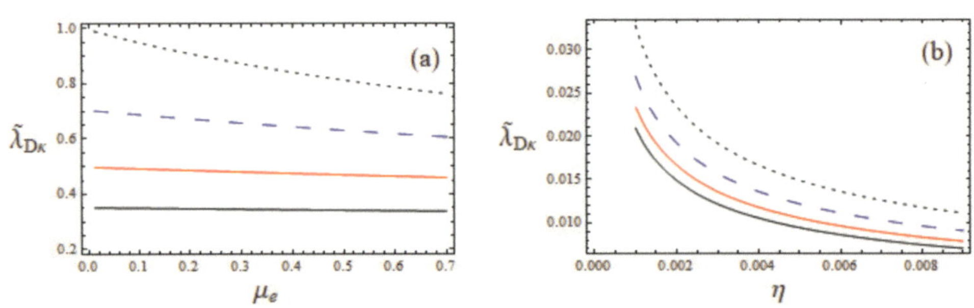

Figure 3.
The shielding length $\tilde{\lambda}_{D\kappa} (= \lambda_{D\kappa}/\lambda_0)$ against the electron and dust concentrations (μ_e, η) for varying the temperature ratios $\left(\frac{T_e}{T_i}, \frac{T_e}{T_d}\right)$ in (a) an electron-ion Maxwellian plasma and (b) superthermal dusty plasma, respectively.

(blue dashed curve), 4 (red solid curve), and 8 (black solid curve), the strength of the effective shielding however reduces in terms of fraction showing no more dominant impact of the electron concentration. In the presence of dust component, the effective shielding length is reduced as compared to the electron-ion plasma and clearly depends on the electron-to-dust temperature ratios $\frac{T_e}{T_d}(= 10, 15, 20, 25)$ at $Z_{d0} = 300$, as can be seen in **Figure 3(b)**.

The variation of the normalized DH potential $\tilde{\phi}_{DH}$ caused by a slowly moving test charge is displayed against the normalized axial distance $\tilde{\xi}$ for varying the dust concentration $\eta = 10^{-3}$, 1.2×10^{-3}, and 1.4×10^{-3} in **Figure 4** for (a) non-Maxwellian electrons and ions, as well as (b) near-Maxwellian electrons and ions. The DH potentials fastly reduce with a variation of dust concentration and attain large magnitudes in the near-Maxwellian case at $\kappa_{e,i} = 100$ compared to non-Maxwellian case at $\kappa_{e,i} = 1.6$.

Figure 5 exhibits how suprathermal electrons and ions modify the profiles of long-range FF potential ($\tilde{\phi}_{FF}$) caused by a slow test charge moving with speed $v_T = 0.02v_{Td}$. For small values of electron spectral index $\kappa_e(= 1.6, 1.8)$, the shielded FF potentials are more localized than the case at near-Maxwellian electrons for $\kappa_e = 100$ (see the red solid curve in **Figure 5(a)**). However, the magnitudes of the FF potential are comparatively decreased in **Figure 5(b)** because of the strong contribution of suprathermal ions.

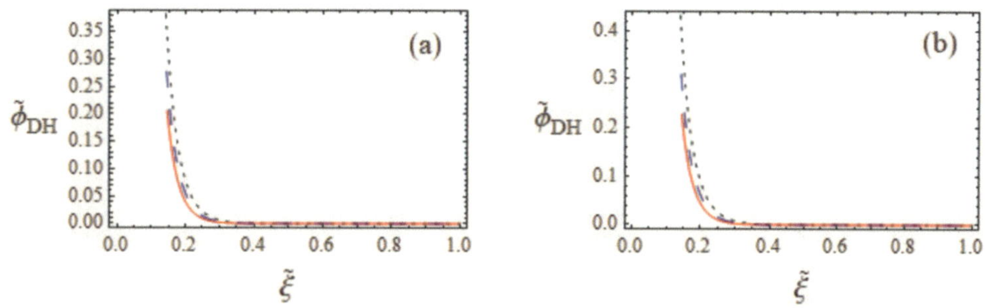

Figure 4.
The DH potential $\tilde{\phi}_{DH}[= \phi_{DH}/(q_T/\lambda_0)]$ vs. the axial distance $\tilde{\xi}(= \xi/\lambda_0)$ for different dust concentrations $\eta = 10^{-3}$ (black dotted curve), 1.2×10^{-3} (blue dashed curve), and 1.4×10^{-3} (red solid curve) at $Z_{do} = 200$ and $\tilde{\rho} = 0$ with (a) $\kappa_{e,i} = 1.6$ and (b) $\kappa_{e,i} = 100$.

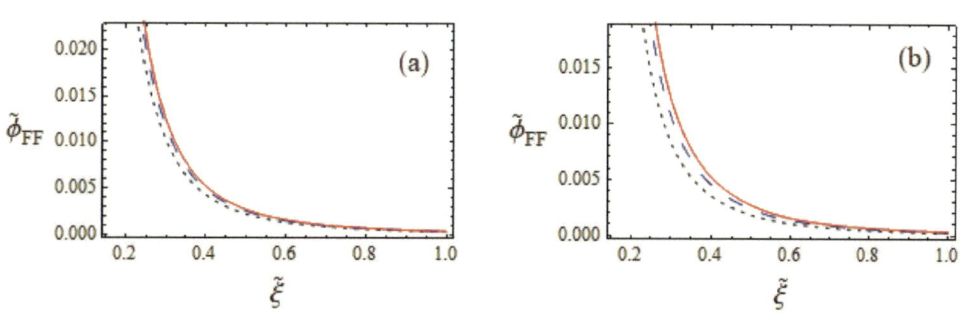

Figure 5.
The FF potential $\tilde{\phi}_{FF}[= \phi_{FF}/(q_T/\lambda_0)]$ vs. $\tilde{\xi}$ for varying (a) $\kappa_e = 1.6$ (black dotted curve), 1.8 (blue dashed curve), and 100 (red solid curve) with $\kappa_i = 100$, and (b) $\kappa_i = 1.6$ (black dotted curve), 1.8 (blue dashed curve), and 100 (red solid curve) with $\kappa_e = 100$. other values are $\eta = 10^{-4}$, $\rho = 0$, and $Z_{do} = 300$.

4.3 Resonating test charge response

To examine the resonant interaction of a test charge with DA waves, we first simply derive the dielectric constant of the DA waves using the limit $v_{Td} \ll \omega/k \ll \theta_{Te}, \theta_{Ti}$ in Eq. (8). We therefore consider the small and large argument expansions $C_s \ll 1$ and $C_d \gg 1$ of the plasma dispersion functions. On DA scales, the inertia is mainly provided by the negatively charged dust grains and restoring force by the pressures of superthermal inertialess electrons and ions for maintaining the propagation of DA waves. In typical laboratory plasmas, the frequency of DA waves is often below the dust plasma oscillation frequency in the range of 1–100 Hz. Thus, the modified dielectric constant for a Lorentzian dusty plasma takes the following form

$$D(k, \mathbf{k}\cdot\mathbf{v}_T) = 1 + \frac{1}{k^2\lambda_D'^2} - \frac{\omega_{pd}^2}{(\mathbf{k}\cdot\mathbf{v}_T)^2}. \tag{12}$$

The effective Debye length now gets a new form $\lambda_D' = \lambda_0/\sqrt{B_\kappa}$ with $B_\kappa = c_{\kappa e}\frac{n_{e0}}{n_0} + c_{\kappa i}\frac{T_e}{T_i}$. Note that dust-charge fluctuations are also ignored here because the characteristic damping rate attributed to dust charge fluctuations is much smaller than the collisional and Landau damping rates. The reciprocal of Eq. (12) can be simplified as

$$D^{-1} = \frac{k^2\lambda_0^2}{B_\kappa + k^2\lambda_0^2}\left\{1 + \frac{\omega_D^2}{(\mathbf{k}\cdot\mathbf{v}_T)^2 - \omega_D^2}\right\}, \tag{13}$$

with

$$\omega_D = \frac{kC_D}{\left(B_\kappa + k^2\lambda_0^2\right)^{1/2}}. \tag{14}$$

This is the DA resonance frequency with DA speed $C_D(=\omega_{pd}\lambda_0)$. For long wavelength limit $k\lambda_0 \ll B_\kappa$, the DA frequency reduces to $\omega_D = kC_D/\sqrt{B_\kappa}$ in the Lorentzian dusty plasma, while in the short wavelength limit $k\lambda_0 \gg B_\kappa$, the frequency simply approaches to the dust plasma oscillation frequency $\omega_D \simeq \omega_{pd}$. The factor $\omega_D^2/\left\{(\mathbf{k}\cdot\mathbf{v}_T)^2 - \omega_D^2\right\}$ in Eq. (13) identifies the dynamical effects of dust grains, which may lead to an oscillatory WF potential strongly depending upon whether the product $\mathbf{k}\cdot\mathbf{v}_T$ is smaller or larger than ω_D.

4.4 Short-range DH and long-range WF potentials

For static or slowly moving test charge in a Lorentzian dusty plasma, its potential distributions are found spherically symmetric both in the axial and radial directions. Consequently, the DH and FF shielded potentials are appropriately solved with spherical polar coordinates. However, if the test charge moves with finite speed in a specific direction along the z-axis, the resonant interaction of test charge with the DA wave leads to the asymmetric distribution of potential in the form of WF behind the test charge. The plasma model is then preferably solved in cylindrical coordinates. Thus, following the standard techniques [37, 38, 40] for DH and WF potentials, we make use of Eq. (13) into Eq. (7) to finally arrive at

$$\phi_1(\mathbf{r},t) = \frac{q_T}{r}\exp\left(-\frac{r\sqrt{B_\kappa}}{\lambda_0}\right) + \frac{2q_T}{\xi}\left(1+\frac{C_D^2}{B_\kappa v_T^2}\right)\left(B_\kappa - \frac{C_D^2}{v_T^2}\right)^{-1}\cos\left(\frac{\omega_{pd}\xi}{\sqrt{B_\kappa}v_T}\right), \tag{15}$$

where $\xi(=r_\parallel - v_T t)$ is the axial distance between the test charge and observer. The first part on the right hand side of Eq. (15) shows the contribution of modified DH potential and the second part corresponds to oscillatory WF potential accounting for the suprathermal electrons and ions in a Lorentzian dusty plasma. For $v_T > C_D$ and $\cos\left(\omega_{pd}\xi/v_T\sqrt{B_\kappa}\right) < 0$, the WF potential becomes attractive [40] and dominates over the repulsive DH potential because the latter decreases rapidly beyond the shielding cloud. An oscillatory WF potential $\tilde{\phi}_{WF}(=\phi_{WF}\lambda_0/q_T)$ of a test charge moving with speed $v_T = 0.2C_D$ is shown along the axial direction $\tilde{\xi}$ as a function of spectral indices κ_e and κ_i in **Figure 6(a)** and **(b)**, respectively. Observe that the amplitude of the WF potential increases as the superthermal indices κ_e and κ_i increase at fixed $\eta = 10^{-3}$ and $Z_{d0} = 300$. Moreover, the wakefield damps behind the test charge a bit earlier as shown in **Figure 6(b)** due to the strong dependence of suprathermal ions on the effective shielding length in comparison with the suprathermal electrons. The impact of electron-to-ion temperature ratio is also examined on the profiles of WF and DH potentials as a function of axial distance $\tilde{\xi}$

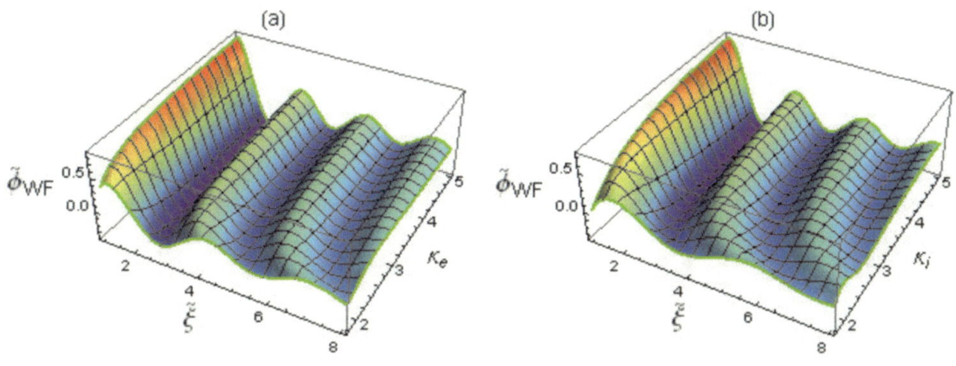

Figure 6.
The normalized WF potential $\tilde{\phi}_{WF}$ as a function of $\tilde{\xi}$ for varying the spectral indices as (a) $1.8 \leq \kappa_e < 5$ with fixed $v_T = 0.2C_D$ and $\kappa_i = 100$, and (b) $1.8 \leq \kappa_i < 5$ with $v_T = 0.2C_D$ and $\kappa_e = 100$.

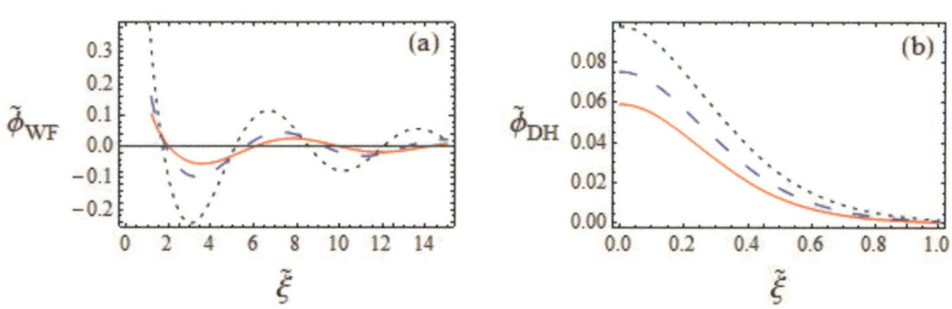

Figure 7.
The WF and DH potentials are shown against the axial distance $\tilde{\xi}$ for changing temperature ratios (a) T_e/T_i =2 (black dotted curve), 2.5 (blue dashed curve), and 3 (red solid curve), and (b) T_e/T_i =2 (black dotted curve), 2.5 (blue dashed curve), and 3 (red solid curve), respectively. Other common parameters are $\kappa_{e,i} = 1.6$, $v_T = 0.2C_D$, $\eta = 10^{-3}$, and $Z_{do} = 100$.

with fixed radial distance $\rho = 0.54\lambda_0$ and $v_T = 0.2C_D$ as can be seen in **Figure 7(a)** and **(b)**, respectively. It is important to examine that electron-to-ion temperature ratio suppresses the magnitudes of the WF and DH potentials at low values of $\kappa_{e,i} = 1.6$ in the Lorentzian dusty plasma (**Figures 6** and 7).

4.5 Fast moving test charge response

In this case, the test charge is assumed to be moving much faster than all the plasma species (viz., the electrons, ions and negatively charged dust grains). Consequently, Eq. (6) can be expressed in 1D form to finally arrive at $D(k, \mathbf{k}.\mathbf{v}_T) \simeq 1$ in the limits $v_T| \gg ku_j$ and $\omega = \mathbf{k}.\mathbf{v}_T$. Thus, the test charge potential (7) simply leads to the Coulomb potential

$$\phi_1(\mathbf{r}, t) = \phi_C \equiv \frac{q_T}{r}. \tag{16}$$

It is now clear that if the test charge is moving very fast, then there is no shielding around it in the Lorentzian dusty plasma.

5. Conclusion

To conclude, we have briefly discussed different plasma diagnostic techniques and specifically investigated the novel features of interaction potentials caused by a test charge moving with constant velocity $\mathbf{v}_T$ along the z-axis in a collisionless unmagnetized Lorentzian dusty plasma. For this purpose, the linearized coupled Vlasov-Poisson equations are employed to model suprathermal electrons and ions with Kappa-df, as well as negatively charged dust grains with Maxwell-df, respectively. After applying the space-time Fourier transformations, an electrostatic potential is obtained with a modified dielectric constant. For taking the test charge speed much smaller than the dust thermal speed in a Lorentzian dusty plasma, we then express the total potential distribution in terms of short-range Debye-Hückel (DH) and long-range far-field potentials. The DH potential exponentially decays with distance, whereas FF potential decreases as the inverse cube of the distance. Both the potentials are substantially influenced by the plasma and superthermality parameters. However, a resonating test charge with DA oscillations introduces the long-range WF potential excitations behind the test charge in Lorentzian dusty plasmas. A Coulomb potential is obtained when the test charge is moving very fast compared to plasma species, and there is no shielding around it in the Lorentzian dusty plasma.

Vladimirov and Nambu [40] have already utilized the idea of WF potential for making new materials by attracting the same polarity dust grains in dusty plasmas. The physics of attractive forces between the negatively charged dust grains is completely analogous to that of Cooper pairing of electrons in superconductors [59]. The dust particle physically polarizes the plasma medium and creates attractive potential regions, where positive ions from collective interaction of DA waves can be focused. This may in turn lead to the possibility for dust crystallization and dust coagulation in both laboratory and space dusty plasmas.

Acknowledgements

Dr. S. Ali dedicates this document to Late Prof. P.K. Shukla and Dr. B. Eliasson (University of Strathclyde, UK) who were very kind to him at many occasions during discussions on dusty plasmas, and acknowledges the partial financial assistance from USTC, Hefei, China, and ICTP, Trieste, Italy, for making his visits feasible in 2018 and 2019, respectively. Professor Y. Al-Hadeethi also acknowledges the technical support of the Deanship of Scientific Research (DSR), King Abdulaziz University, Saudi Arabia.

Author details

Shahid Ali[1]* and Yas Al-Hadeethi[2]

1 National Centre for Physics (NCP), Quaid-e-Azam University Campus, Islamabad, Pakistan

2 Faculty of Sciences, Department of Physics, King AbdulAziz University, Jeddah, Saudia Arabia

*Address all correspondence to: shahid_gc@yahoo.com

IntechOpen

© 2020 The Author(s). Licensee IntechOpen. This chapter is distributed under the terms of the Creative Commons Attribution License (http://creativecommons.org/licenses/by/3.0), which permits unrestricted use, distribution, and reproduction in any medium, provided the original work is properly cited.

References

[1] Al-Hadithi Y, Tallents GJ, Neely D. X-Ray Lasers 1990. Bristol: IOP Publishers; 1991

[2] Al-Hadithi Y, Tallents GJ, Zhang J, Key MH, Norreys PA, Kodama R. Energy transport in plasmas produced by a high brightness Krypton fluoride laser focused to a line. Physics of Plasmas. 1994;**1**:1279. DOI: 10.1063/1.870726

[3] Al-Hadeethi Y, Al-Mujtabi A, Al-Marzouki FM. Laser produced plasma x-ray sources for nanoscale resolution contact microscopy: A candidate in cancerous stem cells imaging. Advances in Molecular Imaging. 2017;7:67. DOI: 10.4236/ami.2017.74004

[4] Chen FF. Introduction to Plasma Physics and Controlled Fusion. New York: Plenum; 1984

[5] Fortov VE, Ivlev AV, Khrapak SA, Morfill GE. Complex (dusty) plasmas: Current status, open issues, perspectives. Physics Reports. 2005;**421**:1. DOI: 10.1016/j.physrep.2005.08.007

[6] Shukla PK, Silin VP. Dust ion-acoustic wave. Physica Scripta. 1992;**45**:508. DOI: 10.1088/0031-8949/45/5/015

[7] Rao NN, Shukla PK, Yu MY. Dust-acoustic waves in dusty plasmas. Planetary Space Science. 1990;**38**:543. DOI: 10.1016/0032-0633(90)90147-I

[8] Barkan A, D'Angelo N, Merlino RL. Experiments on ion-acoustic waves in dusty plasmas. Planetary Space Science. 1996;**44**:239. DOI: 10.1016/0032-0633(95)00109-3

[9] Merlino RL, Barkan A, Thompson C, D'Angelo N. Laboratory studies of waves and instabilities in dusty plasmas. Physics of Plasmas. 1998;**5**:1607. DOI: 10.1063/1.872828

[10] Barkan A, Merlino RL, D'Angelo N. Laboratory observation of the dust-acoustic wave mode. Physics of Plasmas. 1995;**2**:3563. DOI: 10.1063/1.871121

[11] Krall NA, Trivelpiece AW. Principles of Plasma Physics. New York: McGraw-Hill; 1973

[12] Neufeld J, Ritchie RH. Passage of charged particles through plasma. Physical Review. 1955;**98**:1632. DOI: 10.1103/PhysRev.98.1632

[13] Montgomery D, Joyce G, Sugihara R. Inverse third power law for the shielding of test particles. Plasma Physics. 1968;**10**:681. DOI: 10.1088/0032-1028/10/7/304

[14] Stenflo L, Yu MY, Shukla PK. Shielding of a slow test charge in a collisional plasma. Physics of Fluids. 1973;**16**:450. DOI: 10.1063/1.1694361

[15] Shukla PK, Spatschek K-H. Shielding of a moving test charge in a turbulent plasma. Physics Letters A. 1973;**44**:398. DOI: 10.1016/0375-9601(73)90840-2

[16] Yu MY, Stenflo L, Shukla PK. On the electric field of a moving test charge. Radio Science. 1972;**7**:1151. DOI: 10.1029/RS007i012p01151

[17] Shukla PK, Spatschek K-H, Yu MY. Far-field potential of a test charge in an inhomogeneous and magnetized plasma. Canadian Journal of Physics. 1974;**52**:281. DOI: 10.1139/p74-040

[18] Ali S, Nasim MH, Murtaza G. Effects of dust-charge fluctuations on the potential of an array of projectiles in a partially ionized dusty plasma. Physics of Plasmas. 2003;**10**:4207. DOI: 10.1063/1.1619976

[19] Ali S, Nasim MH, Murtaza G. Correlation effects due to an axial

propagation of projectiles in a dusty plasma. Physics of Plasmas. 2005;**12**: 033502. DOI: 10.1063/1.1844496

[20] Ali S, Murtaza G, Nasim MH. Energy loss for the assemblies of charged projectiles in a dusty plasma. Physics of Plasmas. 2005;**12**:072104. DOI: 10.1063/1.1933739

[21] Brandt W, Ratkowski A, Ritchie RH. Energy loss of swift proton clusters in solids. Physical Review Letters. 1974;**33**: 1325. DOI: 10.1103/PhysRevLett.33.1325

[22] Deutsch C, Tahir NA. Fragmentation and stopping of heavy cluster ions in a lithium target—Application to target implosion. Physics of Fluids B. 1992;**4**:3735. DOI: 10.1063/1.860329

[23] Skupsky S. Energy loss of ions moving through high-density matter. Physical Review A. 1977;**16**:727. DOI: 10.1103/PhysRevA.16.727

[24] Li CK, Petrasso RD. Charged-particle stopping powers in inertial confinement fusion plasmas. Physical Review Letters. 1993;**70**:3059

[25] Lindl JD. Inertial Confinement Fusion. New York: Springer Verlag; 1998

[26] Jones ME, Keinigs R. Ion plasma wave wakefield accelerators. IEEE Transactions on Plasma Science. 1987; **15**:203. DOI: 10.1109/TPS.1987.4316686

[27] Mohideen U, Rahman HU, Smith MA, Rosenberg M, Memdis DA. Intergrain coupling in dusty-plasma Coulomb crystals. Physical Review Letters. 1998;**81**:349. DOI: 10.1103/PhysRevLett.81.349

[28] Takahashi K, Oishi T, Shimomai K-I, Hayashi Y, Nishino S. Analyses of attractive forces between particles in Coulomb crystal of dusty plasmas by optical manipulations. Physical Review E. 1998;**58**:7805. DOI: 10.1103/PhysRevE.58.7805

[29] Zwicknagel G, Deutsch C. Correlated ion stopping in plasmas. Physical Review E. 1997;**56**:970. DOI: 10.1103/PhysRevE.56.970

[30] Sanmartin JR, Lam SH. Far-wake structure in rarefield plasma flows past charged bodies. Physics of Fluids. 1971; **14**:62. DOI: 10.1063/1.1693289

[31] Yu MY, Tegeback R, Stenflo L. On test charge potentials in collisional plasmas. Zeitschrift für Physik. 1973; **264**:341. DOI: 10.1007/BF01398860

[32] Peter T. Linearized potential of an ion moving through plasma. Journal of Plasma Physics. 1990;**44**:269. DOI: 10.1017/S0022377800015178

[33] Shivamoggi BK, Mulser P. Dielectric screening and stopping power of a test charge moving in a plasma. Journal of Plasma Physics. 1998;**60**:819

[34] Shukla PK, Singh RN. Total energy loss of a test charge in a collisional magnetoplasma. Physica Scripta. 1971;**4**: 282. DOI: 10.1088/0031-8949/4/6/008

[35] Lakshmi VS, Bharuthram R, Shukla PK. Debye shielding in a dusty plasma. Astrophysics and Space Science. 1993;**209**:213. DOI: 10.1007/BF00627441

[36] Shukla PK. Shielding of a slowly moving test charge in dusty plasmas. Physics of Plasmas. 1994;**1**:1362. DOI: 10.1063/1.870736

[37] Nambu M, Akama H. Attractive potential between resonant electrons. Physics of Fluids. 2300;**1985**:28. DOI: 10.1063/1.865284

[38] Nambu M, Vladimirov SV, Shukla PK. Attractive forces between charged particulates in plasmas. Physics

Letters A. 1995;**203**:40. DOI: 10.1016/0375-9601(95)00380-L

[39] Shukla PK, Rao NN. Coulomb crystallization in colloidal plasmas with streaming ions and dust grains. Physics of Plasmas. 1996;**3**:1770. DOI: 10.1063/1.871695

[40] Vladimirov SV, Nambu M. Attraction of charged particles in plasmas with finite flows. Physical Review E. 1995;**52**:R2172. DOI: 10.1103/PhysRevE.52.R2172

[41] Nambu M, Saikia BJ, Hada T. Wake potential around a test dust particulate in a magnetized plasma with streaming ions. Journal of the Physical Society of Japan. 2001;**70**:1175. DOI: 10.1143/jpsj.70.1175

[42] Freund HP, Smith RA, Papadopoulos K, Palmadesso P. Modulational instability in a plasma with supra-thermal electrons. Physics of Fluids. 1981;**24**:442. DOI: 10.1063/1.863390

[43] Freese KB, Walsh JE, Lohr J. Wave-particle interaction in the late beam-plasma instability. Physics of Fluids. 1979;**22**:2367. DOI: 10.1063/1.862549

[44] Smith RA, Goldstein ML, Papadopoulos K. Nonlinear stability of solar type III radio bursts. I-Theory. The Astrophysical Journal. 1979;**234**:348

[45] Van Compernolle B, Gekelman W, Pribyl P. Generation of supra-thermal electrons and Alfvén waves by a high power pulse at the electron plasma frequency. Physics of Plasmas. 2006;**13**:092112. DOI: 10.1063/1.2261850

[46] Eliasson B, Shao X, Milikh G, Mishin EV, Papadopoulos K. Numerical modeling of artificial ionospheric layers driven by high-power HF heating. Journal of Geophysical Research. 2012;**117**:A10321. DOI: 10.1029/2012JA018105

[47] Vasyliunas VM. A survey of low-energy electrons in the evening sector of the magnetosphere with OGO 1 and OGO 3. Journal of Geophysical Research. 1968;**73**:2839. DOI: 10.1029/JA073i009p02839

[48] Summers D, Thorne RM. The modified plasma dispersion function. Physics of Fluids B. 1991;**3**:1835. DOI: 10.1063/1.859653

[49] Baluku TK, Hellberg MA. Dust acoustic solitons in plasmas with kappa-distributed electrons and/or ions. Physics of Plasmas. 2008;**15**:123705. DOI: 10.1063/1.3042215

[50] Ali S, Eliasson B. Slowly moving test charge in two-electron component non-Maxwellian plasma. Physics of Plasmas. 2015;**22**:084508. DOI: 10.1063/1.4928901

[51] Ali S, Eliasson B. Slow test charge response in a dusty plasma with kappa distributed electrons and ions. Physica Scripta. 2017;**92**:084003. DOI: 10.1088/1402-4896/aa7c09

[52] Burton FD, Conte SD. The Plasma Dispersion Function. New York: Academic Press; 2015

[53] Debye P, Hückel E. On the Debye length in strong electrolytes. Physikalische Zeitschrift. 1923;**24**:185

[54] Rubab N, Murtaza G. Debye length in non-Maxwellian plasmas. Physica Scripta. 2006;**74**:145. DOI: 10.1088/0031-8949/74/2/001

[55] Du J, Guo R, Liu Z, Du S. Slowing down of charged particles in dusty plasmas with power-law kappa-distributions. Contributions to Plasma Physics. 2019;**59**:144. DOI: 10.1002/ctpp.201800046

[56] Wahlund J-E, Andre M, Eriksson AIE, Lundberg M, Morooka MW, Shafiq M, et al. Detection of dusty plasma near the

E-ring of Saturn. Planetary and Space Science. 2009;**57**:1795. DOI: 10.1016/j.pss.2009.03.011

[57] Leisner JS, Russell CT, Dougherty MK, Blanco-Cano X, Strangeway RJ, Bertucci C. Ion cyclotron waves in Saturn's E ring: Initial Cassini observations. Geophysical Research Letter. 2006;**33**:L11101. DOI: 10.1029/2005GL024875

[58] Kurth WS, Averkamp TF, Gurnett DA, Wang Z. Cassini RPWS observations of dust in Saturn's E Ring. Planetary and Space Science. 2006;**54**: 988. DOI: 10.1016/j.pss.2006.05.011

[59] De Gennes PG. Superconductivity of Metals and Alloys. New York: Benjamin; 1966

Section 6

Nonlinear Phenomena in Plasma

Chapter 6

Approximate Analytical Solution of Nonlinear Evolution Equations

Laxmikanta Mandi, Kaushik Roy and Prasanta Chatterjee

Abstract

Analytical solitary wave solution of the dust ion acoustic waves (DIAWs) is studied in the frame-work of Korteweg-de Vries (KdV), damped force Korteweg-de Vries (DFKdV), damped force modified Korteweg-de Vries (DFMKdV) and damped forced Zakharov-Kuznetsov (DFZK) equations in an unmagnetized collisional dusty plasma consisting of negatively charged dust grain, positively charged ions, Maxwellian distributed electrons and neutral particles. Using reductive perturbation technique (RPT), the evolution equations are obtained for DIAWs.

Keywords: solitary wave, soliton, KdV, DKdV, DFZK

1. Introduction

In the field of physics and applied mathematics research getting an exact solution of a nonlinear partial differential equation is very important. The elaboration of many complex phenomena in fluid mechanics, plasma physics, optical fibers, biology, solid-state physics, etc. is possible if analytical solutions can be obtained. Most of the differential equation arises in these field has no explicit solution as popularly known. This problem creates hindrances in the study of nonlinear phenomena and makes it time-consuming in the research of nonlinear models in the plasma and other science. However recent researches in nonlinear differential equations have seen the development of many approximate analytical solutions of partial and ordinary differential equations.

The history behind the discovery of soliton is not only interesting but also significant. In 1834 a Scottish scientist and engineer—John Scott-Russell first noticed the solitary water wave on the Edinburgh Glasgow Canal. In 1844 [1] in "Report on Waves" he accounted his examinations to the British Association. He wrote "I was observing the motion of a boat which was rapidly drawn along a narrow channel by a pair of horses, when the boat suddenly stopped not so the mass of water in the channel which it had put in motion; it accumulated round the prow of the vessel in a state of violent agitation, then suddenly leaving it behind, rolled forward with great velocity, assuming the form of a large solitary elevation, a rounded, smooth and well defined heap of water, which continued its course along the channel apparently without change of form or diminution of speed. I followed it on horseback and overtook it still rolling on at a rate of some eight or nine miles an hour, preserving its original figure some thirty feet long and a foot to a foot and a half in height. Its height gradually diminished and after a chase of one or two miles I lost it in the windings of the channel. Such in the month of August 1834 was my first

 IntechOpen

chance interview with that singular and beautiful phenomenon which I have called the Wave of Translation." He coined the word "solitary wave." The solitary wave is called so because it often occurs as a single entity and is localized. The most important characteristics of solitary waves were unearthed after thorough study along with extensive wave-tank experiments. The following are the properties of solitary waves:

(a) These localized bell-shaped waves travel with enduring form and velocity. The speed of these waves are given by $c^2 = g(h + a)$, where g, a, h are respectively represent the acceleration of the gravity, amplitude of the wave and the undisturbed depth of the water. (b) Solitary waves can cross each other without any alteration.

John Scott-Russell's study created a stir in the scientific community. His study not only initiated a debate with the prevailing knowledge of the theories of waves but also challenged the antecedent knowledge of waves. The previous study claimed that a periodic wave of finite amplitude and permanent shape are feasible only in deep water unlike Russell's observation that the permanent profile is also possible in shallow water. Finally the stable form of solitary waves was received in scientific community with the aid of nonlinearity and dispersion. An ideal equilibrium between nonlinearity and dispersion can generate such waves.

Diederik Johannes Korteweg in 1895 [2] along with his PhD student Gustav De Vries obtained an equation from the primary equation of hydrodynamics. This equation explains shallow water waves where the existence of solitary waves was mathematically recognized. This equation is called KdV equation which is of the form $\frac{\partial u}{\partial t} + Au\frac{\partial u}{\partial x} + B\frac{\partial^3 u}{\partial x^3} = 0$. One of the most popular equations of soliton theory, this equation helps in explaining primary ideas that lie behind the soliton concept. Martin Zabusky and Norman Kruskal [3] in 1965 solved KdV equation numerically and noticed that the localized waves retain their shape and momentum in collisions. These waves were known as "solitons." Soliton are solitary waves with the significant property that the solitons maintain the form asymptotically even when it experiences a collision. The fundamental "microscopic" properties of the soliton interaction; (i) the interaction does not change the soliton amplitudes; (ii) after the interaction, each soliton gets an additional phase shift; (iii) the total phase shift of a soliton acquired during a certain time interval can be calculated as a sum of the elementary phase shifts in pair wise collisions of this soliton with other solitons during this time interval is of importance. Solitons are mainly used in fiber optics, optical computer etc. which has really generated a stir in today's scientific community. The conventional signal dispensation depends on linear system and linear systems. After all in this case nonlinear systems create more well-organized algorithms. The optical soliton is comparatively different from KdV solitons. Unlike the KdV soliton that illustrates the wave in a solitary wave, the optical soliton in fibers is the solitary wave of an envelope of a light wave. In this regard, the optical soliton in a fiber is treated as an envelope soliton.

This chapter will discuss the analytical solitary wave solution of the KdV and KdV-like equations. In the study of nonlinear dispersive waves, these equations are generally seen. The KdV equation, a generic equation, is important in the study of weakly nonlinear long waves. This equation consists of a single humped wave characterized by several unique properties. The Soliton solutions of the KdV equation have been quite popular but it also not devoid of problems. The problems not only restrict to dispersion but also dissipation and interestingly these are not dominated by the KdV equation. The standard KdV equation fails to explain the development of small-amplitude solitary waves in case the particles collide in a plasma system. KdV equation with an additional damping term or the damped

Korteweg-de Vries (DKdV) equation becomes handy in explaining this issue of elaborating the character of the wave. But in the presence of any critical physical situation (critical point) nonlinearity of the KdV equation disappears and the amplitude of the waves reaches infinity. To control this situation, a new nonlinear partial differential equation has to be derived that can explain the system at that critical point. This is known as the modified Korteweg-de Vries (MKdV) equation. In the presence of collisions, this equation is not also adequate and a damped MKdV equation is necessary. Also in the presence of force source term then the equation will be further modified and become DFKdV/DFMKdV.

2. The Korteweg-de Vries equation

Now we will derive the KdV equation from a classic plasma model, in which we consider a collision-free unmagnetized plasma consists of electrons and ions, in which ions are mobile and electrons obey the Maxwell distribution. The basic equation will be given as:

$$\frac{\partial N_i}{\partial T} + \frac{\partial N_i U_i}{\partial X} = 0 \tag{1}$$

$$\frac{\partial U_i}{\partial T} + U_i \frac{\partial U_i}{\partial X} = -\frac{e}{m_i} \frac{\partial \psi}{\partial X} \tag{2}$$

$$\varepsilon_0 \frac{\partial^2 \psi}{\partial X^2} = e(N_e - N_i) \tag{3}$$

where the electrons obey Maxwell distribution, i.e., $N_e = en_0 e^{\frac{e\phi}{K_B T_e}}$. N_i, N_e, U_i, m_i are the ion density, electron density, ion velocity and ion mass, respectively. ψ is the electrostatic potential, K_B is the Boltzmann constant, T_e is the electron temperature and e is the charge of the electrons.

To write Eqs. (1)–(3) in dimensionless from we introduce the following dimensionless variables

$$x = \frac{X}{\lambda_D}, t = \omega_p T, \phi = \frac{e\psi}{KT_e}, n_i = \frac{N_i}{n_0}, u_i = \frac{U_i}{c_s}, \tag{4}$$

where $\lambda_D = \sqrt{\varepsilon_0 K_B T_e / n_0 e^2}$ is the Debye length, $c_s = \sqrt{K_B T_e / m_i}$ is the ion acoustic speed, $\omega_{pi} = \sqrt{n_0 e^2 / \varepsilon_0 m_i}$ is the ion plasma frequency and n_0 is the unperturbed density of ions and electrons. Hence using (4) in (1)–(3) we obtain the normalized set of equations as

$$\frac{\partial n_i}{\partial t} + \frac{\partial (n_i u_i)}{\partial x} = 0 \tag{5}$$

$$\frac{\partial u_i}{\partial t} + u_i \frac{\partial u_i}{\partial x} = -\frac{\partial \phi}{\partial x} \tag{6}$$

$$\frac{\partial^2 \phi}{\partial x^2} = e^{\phi} - n_i \tag{7}$$

To linearized (5)–(7), let us write the dependent variable as sum of equilibrium and perturbed parts, so that we write $n_i = 1 + \overline{n}_i, u_i = \overline{u}_i, \phi = \overline{\phi}$. Putting $n_i = 1 + \overline{n}_i$ where the values of parameters at equilibrium position is given by $n_1 = 1, u_1 = 0$ and $\phi_i = 0$ in Eq. (5), we get

$$\frac{\partial}{\partial t}(1+\overline{n}_i)+\frac{\partial}{\partial x}(\overline{u}_i+\overline{n}_i\overline{u}_i)=0 \tag{8}$$

neglecting the nonlinear term $\frac{\partial(\overline{n}_i\overline{u}_i)}{\partial x}$ from (8), we get

$$\frac{\partial\overline{n}_i}{\partial\overline{t}}+\frac{\partial\overline{u}_i}{\partial\overline{x}}=0 \tag{9}$$

which is the linearized form of Eq. (5).
Putting $u_i=\overline{u}_i, \phi=\overline{\phi}$ in Eq. (6), we get

$$\frac{\partial\overline{u}_i}{\partial t}+\overline{u}_i\frac{\partial\overline{u}_i}{\partial x}=-\frac{\partial\overline{\phi}}{\partial x} \tag{10}$$

Neglecting the nonlinear term from (10), we get

$$\frac{\partial\overline{u}_i}{\partial t}+\frac{\partial\overline{\phi}}{\partial\overline{x}}=0 \tag{11}$$

This is the linearized form of Eq. (6).
Putting $n_i=1+\overline{n}_i, \phi=\overline{\phi}$ in Eq. (7), we get

$$\begin{aligned}\frac{\partial^2\overline{\phi}}{\partial x}&=1+\overline{\phi}-1-\overline{n}_i\\ \Rightarrow\frac{\partial^2\overline{\phi}}{\partial x}&=\overline{\phi}-\overline{n}_i\end{aligned} \tag{12}$$

Hence Eqs. (9), (11), (12) are the linearized form of Eq. (5)–(7) respectively.
To get dispersion relation for low frequency wave let us assume that the perturbation is proportional to $e^{i(kx-\omega t)}$ and of the form

$$\overline{n}=n_0e^{i(kx-\omega t)} \tag{13}$$

$$\overline{u}=u_0e^{i(kx-\omega t)} \tag{14}$$

$$\overline{\phi}=\phi_0e^{i(kx-\omega t)} \tag{15}$$

So,

$$\frac{\partial\overline{n}}{\partial t}=-in_0\omega e^{i(kx-\omega t)} \tag{16}$$

$$\frac{\partial\overline{n}}{\partial x}=ikn_0e^{i(kx-\omega t)} \tag{17}$$

$$\frac{\partial\overline{u}}{\partial t}=-iu_0\omega e^{i(kx-\omega t)} \tag{18}$$

$$\frac{\partial\overline{u}}{\partial x}=iku_0e^{i(kx-\omega t)} \tag{19}$$

$$\frac{\partial\overline{\phi}}{\partial x}=ik\phi_0e^{i(kx-\omega t)} \tag{20}$$

$$\frac{\partial^2\overline{\phi}}{\partial x^2}=(ik)^2\phi_0e^{i(kx-\omega t)} \tag{21}$$

Putting these value in Eqs. (9), (11) and (12), we get,

$$-i\omega n_0 + iku_0 = 0 \tag{22}$$

$$-i\omega u_0 + ik\phi_0 = 0 \tag{23}$$

$$n_0 - \left(k^2+1\right)\phi_0 = 0 \tag{24}$$

Since the system (22)–(24) is a system of linear homogeneous equation so for nontrivial solutions we have

$$\begin{vmatrix} -i\omega & ik & 0 \\ 0 & -i\omega & ik \\ 1 & 0 & -\left(k^2+1\right) \end{vmatrix} = 0 \tag{25}$$

$$\Rightarrow -i^2\omega^2\left(k^2+1\right) + i^2k^2 = 0$$

$$\Rightarrow \omega^2\left(k^2+1\right) = -i^2k^2$$

$$\Rightarrow \omega^2 = \frac{k^2}{\left(k^2+1\right)}$$

This is the **dispersion relation.**

For small k, i.e., for weak dispersion we can expand as

$$\begin{aligned} \omega &= k\left(1+k^2\right)^{-\frac{1}{2}} \\ &= k - \frac{1}{2}K^3 + \cdots \end{aligned} \tag{26}$$

The phase velocity as

$$V_p = \frac{\omega}{k} = \frac{1}{\sqrt{\left(1+k^2\right)}} \tag{27}$$

so that $V_p \to 1$ as $k \to 0$ and $V_p \to 0$ as $k \to \infty$. The group velocity $V_g = \frac{d\omega}{dk}$ is given by

$$V_g = \frac{1}{\left(1+k^2\right)^{3/2}} \tag{28}$$

In this case, we have $V_g < V_p$ for all $k > 0$. The group velocity is more important as energy of a medium transfer with this velocity.

For long-wave as $k \to 0$, the leading order approximation is $\omega = k$, corresponding to non-dispersive acoustic waves with phase speed $\omega/k = 1$. Hence this speed is the same as the speed of the ion-acoustic waves c_s. The long wave dispersion is weak, i.e., $k\lambda_D < <1$. This means that the wavelength is much larger than the Debye length. In these long waves, the electrons oscillate with the ions. The inertia of the wave is provided by the ions and the restoring pressure force by the electrons. At the next order in k, we find that

$$\omega = k - \frac{1}{2}k^3 + O\left(k^5\right) \quad as\ k \to 0 \tag{29}$$

The $O(k^5)$ correction corresponds to weak KdV type long wave dispersion. For short wave ($k \rightarrow \infty$), the frequency $\omega = 1$, corresponding to the ion plasma frequency $\omega_{pi} = \frac{c_s}{\lambda_D}$. Hence the ions oscillate in the fixed background of electrons.

Now the phase of the waves can be written as

$$kx - \omega t = k(x - t) + \frac{1}{2}k^3 t \tag{30}$$

Here $k(x-t)$ and $k^3 t$ have same dynamic status (dimension) in the phase. Assuming k to be small order of $\varepsilon^{1/2}$, ε being a small parameter measuring the weakness of the dispersion, Here $(x-t)$ is the traveling wave form and time t is the linear form.

Let us consider a new stretched coordinates ξ,τ such that

$$\xi = \varepsilon^{1/2}(x - \lambda t), \tau = \varepsilon^{3/2} t \tag{31}$$

where ε is the strength of nonlinearity and λ is the Mach number (phase velocity of the wave). ε may be termed as the size of the perturbation. Let the variables be perturbed from the stable state in the following way (considering $n_i = 1, u_i = 0$, $\phi = 0$ and $n_e = e^{\phi} = e^0 = 1$ at equilibrium)

$$n_i = 1 + \varepsilon n_i^{(1)} + \varepsilon^2 n_i^{(2)} + \varepsilon^3 n_i^{(3)} + \cdots, \tag{32}$$

$$u_i = 0 + \varepsilon u_i^{(1)} + \varepsilon^2 u_i^{(2)} + \varepsilon^3 u_i^{(3)} + \cdots, \tag{33}$$

$$\phi = 0 + \varepsilon \phi^{(1)} + \varepsilon^2 \phi^{(2)} + \varepsilon^3 \phi^{(3)} + \cdots. \tag{34}$$

where x and t are function of ξ and τ so partial derivatives with respect to x and t can be transform into partial derivative in terms of ξ and τ so

$$\frac{\partial}{\partial x} = \frac{\partial}{\partial \xi}\frac{\partial \xi}{\partial x} + \frac{\partial}{\partial \tau}\frac{\partial \tau}{\partial x}, \Rightarrow \frac{\partial}{\partial x} = \varepsilon^{\frac{1}{2}}\frac{\partial}{\partial \xi} \tag{35}$$

$$\frac{\partial}{\partial t} = \frac{\partial}{\partial \xi}\frac{\partial \xi}{\partial t} + \frac{\partial}{\partial \tau}\frac{\partial \tau}{\partial t}, \Rightarrow \frac{\partial}{\partial t} = -\varepsilon^{\frac{1}{2}}\frac{\partial}{\partial \xi} + \varepsilon^{\frac{3}{2}}\frac{\partial}{\partial \tau} \tag{36}$$

$$\frac{\partial^2}{\partial x^2} = \frac{\partial}{\partial x}\left(\varepsilon^{\frac{1}{2}}\frac{\partial}{\partial \xi}\right), \Rightarrow \frac{\partial^2}{\partial x^2} = \varepsilon\frac{\partial^2}{\partial \xi^2} \tag{37}$$

We can express (5)–(7) in terms of ξ and τ as

$$\varepsilon^{3/2}\frac{\partial n_i}{\partial \tau} - \varepsilon^{1/2}\lambda\frac{\partial n_i}{\partial \xi} + \varepsilon^{1/2}\frac{\partial (n_i u_i)}{\partial \xi} = 0 \tag{38}$$

$$\varepsilon^{3/2}\frac{\partial u_i}{\partial \tau} - \varepsilon^{1/2}\lambda\frac{\partial u_i}{\partial \xi} + \varepsilon^{1/2} u_i\frac{\partial u_i}{\partial x} = -\varepsilon^{1/2}\frac{\partial \phi}{\partial x} \tag{39}$$

$$\varepsilon\frac{\partial^2 \phi}{\partial \xi^2} = e^{\phi} - n_i \tag{40}$$

Substituting the Eqs. (31)–(34) in Eqs. (38)–(40) and collecting the lowest order $O(\varepsilon^{3/2})$ terms we get

$$-\lambda\frac{\partial n_i^{(1)}}{\partial \xi} + \frac{\partial u_i^{(1)}}{\partial \xi} = 0, \tag{41}$$

$$-\lambda \frac{\partial u_i^{(1)}}{\partial \xi} = \frac{\partial \phi^{(1)}}{\partial \xi}, \tag{42}$$

$$\phi^{(1)} - n_i^{(1)} = 0. \tag{43}$$

Integrating Eqs. (41)–(43) and all the variables tend to zero as $\xi \to \infty$. We get

$$n_i^{(1)} = \frac{u_i^{(1)}}{\lambda}, \tag{44}$$

$$u_i^{(1)} = \frac{\phi^{(1)}}{\lambda}, \tag{45}$$

$$\phi^{(1)} = n_i^{(1)}. \tag{46}$$

From Eq. (44)–(46) we get the phase velocity as

$$\lambda^2 = \pm 1 \tag{47}$$

Substituting the Eqs.(31)–(34) in Eqs. (38)–(40) and collecting order $O(\varepsilon^{5/2})$, we get

$$\frac{\partial n_i^{(1)}}{\partial \tau} - \lambda \frac{\partial n_i^{(2)}}{\partial \xi} + \frac{\partial n_i^{(1)} u_i^{(1)}}{\partial \xi} + \frac{\partial u_i^{(2)}}{\partial \xi} = 0, \tag{48}$$

$$\frac{\partial u_i^{(1)}}{\partial \tau} - \lambda \frac{\partial u_i^{(2)}}{\partial \xi} + u_i^{(1)} \frac{\partial u_i^{(1)}}{\partial \xi} = -\frac{\partial \phi^{(2)}}{\partial \xi^2}, \tag{49}$$

$$\frac{\partial \phi^{(1)}}{\partial \xi^2} = \phi^{(2)} + \frac{1}{2}\left(\phi^{(1)}\right)^2 - n_i^{(1)}. \tag{50}$$

Differentiating Eq. (50) With respect to ξ and substituting for $\frac{\partial n_i^{(2)}}{\partial \xi}$ from Eq. (48) and for $\frac{\partial u_i^{(2)}}{\partial \xi}$ from Eq. (49), we finally obtain

$$\frac{\partial \phi^{(1)}}{\partial \tau} + \phi^{(1)} \frac{\partial \phi^{(1)}}{\partial \xi} + \frac{1}{2} \frac{\partial^3 \phi^{(1)}}{\partial \xi^3} = 0. \tag{51}$$

Eq. (51) is known as **KdV equation.** $\phi^{(1)} \frac{\partial \phi^{(1)}}{\partial \xi}$ is the nonlinear term and $\frac{1}{2}\frac{\partial^3 \phi^{(1)}}{\partial \xi^3}$ is the dispersive terms. Only nonlinearity can impose energy into the wave and the wave breaks but in presence of both nonlinearity and dispersive a stable wave profile is possible.

The steady-state solution of this KdV equation is obtained by transforming the independent variables ξ and τ to $\eta = \xi - u_0 \tau$ where u_0 is a constant velocity normalized by c_s.

The steady state solution of the KdV Eq. (51) can be written as

$$\phi^{(1)} = \phi_m sech^2\left(\frac{\eta}{\Delta}\right) \tag{52}$$

where $\phi_m = 3u_0$ and Δ are the amplitude and width of the solitary waves. It is clear that height, width and speed of the pulse propotional to u_0, $\frac{1}{\sqrt{u_0}}$, and u_0 respectively. As ϕ_m the amplitude is equal to $3u_0$ so u_0 specify the energy of the

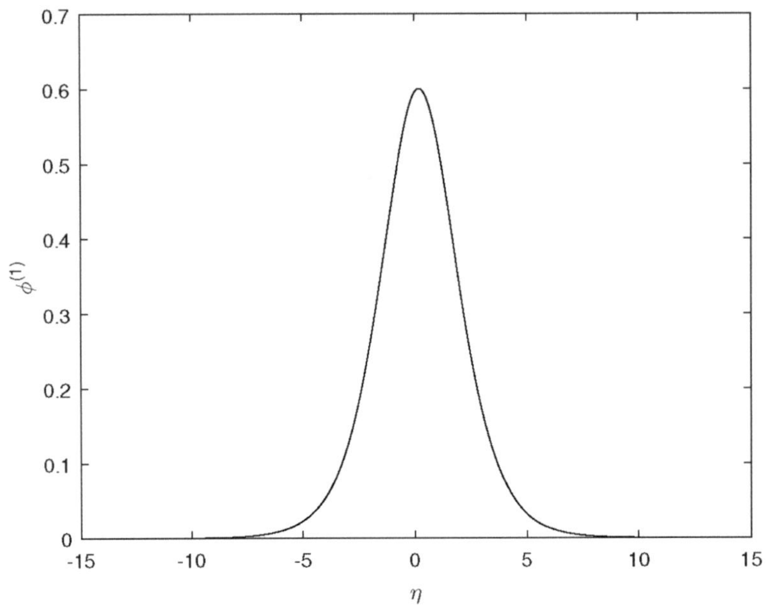

Figure 1.
Solitary wave solution of Eq. (52) for the parameter value $t = 1, u_0 = 0.2$.

solitary waves. So the larger the energy, the greater the speed, larger the amplitude and narrower the width (**Figure 1**).

3. Damped force KdV equation

Let us consider an unmagnetized collisional dusty plasma that contains cold inertial ions, stationary dusts with negative charge and Maxwellian electrons. The normalized ion fluid equations which include the equation of continuity, equation of momentum balance and Poisson's equation, governing the DIAWs, are given by

$$\frac{\partial n_i}{\partial t} + \frac{\partial (n_i u_i)}{\partial x} = 0, \tag{53}$$

$$\frac{\partial u_i}{\partial t} + u_i \frac{\partial u_i}{\partial x} = -\frac{\partial \phi}{\partial x} - \nu_{id} u_i, \tag{54}$$

$$\frac{\partial^2 \phi}{\partial x^2} = (1-\mu) n_e - n + \mu, \tag{55}$$

where n_j (j = i,e for ion, electron), u_i, ϕ are the number density, ion fluid velocity and the electrostatic wave potential respectively. Here $\mu = \frac{Z_d n_{d0}}{n_0}$, ν_{id} is the dust ion collisional frequency and the term $S(x,t)$ [4, 5], is a charged density source arising from experimental conditions for a single definite purpose. n_0, Z_d, n_{d0} are the

3.1 Normalization

$$n_i \to \frac{n_i}{n_0}, u_i \to \frac{u_i}{C_s}, \phi \to \frac{e\phi}{K_B T_e}, x \to \frac{x}{\lambda_D}, t \to \omega_{pi} t \tag{56}$$

where $C_s\left(=\sqrt{\left(\frac{K_B T_e}{m_i}\right)}\right)$ is the ion acoustic speed, T_e as electron temperature, K_B as Boltzmann constant, e as magnitude of electron charge and m_i as mass of ions. $\lambda_D\left(=\left(\frac{T_e}{4\pi n_{e0}e^2}\right)^{\frac{1}{2}}\right)$ is the Debye length and $\omega_{pi}\left(=\left(\frac{4\pi n_{e0}e^2}{m_i}\right)^{\frac{1}{2}}\right)$ as ion-plasma frequency.

The normalized electron density is given by

$$n_e = e^{\phi}. \tag{57}$$

3.2 Phase velocity and nonlinear evolution equation

We introduced the same stretched coordinates use in Eq.(31). The expansion of the dependent variables also considered as (32)–(34) with

$$\nu_{id} \sim \varepsilon^{3/2}\nu_{id0}. \tag{58}$$

$$S \sim \varepsilon^2 S_2. \tag{59}$$

Substituting (31)–(34) and (58)–(59) along with stretching coordinates into Eqs. (53)–(55) and equating the coefficients of lowest order of ε, we get the phase velocity as

$$\lambda = \frac{1}{\sqrt{(1-\mu)}}. \tag{60}$$

Taking the coefficients of next higher order of ε, we obtain the **damped force KdV equation**

$$\frac{\partial \phi^{(1)}}{\partial \tau} + A\phi^{(1)}\frac{\partial \phi^{(1)}}{\partial \xi} + B\frac{\partial^3 \phi^{(1)}}{\partial \xi^3} + C\phi^{(1)} = B\frac{\partial S_2}{\partial \xi}, \tag{61}$$

where $A = \frac{3-\lambda^2}{2\lambda}$, $B = \frac{\lambda^3}{2}$, $C = \frac{\nu_{id0}}{2}$.

It has been noticed that the behavior of nonlinear waves changes significantly in the presence of external periodic force. It is paramount to note that the source term or forcing term due to the presence of space debris in plasmas may be of different kind, for example, Gaussian forcing term [4], hyperbolic forcing term [4], (in the form of $sech^2(\xi,\tau)$ and $sech^4(\xi,\tau)$ functions) and trigonometric forcing term [6] (in the form of $\sin(\xi,\tau)$ and $\cos(\xi,\tau)$ functions). Motivated by these work we assume that S_2 is a linear function of ξ such as $S_2 = \frac{f_0\xi}{B}\cos(\omega\tau) + P$, where P is some constant and f_0, ω denote the strength and the frequency of the source respectively. Put the expression of S_2 in Eq. (61) we get,

$$\frac{\partial \phi^{(1)}}{\partial \tau} + A\phi^{(1)}\frac{\partial \phi^{(1)}}{\partial \xi} + B\frac{\partial^3 \phi^{(1)}}{\partial \xi^3} + C\phi^{(1)}1 = f_0\cos(\omega\tau), \tag{62}$$

which is termed as **damped and forced KdV (DFKdV) equation**.

In absence of C and f_0, i.e., for $C = 0$ and $f_0 = 0$ the Eq.(62) takes the form of well-known KdV equation with the solitary wave solution

$$\phi_1 = \phi_m sech^2\left(\frac{\xi - M\tau}{W}\right), \tag{63}$$

where $\phi_m = \frac{3M}{A}$ and $W = 2\sqrt{\frac{B}{M}}$, with M as the Mach number.

In this case, it is well established that

$$I = \int_{-\infty}^{\infty} \phi_1^2 \, d\xi, \tag{64}$$

is a conserved. For small values of C and f_0, let us assume that the solution of Eq. (62) is of the form

$$\phi_1 = \phi_m(\tau) sech^2\left(\frac{x - M(\tau)\tau}{W(\tau)}\right), \tag{65}$$

where $M(\tau)$ is an unknown function of τ and $\phi_m(\tau) = \frac{3M(\tau)}{A}$, $W(\tau) = 2\sqrt{B/M(\tau)}$. Differentiating Eq. (64) with respect to τ and using Eq. (62), one can obtain

$$\begin{aligned} &\frac{dI}{d\tau} + 2CI = 2f_0 \cos(\omega\tau) \int_{-\infty}^{\infty} \phi_1 \, d\xi, \\ &\Rightarrow \frac{dI}{d\tau} + 2CI = \frac{24 f_0 \sqrt{B}}{A} \sqrt{M(\tau)} \cos(\omega\tau). \end{aligned} \tag{66}$$

Again,

$$\begin{aligned} I &= \int_{-\infty}^{\infty} \phi_1^2 \, d\xi, \\ I &= \int_{-\infty}^{\infty} {\phi_m}^2(\tau) sech^4\left(\frac{\xi - M(\tau)\tau}{W(\tau)}\right) d\xi, \\ I &= \frac{24\sqrt{B}}{A^2} M^{3/2}(\tau). \end{aligned} \tag{67}$$

Using Eq. (66) and (67) the expression of $M(\tau)$ is obtained as

$$M(\tau) = \left(M - \frac{8ACf_0}{16C^2 + 9\omega^2}\right) e^{-\frac{4}{3}C\tau} + \frac{6Af_0}{16C^2 + 9\omega^2}\left(\frac{4}{3} C cos(\omega\tau) + \omega sin(\omega\tau)\right).$$

Therefore, the solution of the Eq. (62) is

$$\phi_1 = \phi_m(\tau) sech^2\left(\frac{\xi - M(\tau)\tau}{W(\tau)}\right), \tag{68}$$

where $\phi_m(\tau) = \frac{3M(\tau)}{A}$ and $W(\tau) = 2\sqrt{\frac{B}{M(\tau)}}$. The effect of the parameters, i.e., ion collision frequency parameter(ν_{id0}), strength of the external force (f_0)on the solitary wave solution of the damp force KdV Eq. (62) have been numerically studied. In **Figure 2**, the soliton solution of (62) is plotted from (63)in the absence of external periodic force and damping.

In **Figure 3**, the soliton solution of the damp force KdV equation is plotted from Eq. (65) for different values of the strength of the external periodic force (f_0). The values of other parameters are $M_0 = 0.2, \omega = 1, \tau = 1, \mu = 0.2, \nu_{id0} = 0.01$. It is observed that the solution produces solitary waves and the amplitude of the solitary waves increases as the value of the parameter f_0 increases. In **Figure 4**, damp force KdV equation is plotted from Eq. (65) for different values of the dust ion collision

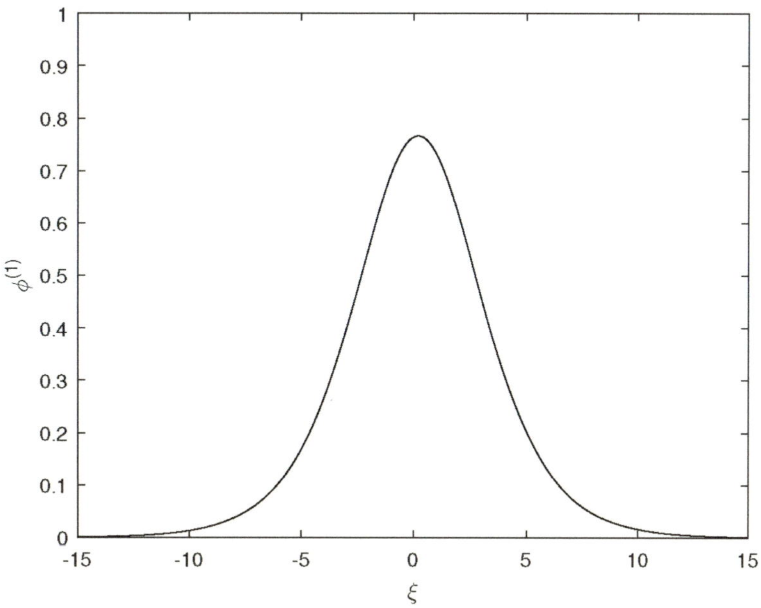

Figure 2.
Solitary wave solution of Eq. (62) in the absence of damping$(\nu_{ido} = 0)$ *and external force*$(f_o = 0)$ *with the parameter value* $M_o = 0.2, \omega = 1, \tau = 1, \mu = 0.2$.

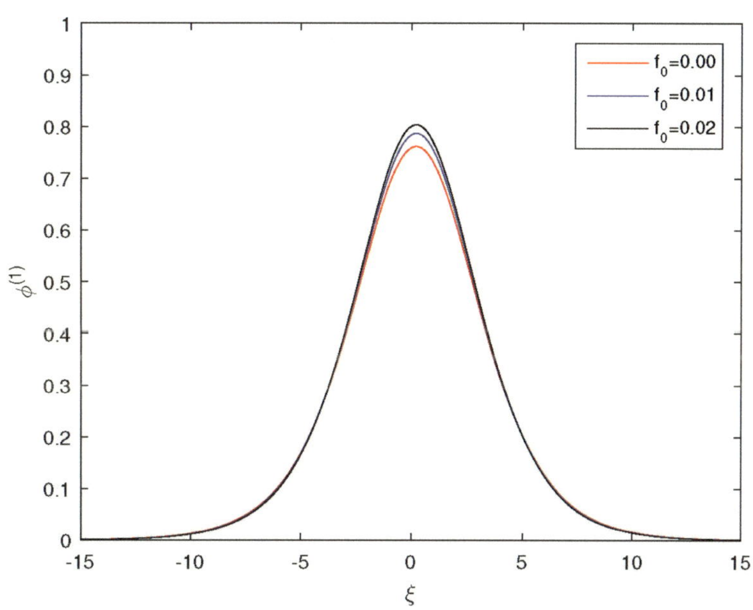

Figure 3.
Variation of solitary wave from Eq. (62) for the different values of f_o *with* $M_o = 0.2, \omega = 1, \tau = 1, \mu = 0.2$, $\nu_{ido} = 0.01$.

frequency parameter (ν_{id0}). The values of other parameters are $M_0 = 0.2, \omega = 1$, $\tau = 1, \mu = 0.2, f_0 = 0.01$. It is observed that the solution produces solitary waves and the amplitude of the solitary waves decreases as the value of the parameter ν_{id0} increases and width of the solitary waves increases for increasing value of ν_{id0}.

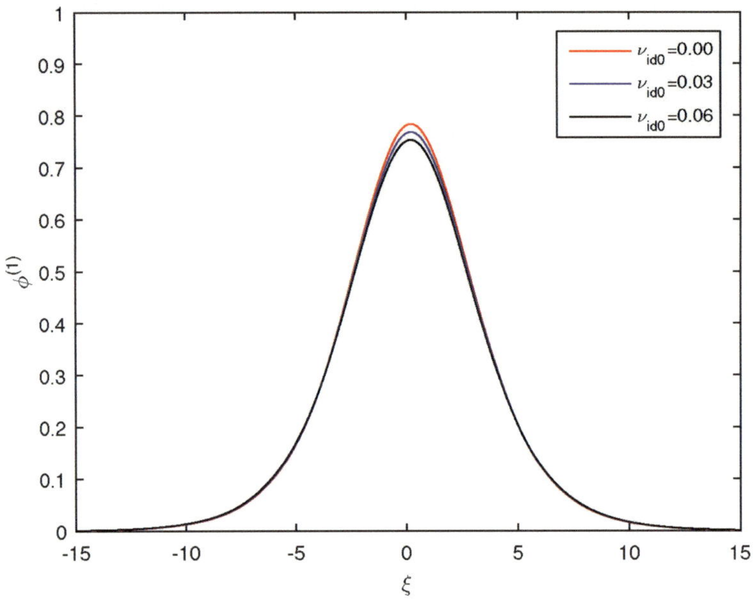

Figure 4.
Variation of solitary wave from Eq. (62) for the different values of collisional frequency ν_{id0} *with* $M_0 = 0.2$, $\omega = 1, \tau = 1, \mu = 0.2, f_0 = 0.01$.

4. Damped KdV Burgers equation

To obtain damped KdV Burgers equation we considered an unmagnetized collisional dusty plasma which contains cold inertial ions, stationary dusts with negative charge and Maxwellian distributed electrons. The normalized ion fluid equations are as follows

$$\frac{\partial n_i}{\partial t} + \frac{\partial (n_i u_i)}{\partial x} = 0, \tag{69}$$

$$\frac{\partial u_i}{\partial t} + u_i \frac{\partial u_i}{\partial x} = -\frac{\partial \phi}{\partial x} + \eta \frac{\partial^2 u_i}{\partial x^2} - \nu_{id} u_i, \tag{70}$$

$$\frac{\partial^2 \phi}{\partial x^2} = (1-\mu) n_e - n_i + \mu, \tag{71}$$

$$n_e = e^{\phi}, \tag{72}$$

where $n_i, n_e, u_i, \phi,$ are the number density of ions, the number density of electrons, the ion fluid velocity and the electrostatic wave potential, respectively.

Here normalization is taken as follows

$$n_i \to \frac{n_i}{n_0}, u_i \to \frac{u_i}{C_s}, \phi \to \frac{e\phi}{K_B T_e}, x \to \frac{x}{\lambda_D}, t \to \omega_{pi} t$$

$C_s\left(=\sqrt{\left(\frac{K_B T_e}{m_i}\right)}\right)$ is the ion acoustic speed, T_e as electron temperature, K_B as Boltzmann constant and m_i as mass of ions, e as magnitude of electron charge.

$\lambda_D\left(=\left(\frac{T_e}{4\pi n_{e0} e^2}\right)^{\frac{1}{2}}\right)$ is the Debye length and $\omega_{pi}\left(=\left(\frac{m_i}{4\pi n_{e0} e^2}\right)^{\frac{1}{2}}\right)$ as ion-plasma frequency. Here, ν_{id} is the dust-ion collisional frequency and $\mu = \frac{n_{0e}}{n_{0i}}$, where n_{0e} and n_{0i} are the unperturbed number densities of electrons and ions, respectively.

4.1 Perturbation

To obtain damped KdV burger we introduced the same stretched coordinates use in Eq.(31). The expansion of the dependent variables are also considered same as (32)–(34) with

$$\eta = \varepsilon^{1/2}\eta_0, \tag{73}$$

$$\nu_{id} \sim \varepsilon^{3/2}\nu_{id0}. \tag{74}$$

4.2 Phase velocity and nonlinear evolution equation

Substituting the above expansions (32)-(34) and (73)–(74) along with stretching coordinates (31) into Eqs. (69)–(71) and equating the coefficients of lowest order of ε, the phase velocity is obtained as

$$\lambda = \frac{1}{\sqrt{(1-\mu)}}. \tag{75}$$

Taking the coefficients of next higher order of ε, we obtain the **DKdVB equation**

$$\frac{\partial \phi^{(1)}}{\partial \tau} + A\phi^{(1)}\frac{\partial \phi^{(1)}}{\partial \xi} + B\frac{\partial^3 \phi^{(1)}}{\partial \xi^3} + C\frac{\partial^2 \phi^{(1)}}{\partial \xi^2} + D\phi^{(1)} = 0, \tag{76}$$

where $A = \frac{3-\lambda^2}{2\lambda}$, $B = \frac{v^3}{2}$, $C = -\frac{\eta_{10}}{2}$ and $D = \frac{\nu_{id0}}{2}$.

In absence of C and D, i.e., for $C = 0$ and $D = 0$ the Eq.(76) takes the form of well-known KdV equation with the solitary wave solution

$$\phi_1 = \phi_m sech^2\left(\frac{\xi - M_0\tau}{W}\right), \tag{77}$$

where amplitude of the solitary waves $\phi_m = \frac{3M_0}{A}$ and width of the solitary waves $W = 2\sqrt{\frac{B}{M_0}}$, with M_0 is the speed of the ion-acoustic solitary waves or Mach number.

It is well established for the KdV equation that,

$$I = \int_{-\infty}^{\infty} \phi_1^2 \, d\xi, \tag{78}$$

is a conserved quantity [7].

For small values of C and D, let us assume that amplitude, width and velocity of the dust ion acoustic waves are dependent on τ and the slow time dependent solution of Eq. (76) is of the form

$$\phi^{(1)} = \phi_m(\tau) sech^2\left(\frac{\xi - M(\tau)\tau}{W(\tau)}\right), \tag{79}$$

where the amplitude $\phi_m(\tau) = \frac{3M(\tau)}{A}$, width $W(\tau) = 2\sqrt{B/M(\tau)}$ and velocity $M(\tau)$ have to be determined.

Differentiating Eq. (78) with respect to τ and using Eq. (76), one can obtain

$$\begin{aligned} &\frac{dI}{d\tau} + 2DI = 2C\int_{-\infty}^{\infty}\left(\frac{\partial\phi_1}{\partial\xi}\right)^2 d\xi, \\ &\Rightarrow \frac{dI}{d\tau} + 2DI = 2C \times \frac{24}{5}\frac{M^{5/2}(\tau)}{A^2\sqrt{B}}. \end{aligned} \tag{80}$$

where,

$$\int_{-\infty}^{\infty}\left(\frac{\partial\phi_1}{\partial\xi}\right)^2 d\xi = \frac{24}{5}\frac{M^{5/2}(\tau)}{A^2\sqrt{B}} \tag{81}$$

and

$$\begin{aligned} I &= \int_{-\infty}^{\infty}\phi_1^2\, d\xi, \\ I &= \int_{-\infty}^{\infty}\phi_m{}^2(\tau)sech^4\left(\frac{\xi - M(\tau)\tau}{W(\tau)}\right) d\xi, \\ I &= \frac{24\sqrt{B}}{A^2}M^{3/2}(\tau). \end{aligned} \tag{82}$$

Substituting Eq. (81) and (82) into Eq. (80), we obtain

$$\frac{dM(\tau)}{d\tau} + PM(\tau) = QM^2(\tau), \tag{83}$$

which is the Bernoulli's equation, where $P = \frac{4}{3}D$ and $Q = \frac{4}{15}\frac{C}{B}$.
The solution of the Eq. (83) is

$$M(\tau) = \frac{PM_0}{M_0Q(1 - e^{P\tau}) + Pe^{P\tau}}$$

Therefore, the slow time dependence form of the ion acoustic solitary wave solution of the DKdVB Eq. (76) is given by (79)where.
$M(\tau) = \frac{PM_0}{M_0Q(1-e^{P\tau})+Pe^{P\tau}}$ and $M(0) = M_0$ for $\tau = 0$.

5. Damped force MKdV equation

Let us consider an unmagnetized collisional dusty plasma that contains cold inertial ions, stationary dusts with negative charge and Maxwellian distributed electrons. The normalized ion fluid equations which include the equation of continuity, equation of momentum balance and Poisson's equation, governing the DIAWs, are given by

$$\frac{\partial n_i}{\partial t} + \frac{\partial(n_i u_i)}{\partial x} = 0, \tag{84}$$

$$\frac{\partial u_i}{\partial t} + u_i\frac{\partial u_i}{\partial x} = -\frac{\partial\phi}{\partial x} - \nu_{id}u, \tag{85}$$

$$\frac{\partial^2\phi}{\partial x^2} = (1-\mu)n_e - n_i + \mu + S(x,t) \tag{86}$$

where n_j ($j = i,e$ for ion, electron), u_i, ϕ are the number density, ion fluid velocity and the electrostatic wave potential respectively. Here $\mu = \frac{Z_d n_{d0}}{n_0}$, ν_{id} is the dust-ion collisional frequency and the term $S(x,t)$ [4, 5], is a charged density source arising from experimental conditions for a single definite purpose. n_0, Z_d, n_{d0} are the normalization:

$$n_i \to \frac{n_i}{n_0}, u_i \to \frac{u_i}{C_s}, \phi \to \frac{e\phi}{K_B T_e}, x \to \frac{x}{\lambda_D}, t \to \omega_{pi} t \tag{87}$$

where $C_s\left(=\sqrt{\left(\frac{K_B T_e}{m_i}\right)}\right)$ is the ion acoustic speed, T_e as electron temperature, K_B as Boltzmann constant, e as magnitude of electron charge and m_i as mass of ions. $\lambda_D\left(=\left(\frac{T_e}{4\pi n_{e0} e^2}\right)^{\frac{1}{2}}\right)$ is the Debye length and $\omega_{pi}\left(=\left(\frac{4\pi n_{e0} e^2}{m_i}\right)^{\frac{1}{2}}\right)$ as ion-plasma frequency.

The normalized q-nonextensive electron number density takes the form [8]:

$$n_e = n_{e0}\{1 + (q-1)\phi\}^{\frac{q+1}{2(q-1)}} \tag{88}$$

Phase velocity and nonlinear evolution equation

We introduced the same stretched coordinates use in Eq. (31). The expansion of the dependent variables also considered same as (32)–(34) and (58)–(59). Substituting (31)–(34) and (58)–(59) along with stretching coordinates into Eqs. (84)–(86) and equating the coefficients of lowest order of ε, we get the phase velocity as

$$\lambda = \frac{1}{\sqrt{a(1-\mu)}}, \tag{89}$$

with $a = \frac{q+1}{2}$. Now taking the coefficients of next higher order of ε [i.e., coefficient of $\varepsilon^{5/2}$ from Eqs. (84) and (85) and coefficient of ε^2 from Eq. (86)], we obtain the **DFKdV** equation

$$\frac{\partial \phi^{(1)}}{\partial \tau} + A\phi^{(1)}\frac{\partial \phi^{(1)}}{\partial \xi} + B\frac{\partial^3 \phi^{(1)}}{\partial \xi^3} + C\phi^{(1)} = B\frac{\partial S_2}{\partial \xi}, \tag{90}$$

where $A = \left(\frac{3}{2\lambda} - \frac{b\lambda}{a}\right)$, $B = \frac{\lambda^3}{2}$ and $C = \frac{\nu_{id0}}{2}$, with $b = \frac{(q+1)(3-q)}{8}$.

Now at the certain values, for example $q = 0.6$ and $\mu = 0.5$, there is a critical point at which $A = 0$, which imply the infinite growth of the amplitude of the DIASW solution as nonlinearity goes to zero. Therefore, at the critical point at which $A = 0$ the stretching (31) is not valid. For describing the evolution of the nonlinear system at or near the critical point we introduce the new stretched coordinate as

$$\xi = \varepsilon(x - \lambda t), \tau = \varepsilon^3 t, \tag{91}$$

and expand of the dependent variables same as Eqs. (32)–(34) with

$$\nu_{id} \sim \varepsilon^3 \nu_{id0}, \tag{92}$$

$$S \sim \varepsilon^3 S_2. \tag{93}$$

Now substituting Eq. (32)–(34) and (91)–(93) into the basic Eqs. (84)–(86) and equating the coefficients of lowest order of ε, [i.e., coefficients of ε^2 from Eq. (84) and (85) and coefficients of ε from Eq. (86)], we obtain the following relations:

$$n_i^{(1)} = \frac{u_i^{(1)}}{\lambda}, \tag{94}$$

$$u_i^{(1)} = \frac{\phi^{(1)}}{\lambda}, \tag{95}$$

$$n_i^{(1)} = a(1-\mu)\phi^{(1)}. \tag{96}$$

Equating the coefficients of next higher order of ε, [i.e., coefficients of ε^3 from Eq. (84) and (85) and coefficients of ε from Eq. (86)],we obtain the following relations:

$$n_i^{(2)} = \frac{1}{\lambda}\left(u_i^{(2)} + n_i^{(1)}u_i^{(1)}\right) \tag{97}$$

$$\frac{\partial u_i^{(1)}}{\partial \xi} = \frac{1}{\lambda}\left(u_i^{(1)}\frac{\partial u_i^{(1)}}{\partial \xi} + \frac{\partial \phi^{(2)}}{\partial \xi}\right) \tag{98}$$

$$n_i^{(2)} = a(1-\mu)\left(a\phi^{(2)} + b\left(\phi^{(1)}\right)^2\right) \tag{99}$$

Equating the coefficients of next higher order of ε, [i.e., coefficients of ε^4 from Eq. (84) and (85) and coefficients of ε from Eq. (86)], we obtain the following relations:

$$\frac{\partial n_i^{(1)}}{\partial \tau} - \lambda\frac{\partial n_i^{(3)}}{\partial \xi} + \frac{\partial u_i^{(3)}}{\partial \xi} + \frac{\partial\left(n_i^{(1)}u_i^{(2)}\right)}{\partial \xi} + \frac{\partial\left(n_i^{(2)}u_i^{(1)}\right)}{\partial \xi} = 0 \tag{100}$$

$$\frac{\partial u_i^{(1)}}{\partial \tau} - \lambda\frac{\partial u_i^{(3)}}{\partial \xi} + \frac{\partial \phi^{(3)}}{\partial \xi} + \frac{\partial\left(u_i^{(1)}u_i^{(2)}\right)}{\partial \xi} + \nu_{id0}u_i^{(1)} = 0 \tag{101}$$

$$\frac{\partial^2 \phi^{(1)}}{\partial \xi^2} = (1-\mu)\left(a\phi^{(3)} + 2b\phi^{(1)}\phi^{(2)} + c\left(\phi^{(1)}\right)^3\right) - n_i^{(3)} + S_2 \tag{102}$$

where $a = \frac{(1+q)}{2}$, $b = \frac{(1+q)(3-q)}{8}$ and $c = \frac{(1+q)(3-q)(5-3q)}{48}$.

From Eq. (94)–(96), one can obtain the Phase velocity as $\lambda^2 = \frac{1}{a(1-\mu)}$ and from Eqs. (94)–(102), one can obtain the following nonlinear evaluation equation as:

$$\frac{\partial \phi^{(1)}}{\partial \tau} + A_1\left(\phi^{(1)}\right)^2\frac{\partial \phi^{(1)}}{\partial \xi} + B_1\frac{\partial^3 \phi^{(1)}}{\partial \xi^3} + C_1\phi^{(1)} = B_1\frac{\partial S_2}{\partial \xi}, \tag{103}$$

where $A_1 = \frac{15}{4\lambda^3} - \frac{3\lambda^3 c(1-\mu)}{2}$, $B_1 = \frac{\lambda^3}{2}$ and $C_1 = \frac{\nu_{id0}}{2}$.

It has been noticed that the behavior of nonlinear waves changes significantly in the presence of external periodic force. For simplicity, we assume that S_2 is a linear function of ξ such as $S_2 = f_0\xi\cos(\omega\tau) + P$, where P is some constant and f_0, ω denote the strength and the frequency of the source respectively. Put the expression of S_2 in the Eq. (103) we get,

$$\frac{\partial \phi^{(1)}}{\partial \tau} + A_1\left(\phi^{(1)}\right)^2 \frac{\partial \phi^{(1)}}{\partial \xi} + B_1 \frac{\partial^3 \phi^{(1)}}{\partial \xi^3} + C_1 \phi^{(1)} = B_1 f_0 \cos(\omega\tau). \tag{104}$$

Such a form of this source function is observed in experimental situations or conditions for a particular device. Eq. (104) is termed as **damped force modified Korteweg-de Varies (DFMKdV)** equation.

In absence of C_1 and f_0, i.e., for $C_1 = 0$ and $f_0 = 0$ the Eq.(104) takes the form of well-known **MKdV equation**.

The slow time dependence form of the ion acoustic waves solution of the DFMKdV Eq. (104) is given by,

$$\phi^{(1)} = \phi_m(\tau) sech\left(\frac{\xi - M(\tau)\tau}{W(\tau)}\right), \tag{105}$$

where $M(\tau)$ is given by equation

$$M(\tau) = \left[\frac{\pi f_0 B_1 \sqrt{A_1/6}}{2}\left(\frac{\omega}{\omega^2 + 4C_1^2}\right)\left\{sin(\omega\tau) + \frac{2C_1}{\omega} cos(\omega\tau)\right\}\right.$$
$$\left. + \left\{\sqrt{M} - \pi f_0 B_1 \sqrt{A_1/24}\left(\frac{2C_1}{\omega^2 + 4C_1^2}\right)\right\} e^{-\nu_{id0}\tau}\right]^2.$$

The amplitude and width are as follows:

$$\phi_m(\tau) = \frac{1}{\sqrt{A}}\left(\sqrt{6}\left[\frac{\pi f_0 B_1 \sqrt{A_1/6}}{2}\left(\frac{\omega}{\omega^2 + 4C_1^2}\right)\left\{\sin(\omega\tau) + \frac{2C_1}{\omega}\cos(\omega\tau)\right\}\right.\right.$$
$$\left.\left. + \left\{\sqrt{M} - \pi f_0 B_1 \sqrt{A_1/24}\left(\frac{2C_1}{\omega^2 + 4C_1^2}\right)\right\} e^{-\nu_{id0}\tau}\right]\right)$$

$$W(\tau) = \frac{\sqrt{B_1}}{W_1 + W_2}$$

where

$$W_1 = \frac{\pi f_0 B_1 \sqrt{A_1/6}}{2}\left(\frac{\omega}{\omega^2 + 4C_1^2}\right)\left\{sin(\omega\tau) + \frac{2C_1}{\omega} cos(\omega\tau)\right\}$$

$$W_2 = \left\{\sqrt{M} - \pi f_0 B_1 \sqrt{A_1/24}\left(\frac{2C_1}{\omega^2 + 4C_1^2}\right)\right\} e^{-\nu_{id0}\tau}$$

6. Damped force Zakharov-Kuznetsov equation

Let us consider a plasma model [9] consisting of cold ions, Maxwellian electrons in the presence of dust particles and the external static magnetic field $B = \hat{y}B_0$ along the y-axis. The normalized continuity, momentum and Poisson's equations are as follows

$$\frac{\partial n}{\partial t} + \frac{\partial(nu)}{\partial x} + \frac{\partial(nv)}{\partial y} + \frac{\partial(nw)}{\partial z} = 0, \tag{106}$$

$$\frac{\partial u}{\partial t}+\left(u\frac{\partial}{\partial x}+v\frac{\partial}{\partial y}+w\frac{\partial}{\partial z}\right)u=-\frac{\partial\phi}{\partial x}-\frac{\Omega_i}{\omega_{pi}}w, \tag{107}$$

$$\frac{\partial v}{\partial t}+\left(u\frac{\partial}{\partial x}+v\frac{\partial}{\partial y}+w\frac{\partial}{\partial z}\right)v=-\frac{\partial\phi}{\partial y}-\nu_{id}v, \tag{108}$$

$$\frac{\partial w}{\partial t}+\left(u\frac{\partial}{\partial x}+v\frac{\partial}{\partial y}+w\frac{\partial}{\partial z}\right)w=-\frac{\partial\phi}{\partial z}+\frac{\Omega_i}{\omega_{pi}}u, \tag{109}$$

$$\frac{\partial^2\phi}{\partial x^2}+\frac{\partial^2\phi}{\partial y^2}+\frac{\partial^2\phi}{\partial z^2}=\delta_1+\delta_2 n_e-n \tag{110}$$

The normalized electron density is given by

$$n_e=e^{\phi}, \tag{111}$$

where $n, n_e, u_i(=u,v,w), T_e, m_i, e, \phi, \Omega_i, \omega_{pi}, \nu id$ and λ_D are the ion number density, electron number density, ion velocity, electron temperature, ion mass, electron charge, electrostatic potential, ion cyclotron frequency, ion plasma frequency, dust ion collision frequency and Debye length respectively.

Here the normalization is done as follows:

$$n\rightarrow\frac{n}{n_0}, n_e\rightarrow\frac{n_e}{n_{e0}}, u_i\rightarrow\frac{u_i}{C_s}, \phi\rightarrow\frac{e\phi}{T_e}, x\rightarrow\frac{x}{\lambda_D}, t\rightarrow\omega_{pi}t$$

Here $\delta_1=\frac{n_{d0}}{n_{i0}}, \delta_2=\frac{n_{e0}}{n_{i0}}$ with the condition $\delta_1+\delta_2=1$. $\lambda_D=\left(\frac{T_e}{4\pi n_{e0}e^2}\right)^{1/2}$, $\omega_{pi}^{-1}=\left(\frac{m_i}{4\pi n_{e0}e^2}\right)^{1/2}, C_s=\sqrt{\frac{T_e}{m_i}}$.

To obtain the DFZK equation we introduce the new stretched coordinates as

$$\begin{aligned}\xi&=\varepsilon^{1/2}x\\ \zeta&=\varepsilon^{1/2}(x-\lambda t),\\ \eta&=\varepsilon^{1/2}y,\\ \tau&=\varepsilon^{3/2}t\end{aligned} \tag{112}$$

where ε is the strength of nonlinearity and λ is the phase velocity of waves. The expression of the dependent variables as follows:

$$n=1+\varepsilon n_1+\varepsilon^2 n_2+\cdots \tag{113}$$

$$u=0+\varepsilon^{3/2}u_1+\varepsilon^2 u_2+\cdots \tag{114}$$

$$v=0+\varepsilon v_1+\varepsilon^2 v_2+\cdots \tag{115}$$

$$w=0+\varepsilon^{3/2}w_1+\varepsilon^2 w_2+\cdots \tag{116}$$

$$\phi=0+\varepsilon\phi_1+\varepsilon^2\phi_2+\cdots \tag{117}$$

$$\nu_{id}\sim\varepsilon^{3/2}\nu_{id0} \tag{118}$$

$$S(x,y,z)\sim\varepsilon^2 S_2(x,y,z) \tag{119}$$

Substituting the equations (112)-(119) into the system of Eqs. (106)-(110) equating the coefficient of ε, we get

$$\phi_1 = \frac{n_1}{\delta_2}. \tag{120}$$

Equating the coefficient of $\varepsilon^{3/2}$, we get

$$n_1 = \frac{v_1}{\lambda} \tag{121}$$

$$w_1 = -\frac{\omega_{pi}}{\Omega_i}\frac{\partial \phi_1}{\partial \xi}, \tag{122}$$

$$v_1 = \frac{\phi_1}{\lambda}, \tag{123}$$

$$u_1 = \frac{\omega_{pi}}{\Omega_i}\frac{\partial \phi_1}{\partial \eta}. \tag{124}$$

Considering the coefficient of ε^2, the following relationships are obtained

$$w_2 = \lambda \frac{\omega_{pi}}{\Omega_i}\frac{\partial u_1}{\partial \zeta}, \tag{125}$$

$$u_2 = -\lambda \frac{\omega_{pi}}{\Omega_i}\frac{\partial w_1}{\partial \zeta}, \tag{126}$$

$$\frac{\partial^2 \phi_1}{\partial \xi^2} + \frac{\partial^2 \phi_1}{\partial \zeta^2} + \frac{\partial^2 \phi_1}{\partial \eta^2} = \delta_1\left(1 - \phi_2 + \frac{\phi_1^2}{2}\right) - n_2 + S_2. \tag{127}$$

Comparing the coefficients of $\varepsilon^{5/2}$, we obtain

$$\frac{\partial n_1}{\partial \tau} - \lambda \frac{\partial n_2}{\partial \zeta} + \frac{\partial u_2}{\partial \xi} + \frac{\partial}{\partial \zeta}(n_1 v_1) + \frac{\partial v_2}{\partial \zeta} + \frac{\partial w_2}{\partial \eta} = 0 \tag{128}$$

$$\frac{\partial v_1}{\partial \tau} - \lambda \frac{\partial v_2}{\partial \zeta} + v_1 \frac{\partial v_1}{\partial \zeta} + \frac{\partial \phi_2}{\partial \zeta} - \nu_{id0} v_1 = 0. \tag{129}$$

Using the relationships (120)–(124), one can obtain the linear dispersion relation as

$$1 - \lambda^2 \delta_2 = 0 \tag{130}$$

Expressing all the perturbed quantities in terms of ϕ_1 from Eq. (125)–(129), the **damped forced ZK equation** is obtained as

$$\frac{\partial \phi_1}{\partial \tau} + A\phi_1 \frac{\partial \phi_1}{\partial \zeta} + B\frac{\partial^3 \phi_1}{\partial \zeta^3} + D\phi_1 + C\frac{\partial}{\partial \zeta}\left(\frac{\partial^2 \phi_1}{\partial \xi^2} + \frac{\partial^2 \phi_1}{\partial \eta^2}\right) + B\frac{\partial S_2}{\partial \zeta} = 0 \tag{131}$$

where

$$A = \frac{3}{2\lambda} - \frac{\lambda}{2}, \; B = \frac{\lambda}{2\delta_2}, \; C = \frac{\lambda}{2\delta_2}\left(1 + \frac{\omega_{pi}^2}{\Omega_i^2}\right), \; D = \frac{\nu_{id0}}{2}.$$

Choudhury et al. [5] studied analytical electron acoustic solitary wave (EASW) solution in the presence of periodic force for an unmagnetized plasma consisting of cold electron fluid, superthermal hot electrons and stationary ions. Motivated by the these works, here we consider the source term as $S_2 = \frac{f_0}{B}(e\zeta + f\xi + g\eta)\cos(\omega\tau)$, where f_0 and ω denote the strength and frequency of the source term respectively. Then Eq. (131) is of the form,

$$\frac{\partial\phi_1}{\partial\tau} + A\phi_1\frac{\partial\phi_1}{\partial\zeta} + B\frac{\partial^3\phi_1}{\partial\zeta^3} + D\phi_1 + C\frac{\partial}{\partial\zeta}\left(\frac{\partial^2\phi_1}{\partial\xi^2} + \frac{\partial^2\phi_1}{\partial\eta^2}\right) = F_0\,cos\,(\omega\tau) \tag{132}$$

where $F_0 = -\frac{ef_0}{B}$. To find the analytical solution of Eq. (132), we transform the damped-forced ZK equation to the KdV equation. We introduce new variable:

$$\xi = (l\zeta + m\xi + n\eta), \tag{133}$$

where l, m, n are the direction cosines of the line of wave propagation, with $l^2 + m^2 + n^2 = 1$. Substituting Eqs. (133) into the Eq. (132), we get

$$\begin{aligned} &\frac{\partial\phi_1}{\partial\tau} + Al\phi_1\frac{\partial\phi_1}{\partial\xi} + Bl^3\frac{\partial^3\phi_1}{\partial\xi^3} + Cl\left(m^2+n^2\right)\frac{\partial^3\phi_1}{\partial\xi^3} + D\phi_1 = F_0\,cos\,(\omega\tau) \\ &\Rightarrow \frac{\partial\phi_1}{\partial\tau} + P\phi_1\frac{\partial\phi_1}{\partial\xi} + Q\frac{\partial^3\phi_1}{\partial\xi^3} + D\phi_1 = F_0\,cos\,(\omega\tau) \end{aligned} \tag{134}$$

where, $P = Al$, $Q = Bl^3 + Cl(m^2 + n^2)$,
The analytical solitary wave solution of the Eq. (134) as obtained in (68), is

$$\phi_1 = \phi_m(\tau)sech^2\left(\frac{\xi - M(\tau)\tau}{W(\tau)}\right) \tag{135}$$

where $\phi_m(\tau) = \frac{3M(\tau)}{P}$ and $W(\tau) = 2\sqrt{\frac{Q}{M(\tau)}}$, with

$$M(\tau) = \left(M - \frac{8PF_0}{16D^2 + 9\omega^2}\right)e^{-\frac{4}{3}D\tau} + \frac{6PF_0}{16D^2 + 9\omega^2}\left(\frac{4}{3}Dcos(\omega\tau) + \omega sin(\omega\tau)\right). \tag{136}$$

7. Conclusions

It is clear from the structure of the solitary wave solution of the DFKdV, DFMKdV and DFZK that the soliton amplitude and width depends on the nonlinearity and dispersion of the evolution equations, which are the function of different plasma parameter involve in the consider plasma system. Also evident from the structure of the approximate analytical solution that the amplitude and the width of the soliton depends on the Mach number $(M(\tau))$ which involve the forcing term $F_0\cos(\omega\tau)$ and the damping parameter. Thus the amplitude and the width of the solitary wave structure changes with the different plasma parameters. Also they are changes with the change of strength of external force F_0, frequency of the external force ω and the collisional frequency between the different plasma species. The effect of these parameter can be studied through numerical simulation.

Author details

Laxmikanta Mandi[1,2†], Kaushik Roy[3†] and Prasanta Chatterjee[1*]

1 Department of Mathematics, Visva-Bharati, Santiniketan, India

2 Department of Mathematics, Gushkara Mahavidyalaya, Guskhara, India

3 Beluti M.K.M. High School, Beluti, Birbhum, India

*Address all correspondence to: prasantacvb@gmail.com

† These authors contributed equally.

IntechOpen

© 2020 The Author(s). Licensee IntechOpen. This chapter is distributed under the terms of the Creative Commons Attribution License (http://creativecommons.org/licenses/by/3.0), which permits unrestricted use, distribution, and reproduction in any medium, provided the original work is properly cited.

References

[1] Russell JS. Report on waves. In: Murray J, editor. Report of the British Association for the Advancement of Science. 1944. pp. 311-390

[2] Korteweg DJ, de Vries G. On the change of form of long waves advancing in a rectangular canal, and on a new type of long stationary waves. Philosophical Magazine. 1985;**39**:422

[3] Zabusky NJ, Kruskal MD. Interaction of "solitons" in a collisionless plasma and the recurrence of initial states. Physical Review Letters. 1965;**15**: 240-243

[4] Sen A, Tiwary S, Mishra S, Kaw P. Nonlinear wave excitations by orbiting charged space debris objects. Advances in Space Research. 2015;**56**(3):429

[5] Chowdhury S, Mandi L, Chatterjee P. Effect of externally applied periodic force on ion acoustic waves in superthermal plasmas. Physics of Plasmas. 2018;**25**:042112

[6] Aslanov VS, Yudintsev VV. Dynamics, analytical solutions and choice of parameters for towed space debris with flexible appendages. Advances in Space Research. 2015; **55**:660

[7] Israwi S, Kalisch H. Approximate conservation laws in the KdV equation. Physics Letters A. 2019;**383**:854

[8] Tsallis CJ. Possible generalization of Boltzmann-Gibbs statistics. Journal of Statistical Physics. 1988;**52**:479

[9] Zakharov VE, Kuznetsov EA. Three-dimensional solitons. Soviet Physics JETP. 1974;**39**:285

Accounting and Finance - New Perspectives on Banking, Financial Statements and Reporting

Edited by Reza Gharoie Ahangar and Can Öztürk

Published in London, United Kingdom

IntechOpen

Supporting open minds since 2005

Accounting and Finance - New Perspectives on Banking, Financial Statements and Reporting
http://dx.doi.org/10.5772/intechopen.79593
Edited by Reza Gharoie Ahangar and Can Öztürk

Contributors
Maximo Paulino Iii Sison, Gloria Rodriguez-Lozano, Yuriy Zaychenko, Michael Zgurovsky, Birsel Sabuncu, Elya Nabila Abdul Bahri, Nor Hakimah Haji Mohd Nor, Maria Gabriella Baldarelli, Matthias Nnadi, Oliver Neoh, Reza Gharoie Ahangar, Can Ozturk

© The Editor(s) and the Author(s) 2019
The rights of the editor(s) and the author(s) have been asserted in accordance with the Copyright, Designs and Patents Act 1988. All rights to the book as a whole are reserved by INTECHOPEN LIMITED. The book as a whole (compilation) cannot be reproduced, distributed or used for commercial or non-commercial purposes without INTECHOPEN LIMITED's written permission. Enquiries concerning the use of the book should be directed to INTECHOPEN LIMITED rights and permissions department (permissions@intechopen.com).
Violations are liable to prosecution under the governing Copyright Law.

CC BY

Individual chapters of this publication are distributed under the terms of the Creative Commons Attribution 3.0 Unported License which permits commercial use, distribution and reproduction of the individual chapters, provided the original author(s) and source publication are appropriately acknowledged. If so indicated, certain images may not be included under the Creative Commons license. In such cases users will need to obtain permission from the license holder to reproduce the material. More details and guidelines concerning content reuse and adaptation can be found at http://www.intechopen.com/copyright-policy.html.

Notice
Statements and opinions expressed in the chapters are these of the individual contributors and not necessarily those of the editors or publisher. No responsibility is accepted for the accuracy of information contained in the published chapters. The publisher assumes no responsibility for any damage or injury to persons or property arising out of the use of any materials, instructions, methods or ideas contained in the book.

First published in London, United Kingdom, 2019 by IntechOpen
IntechOpen is the global imprint of INTECHOPEN LIMITED, registered in England and Wales, registration number: 11086078, 7th floor, 10 Lower Thames Street, London, EC3R 6AF, United Kingdom
Printed in Croatia

British Library Cataloguing-in-Publication Data
A catalogue record for this book is available from the British Library

Additional hard and PDF copies can be obtained from orders@intechopen.com

Accounting and Finance - New Perspectives on Banking, Financial Statements and Reporting
Edited by Reza Gharoie Ahangar and Can Öztürk
p. cm.
Print ISBN 978-1-83968-197-4
Online ISBN 978-1-83968-198-1
eBook (PDF) ISBN 978-1-83968-199-8

For EU product safety concerns:
IN TECH d.o.o., Prolaz Marije Krucifikse Kozulić 3, 51000 Rijeka, Croatia, info@intechopen.com or visit our website at intechopen.com.

We are IntechOpen,
the world's leading publisher of
Open Access books
Built by scientists, for scientists

4,500+
Open access books available

118,000+
International authors and editors

130M+
Downloads

151
Countries delivered to

Our authors are among the
Top 1%
most cited scientists

12.2%
Contributors from top 500 universities

BOOK CITATION INDEX
CLARIVATE ANALYTICS
INDEXED
WEB OF SCIENCE™

Selection of our books indexed in the Book Citation Index
in Web of Science™ Core Collection (BKCI)

Interested in publishing with us?
Contact book.department@intechopen.com

Numbers displayed above are based on latest data collected.
For more information visit www.intechopen.com

Meet the editors

Reza Gharoie Ahangar is a PhD student in Management Science with a minor in Finance, and a Teaching and Research Assistant in G. Brint Ryan College of Business at the University of North Texas, USA. He holds a BSc in Business Management, an MBA in Finance, an MSc in Finance, and an MSc in Business Analytics. He is a member of the American Finance Association (AFA), Decision Sciences Institute (DSI), and the Institute for Operations Research and the Management Sciences (INFORMS) as well as the Association of Scientists, Developers, and Faculties (ASDF). Mr. Gharoie has authored a series of books and journal and conference publications in finance and business analytics, with more than 30 technical publications.
He serves as an editorial board member for several scientific journals. His current research interests include asset pricing, investment, banking, forecasting, machine learning, artificial intelligence, and optimization.

Dr. Can Öztürk is an associate professor of financial, international and comparative accounting at the Department of Management, Çankaya University, Turkey. He received a bachelor's degree in Banking and Finance (2001) from Bilkent University and a second bachelor's degree in Finance (2003) and a master's in Business Administration (2004) from the University of Massachusetts Dartmouth, USA. He obtained a PhD in Accounting and Finance in 2011 from Başkent University, Turkey. His primary research interests are International Financial Reporting Standards (IFRS), IFRS for small and medium-sized enterprises (SMEs), ISAs, and corporate social responsibility (CSR) reporting. He has published several articles in national and international journals such as *Accounting in Europe*, *Accounting and Management Information Systems*, *Emerging Markets Journal*, and *The World of Accounting Science*. Dr. Öztürk speaks English, French, and Turkish.

Contents

Preface

This book focuses on issues in accounting and finance. The volume contains an introductory chapter from editors and seven chapters from a global perspective written by authors from Turkey, Italy, Ukraine, Colombia, ASEAN-5 (Malaysia, Thailand, Indonesia, Singapore, and the Philippines), and the United States. It is organized into four sections: "Introduction," "Financial and Non-Financial Reporting," "Accounting and Financial Statements," and "Banking." The target audience is scholars and specialists in the fields of accounting, reporting, and banking.

The first section contains the introductory chapter.

The second section, "Financial and Non-Financial Reporting," includes two chapters. In Chapter 2, "National Accounting Standards in Turkey," Birsel Sabuncu describes the accounting standards being implemented in Turkey, which include Uniform Accounting System and Uniform Chart of Accounts, Turkish Financial Reporting Standards (TFRS), and Financial Reporting Standards for Large- and Medium-Sized Enterprises (FRS for LMEs). In Chapter 3, "CSR Training and Financial Statement Disclosure: The Case of Italy," Maria-Gabriella Baldarelli analyses the impact that corporate social responsibility (CSR) training had on non-financial information that is contained in financial statements following the EU directive 2014/95/EU and the Italian law 254/2016.

The third section, "Accounting and Financial Statements," includes two chapters. In Chapter 4, "Nonlinear Effect of Financial Development and Foreign Direct Investment in Integration Economies Among ASEAN-5 Countries Following IFRS Adoption," Elya Nabila Abdul Bahri and Nor Hakimah Haji Mohd Nor examine the role of financial development on foreign direct investment (FDI) inflows in ASEAN-5 countries following adoption of International Financial Reporting Standards (IFRS) over the period 1980–2017. In Chapter 5, "The Roles of Accounting Valuations and Earnings Management in the Survivorship of Technology Firms during the Global Financial Crisis," Oliver Neoh and Matthias Nnadi examine the survivorship of technology firms listed on the NASDAQ market during the immediate and post-2008 global financial crisis period and demonstrate the varying roles of accounting valuation and earnings management metrics in the technology industry.

The fourth section, "Banking," includes three chapters. In Chapter 6, "Bank Financial State Analysis and Bankruptcy Risk Forecasting with Application of Fuzzy Neural Networks," Zaychenko Yuriy, Michael Zgurovsky, and Galib Hamidov forecast the bankruptcy risk of Ukrainian and European Banks with artificial neural networks. In Chapter 7, "The Colombian Banking Sector, Analysis from Relative Efficiency," Gloria Rodriguez-Lozano examines the efficiency of Colombian banks with data envelopment analysis. In Chapter 8, "Sustainability in Indebtedness: A Proposal for a Treaty-Based Framework in Sovereign Debt Restructuring," Sison Maximo Paulino T. Sison III proposes a framework for debt restructuring in United Nations General Assembly Resolution 69/319.

We would like to express our sincere gratitude to all the authors for their contributions. The successful completion of this book is the result of cooperation among many people. We would like to thank Ms. Rozmari Marijan and Mr. Edi Lipović, Author Service Managers at IntechOpen, for their support during the publishing process.

Reza Gharoie Ahangar
G. Brint Ryan College of Business,
University of North Texas,
USA

Can Öztürk
Associate Professor,
Department of Management,
Faculty of Economics and Administrative Sciences,
Cankaya University,
Turkey

Section 1

Introduction

Chapter 1

Introductory Chapter: Some Current Issues on Accounting and Finance

Reza Gharoie Ahangar and Can Öztürk

1. Introduction

This book deals with new perspectives on banking, financial statements, and reporting. The introductory chapter focuses on the following overviews regarding accounting and finance, and it gives the target audience the opportunity to be familiar with recent developments in accounting and finance.

2. Accounting overview

Accounting is an information that collects financial data from the operations of a business entity to create financial statements which consist of an integral part of financial communication so that related parties are informed about the financial situation, financial performance, and cash flows of a business. In recent years, financial reporting side of accounting became very important rather than traditional bookkeeping side of accounting. Within the framework of transparency and accountability concepts of corporate governance, countries started to integrate International Financial Reporting Standards (IFRS) into their regulations so that publicly accountable entities can prepare understandable, reliable, transparent, comparative financial statements particularly to investors and creditors in the global context. In addition, some countries started to adopt "think small first approach" in terms of nonpublicly accountable entities focusing on financial reporting based on the size of a business enterprise. That is why either they adopted International Financial Reporting Standard for Small and Medium-Sized Entities (IFRS for SMEs) or they prepared a new financial reporting standard under the influence of international practices and national accounting rules by considering some exemptions in order not to create additional reporting costs for small and micro-sized entities [1]. Even if this is the case, there are still some countries that continue to use their national GAAP.

On the other hand, another reporting issue that complements financial reporting refers to corporate social responsibility (CSR) or sustainability reporting. Sustainability reporting gives (1) economics data, (2) environmental data, and (3) social data about a business entity [2]. Therefore, financial reporting data of a business enterprise is supported by other three aspects of a business so that related parties are informed about the business entity as a whole.

No matter which accounting practice is adopted by a business enterprise, financial statements create the basis and provide the necessary infrastructure so that investors and creditors can financially analyze the business entity using different scenarios such as financial development and survivorship under different accounting metrics.

3. Finance overview

Financial institutions are public or private organizations responsible for the supply of money in a marketplace. They do this by collecting money and investing in assets such as bank deposits, stock, bonds, and loans. Consequently, they act as an intermediary between savers and borrowers; they play a very important role in society. Financial institutions are not just commercial banks; they can also be national banks (e.g., Bank of England), brokerage firms, money services, insurance companies, and even casinos [3, 4].

The financial institutions are the most critical institutions in the world, since they play a very important role in society. Financial institutions are among the largest employers in many countries, and the development of a nation's economy is closely related to the performance of its financial institutions. The banking system among financial institutions has such a heavy responsibility that is considered as one of the most important parts of a country's economy. The bank represents services, in which without them the economic system of a country would stop. The available capital in banks is the main resource of purchasing products and services; moreover, their donated loans are resources of credit for all units of the economy like families, occupations, companies, and government. Therefore, banks' optimum activity and efficient use of their available facilities are important to help them to reach their goals. Appropriate allocation of capital and resources of different financial activities will ultimately conclude to the improvement of the whole economic situation of a country [5].

Threats and pressures of globalization and rapid development of non-bank credit and financial institutions in recent years have forced the managers of banks to establish research centers and launch research activities in their own status field to stay afloat in the market and compete with other financial institutions in domestic and international markets. In this respect, banks need to determine their weaknesses and strengths in the markets and improve productivity by finding a logical and efficient solution for their activities. A bank or financial institute would be successful in this competitive world if high efficiency can be obtained in its activity. The final goal of a bank is to reach the highest efficiency with the lowest cost. The cost phenomenon is one of the most important issues that will have an effect on the efficiency of banks. An economic project with high efficiency can be achieved when you get a better and economic result toward your goals with lower costs [6, 7].

4. Summary

In this respect, some concepts raise particularly for investors and creditors such as bankruptcy, efficiency of financial markets, transparency, accountability, financial information quality, and survivorship. The overall objective of this book is to discuss these concerns to help the investors, creditors, and the managers of financial institutions to better recognize and analyze particular financial problems and solve them successfully. Moreover, the main purpose of this book is to help the managers of the financial institutions and the users of accounting information, especially bank, investors, and creditors, to find the possible solution to improve the performance of firms and financial institutions.

Author details

Reza Gharoie Ahangar[1*] and Can Öztürk[2]

1 G. Brint Ryan College of Business, University of North Texas, USA

2 Faculty of Economics and Administrative Sciences, Department of Management, Cankaya University, Turkey

*Address all correspondence to: rg0381@unt.edu

IntechOpen

© 2019 The Author(s). Licensee IntechOpen. This chapter is distributed under the terms of the Creative Commons Attribution License (http://creativecommons.org/licenses/by/3.0), which permits unrestricted use, distribution, and reproduction in any medium, provided the original work is properly cited.

References

[1] André P. The role and current status of IFRS in the completion of National Accounting Rules—Evidence from European countries. Accounting in Europe. 2017;**14**(1-2):1-12. DOI: 10.1080/17449480.2017.1319965

[2] Okan Gökten P, Marşap B. Paradigm shift in corporate reporting, Accounting and Corporate Reporting. In: Gokten S, editor. Today and Tomorrow. Rijeka, Croatia: IntechOpen; 2017. DOI: 10.5772/intechopen.68832

[3] Kunt AD, Maksimovic V. Institutions, financial markets, and firm debt maturity. Journal of Banking and Finance. 1999;**54**:295-336

[4] Berger AN, Herring RJ, Szego GP. The role of capital in financial institutions. Journal of Banking and Finance. 1995;**19**:393-430

[5] Morck R, Yavuz MD, Yeung MB. Banking system control, capital allocation, and economy performance. Journal of Financial Economics. 2011;**100**:264-283

[6] Richard D, Villanueva D. Relative economic efficiency of banking systems in a developing country. Journal of Banking and Finance. 1980;**4**:315-334

[7] Feng G, Wang C. Why European banks are less profitable than U.S. banks: A decomposition approach. Journal of Banking and Finance. 2018;**90**:1-16

Section 2

Financial and Non-Financial Reporting

Chapter 2

National Accounting Standards in Turkey

Birsel Sabuncu

Abstract

This study describes the accounting standards being implemented in Turkey. Uniform Accounting System and Uniform Chart of Accounts, Turkish Financial Reporting Standards (TFRS), and Financial Reporting Standards for Large- and Medium-Sized Enterprises (FRS for LMEs) are simultaneously applied in Turkey. TFRS are a direct translation of the International Financial Reporting Standards (IFRS). Enterprises that are subject to independent audit apply TFRS. FRS for LMEs is a financial reporting framework that is subject to independent audit and is valid for financial statements presented to general assemblies of enterprises that do not apply the TFRS. Enterprises that are not subject to independent audit apply Uniform Accounting System. Uniform Chart of Accounts is used by all these businesses. Despite these new practices, a revision has not been carried out in Uniform Chart of Accounts. Because of these applications, it is necessary to define the new accounts needed and to review Uniform Chart of Accounts. Banks, insurance and pension companies, finance companies, financial leasing and factoring companies, and asset management companies are required to use a different account plan despite the application of TFRS.

Keywords: Uniform Accounting System, Uniform Chart of Accounts, Turkish Accounting Standards (TAS), Turkish Financial Reporting Standards (TFRS), Financial Reporting Standards for Large- and Medium-Sized Enterprises (FRS for LMEs)

1. Introduction

Every country initially established its own accounting techniques to report the results of the activities of the enterprises to the relevant interest groups. Due to economic, historical, and traditional differences, there are differences between the accounting techniques. A need to standardize accounting techniques was born both locally and internationally in order to eliminate these differences, which caused negativity about the consistency and comparability of the information in the financial statements. The existence of different accounting systems for different countries can cause various difficulties for international money flow, while a Uniform Accounting System makes it easier for businesses. The existence of a single accounting system to be implemented by all enterprises increases the speed of international transactions, facilitates transparency, and supervision in financial statements and also facilitates transactions.

To eliminate the differences in accounting practices; to perform uniformity in the accounting principles; to ensure that the financial statements are clear, appropriate, understandable, objective, and comparable; to prevent misappropriation of the individuals and institutions related to enterprises; to prevent wrong decisions; and to create a common language during the production and presentation of financial information, Uniform Accounting System and Uniform Chart of Accounts came into force in Turkey in 1994 and still continue today.

As with financial reports prepared in accordance with the different legal and financial structures of countries, International Accounting Standards (IAS) were made compatible which resulted in an exact translation of the International Accounting Standards (IAS), and International Financial Reporting Standards (IFRS) were entered into force as Turkish Financial Reporting Standards (TFRS) in Turkey. Public Oversight, Accounting and Auditing Standards Authority (POAASA) is authorized for the creation, publishing, and monitoring of changes of the accounting standards that are in compliance with international standards. POAASA aims to ensure the transparency, reliability, comprehensibility, comparability, and consistency of the financial statements of the firms. Currently, there are various accounting practices in Turkey, and the preference of which one to be used depends on being subject to independent audit by the enterprises. Companies whose shares are traded on the stock exchange and that are subject to independent audit apply TFRS, while Large- and Medium-Sized Enterprises which are subject to independent audit but whose shares are not traded on the stock exchange and do not apply TFRS, apply Financial Reporting Standards for Large- and Medium-Sized Enterprises (FRS for LMEs). Enterprises that are not subject to independent audit apply Uniform Accounting System.

Uniform Chart of Accounts is used by all businesses. Despite the application of new accounting standards in Turkey, there are no new regulations regarding the needs of Uniform Chart of Accounts, which are used by the companies that perform accounting and reporting in accordance with these new standards. In Uniform Chart of Accounts, there is a need for a revision of the needs of enterprises.

Banks and insurance companies: private financial institutions: financial leasing, factoring, and financing companies: security mutual funds, brokerage houses, and investment trusts are subject to Banking Regulation and Supervision Agency (BRSA) legislation and have a different chart of accounts, and a revision has been performed for this chart of accounts. The mentioned company activities are accounted in accordance with the declarations of BRSA and TFRS.

2. Uniform Accounting System

Before 1994, there was no standard in accounting applications and in accounting principles in Turkey. As of 1.1.1994, there has been a transition to the standard Uniform Accounting System in accordance with the Accounting System Application General Communiqué, Order No.1 published in the Official Gazette dated 26.12.1992 and numbered 21,447 (repeated). With the regulation, it was aimed to provide uniformity in all accounting applications in Turkey and to have the same operating language for all segments. Accordingly, natural and legal persons holding books according to balance sheet basis are obliged to apply the accounting procedures and principles stated in the General Communiqué on Accounting System.

Tax regulations in Turkey may directly interfere with accounting practices and education in time. This relationship between tax regulations and accounting causes accounting in Turkey to be intended for tax and focused on tax regulations. Tax Procedure Law obligates application of the conditions required by the General Communique on Accounting System Application and Tax Procedure Law simultaneously. The arrangements made with the General Communique on Accounting System

Application do not regulate or modify any matter related to tax legislation. The difference between period income and taxable income shall not be settled on balance sheets or income statements; taxable profit shall be calculated outside of financial statements.

Uniform Accounting System is defined as the use of the same chart of accounts in accounting applications, adopting the same accounting principles, benefiting from the same accounting concepts, and providing accounting information in uniform financial statements and reports.

Uniform Accounting System is a system that includes financial statements, charts of accounts, account codes, and similar details. This regulation related to the transition to Uniform Accounting System was made in order to facilitate and provide a sound and reliable follow-up of the results of the undertakings and enterprises belonging to real and legal persons holding books on a balance sheet basis and to ensure that the information presented to the related parties through the financial statements reflects the real situation while maintaining the consistency and comparability characteristics of the audit.

It is the responsibility of the business to be in compliance with the procedures and principles seen in changes made in the name of Uniform Accounting System and organization of reports and declarations for related legislation provisions. Regulation [1]:

- Full and direct delivery of accounting information to respective persons making a decision.
- To compare different periods of the same enterprise with different enterprises.
- Account names in the financial statements give the same meaning to all segments.
- Retaining accounting term uniformity while remaining understandable.
- Aims to establish trust between businesses and those concerned.

Uniformity of the financial statements to be prepared within the framework of the procedures and principles related to the regulation and presentation of financial statements in Uniform Accounting System cannot be changed. In the determination of the taxable income, companies are required to make the necessary adjustments and calculations in accordance with tax legislation. The accounting principles and procedures of the Uniform Accounting System are explained in five sections as follows:

- Basic concepts of accounting
- Explanation of accounting policies
- Principles of financial statements
- Arrangement and presentation of financial statements

Uniform account framework, chart of accounts, and descriptions of accounts. Businesses must establish their accounting systems in accordance with the uniform accounting framework and the chart of accounts. The basic concepts of accounting in Turkey are the concept of social responsibility, the concept of business entity, the going-concern concept, the concept of periodicity, the concept of monetary unit, the concept of cost basis, the concept of objectivity and documentation, the consistency concept, the full disclosure concept, the concept of prudence, the concept of materiality, and the concept of substance over form.

In the Uniform Accounting System, it is not required to disclose them if they are prepared on the basis of the concepts of going-concern, consistency, and periodicity in the preparation of financial statements. However, in the case of deviation from these concepts, the disclosures of the financial statements should explain this deviation together with their reasons. All significant accounting policies included in the financial statements should be clarified and explained briefly. Financial statements should be comparable in terms of periods. If a change is made in the current period or will occur in future periods, which has a significant effect on financial policies, it should be explained together with the reasons and the effects of these changes on financial statements. The financial statements consist of the following ones [1]:

- Balance sheet
- Income statement
- Statement of cost of sales
- Fund flow statement
- Cash flow statement
- Statement of profit distribution
- Statement of changes in equity

Providing useful information in making decisions for investors, lenders, and other interested parties; providing useful information on assessing future cash flows; providing information about assets, resources and changes in these items; and obtaining information about the business activities, these statements should be understandable, appropriate, reliable, comparable, and promptly arranged so that the information in the financial statements can be used in the best possible way by decision-makers.

Uniform Accounting System includes real and legal persons holding books according to the balance sheet basis. However, there are institutions that need to use different accounting techniques in terms of their activities. These include bank and insurance companies; private financial institutions; financial leasing, factoring, and financing companies; security investment funds; brokerage houses; and investment trusts. These institutions are subject to BRSA's legislation, and they are required to comply with the basic principles of accounting, explanation of accounting policies, and principles of financial statements in Uniform Accounting System. They have a different chart of accounts and do not have to fulfill other obligations.

Banks are required to account for their activities in accordance with the communiqués published by BRSA and within the framework of the conceptual provisions of POAASA. However, the accounting for subsidiaries, jointly controlled entities and associates and consolidated financial statements, financial statements to be disclosed to public, and procedures and principles regarding the disclosure and footnotes are determined by the communiqués issued by BRSA.

In accordance with Banking Law, banks are obliged to implement a uniform order in accounting systems in accordance with the procedures and principles determined by BRSA by taking the opinion of the POAASA. Banks must be able to account for all transactions in accordance with their real nature in accordance with the accounting and Financial Reporting Standards issued by POAASA in a manner that is capable of meeting the need to obtain financial reports in an understandable, reliable, and comparable style, suitable for auditing, analysis, and interpretation. Banks are required to achieve transparency and uniformity in accounting and reporting systems, by preventing transactions from being left out of record and by accounting for all activities in a timely, healthy, and secure manner within the framework of correct and timely preparation, reporting, and publishing of financial statements showcasing financial performance and management.

The "Declaration in Reference to the Uniform Chart of Accounts and Prospectus" published in 26.01.2007 and 26,415 (repeated) numbered Official Gazette based on the regulation of procedures and principles for the accounting of bank's accounting practices and documents provides uniformity in terms of accounting and financial reporting for banks and direct acquisition of a single type of balance sheet and income statement, obtaining direct and healthy information in a verifiable and auditable manner in order to ensure supervision and monitoring and to provide the information needed for various analyses and interpretations in a standardized way.

2.1 Uniform Chart of Accounts

The chart of accounts is expressed as a list of accounts to be used in accounting. Uniform Chart of Accounts is a system that is organized for the systematic execution of all accounts held in an enterprise and to ensure that the same events are always recorded in the same accounts.

In Turkey, Uniform Chart of Accounts mentioned in the Accounting System Application General Communiqué, Order No.1 published in the Official Gazette dated 26.12.1992 and numbered 21,447 (repeated) became a necessity for all businesses and accounting activities as of 01.01.1994. Thus, it is aimed to provide unity in accounting practices in enterprises. Uniform Chart of Accounts is a scale that shows the accounts of the activities of the enterprises, and the enterprises comply with the chart of accounts and transfer the same type of documents to the accounting records and financial statements according to the same chart of accounts and the same accounting system. The purpose of the Uniform Chart of Accounts is to use the same chart of accounts for companies holding books on the balance sheet basis and to include the economic events occurring in enterprises in the same accounts within the framework of the same systems and principles. Uniform Account Chart framework [1] is as follows:

1. Current assets
2. Noncurrent assets
3. Short-term liabilities
4. Long-term liabilities
5. Equity
6. Income statement accounts
7. Cost accounting accounts
8. Free
9. Off-balance sheet accounts

Uniform Chart of Accounts is divided into nine account classes, from 1 to 9. Likewise, each account class is divided into ten account groups, ranging from 0 to 9. There are a total of ten main accounts in each account group, from 0 to 9. The main accounts are given three-digit numbers. For instance; 1 Current assets (account class), 10 Cash and Cash Equivalents (account group), 100 Cash (main account). In the Uniform Chart of Accounts coding, various code systems are used while numbering the accounts. Account codes make it easy to distinguish between account groups and accounts.

According to the Uniform Accounting System, enterprises have to establish their accounting systems in accordance with the uniform accounting framework and chart of accounts. Businesses plan their accounts in advance and make a list of them. With the introduction of new standards in Turkey, there was a need for new account codes. There are a number of problems related to the definition of new account codes due to new applications of the Uniform Chart of Accounts related to the application of TFRS and FRS for LMEs such as account of investment property and account of qualified assets held for sale purpose. Therefore, a serious review and revision is needed.

The draft chart of account has been drawn up by POAASA that will enable the companies to establish their accounting systems so that they can prepare their financial statements. In the drafting of the chart of account, account groups and accounts have added to ensure that all companies have access to financial statement data directly. This draft chart of account was submitted to the public on 25.12.2018. After taking the comments of the public and related institutions and organizations, the final version of the new chart of account will be given. These accounts have added in the draft chart of account [2]:

- Financial investments whose fair value difference is measured at amortized cost by financial investments reflected in other comprehensive incomes (accounts 111, 112, 201, and 202)

- Financial assets from ongoing construction (undertaking), project or service contracts, and obligations arising from these contracts and construction works performed in advance (groups 14 and 24 with the exception of accounts 142 and 242 and groups 34 and 44)

- Live assets and gains and losses related to them (account numbers 17 and 27 and account numbers 643 and 653)

- Liabilities related to noncurrent assets classified for sale and groupings classified as held for sale (account numbers 196–199 and account numbers 396–399)

- Investment properties and related income and expenses (accounts numbers 245–249 and account numbers 647 and 657)

- Deferred tax assets and liabilities and deferred tax income and expense effects (account numbers 289 and 489 and account numbers 692 and 696)

- The effect of mergers involving enterprises or enterprises subject to joint control (account number 516)

- Income and expenses accumulated in shareholders' equity (groups 55 and 56)

- Other comprehensive income and expenses for the period (groups 80, 81, 82, and 83)

- Income and expenses relating to discontinued operations and taxes on these income and expenses (accounts 684 and 697)

2.2 Bank Uniform Chart of Accounts

Banks and leasing and factoring companies have a Uniform Chart of Accounts different from the other entities due to the different types of activities and operations of the financing companies. The operations of such institutions are recorded by using Bank Uniform Chart of Accounts that is recognized under the provisions of the Banks Association of Turkey. Banks are subject to the Bank Uniform Chart of Accounts and the regulatory authority, BRSA. Bank Uniform Chart of Accounts consists of the following groups [3]:

0. Return values

1. Credits

2. Investment values and other assets

3. Deposits and other foreign resources

4. Equity

5. Interest income

6. Interest expenses

7. Noninterest income

8. Noninterest expenses

9. Off-balance sheet accounts

The accounts in each group are divided into two as Turkish currency and foreign currency. The foreign currency-denominated accounts are the accounts in which the amounts arising from the foreign currency transactions of the bank are recorded. Foreign currency balances in these accounts are valued at the end of the period and transferred to related accounts. Foreign currency interest, commissions, and income received from foreign currency accounts and transactions are translated into Turkish Lira at the exchange rates prevailing at the transaction date and recorded in the related foreign currency profit/loss accounts. Even if the revenues are collected in Turkish currency instead of foreign currency, they are recorded in the related foreign currency profit and loss accounts. In the Uniform Chart of Accounts, the accounts working for Turkish Lira and foreign currency transactions are separated at the general ledger level. If the last digit of the general ledger accounts is an even number, then it is in Turkish Liras, and if the last digit is an odd number, then it indicates a foreign currency account.

A new account cannot be opened in the book at a new level without the permission of the BRSA. For the accounts that are opened at the level of general ledger and do not have any subaccounts, banks can open auxiliary, sub-, and subordinate accounts if needed. If there is no special account in which to record a transaction, a special account must be opened with permission from the BRSA for such transactions.

3. Turkish Financial Reporting Standards (TFRS)

In the POAASA's decision dated 13.09.2018 numbered (03/161), TFRS are redefined. They cover the standards published under the names of TAS, TFRS, TAS Interpretation, and TFRS Interpretation [4].

The financial statements are prepared and presented by their preparers across the globe for their users. There are some differences due to the use of different financial statements by different countries during the creation of national regulations for social, economic, and legal reasons. The efforts of the International Accounting Standards Board (IASB) to form a common language in the process of globalization are in question. IASB aims to reduce the differences by harmonizing the rules and accounting standards and application procedures for the preparation and presentation of financial statements.

As the common language used by businesses in terms of financial reporting, IFRS are the standard set which directs financial reporting in the world and are formed by the IASB that is the authority for financial reporting in terms of the level of implementation and adoption by the world. The purpose of the IFRS is to increase international investments and trade through a comparable, reliable, understandable, and transparent presentation of financial statements. TAS/TFRS and TAS/TFRS Interpretations are literal translations into the International Accounting Standards/International Financial Reporting Standards/International Financial Reporting Interpretation Committee (IAS/IFRS/IFRIC) as issued by the IASB.

A set of financial statements prepared under the TFRS [5] are as follows:

- Statement of financial position.
- Statement of profit or loss and other comprehensive incomes.
- Statement of cash flows.
- Statement of changes in equity.
- Financial statements are composed of footnotes.

TFRS are prepared to meet the needs of the users of the securities and their financial statements.

The accounting records of companies of the specified nature and scale determined by the decision of the Council of Ministers which was formed around the decision of the POAASA following 01.01.2013 continue to be done in accordance with Tax Procedure Law with compatibility to TFRS [6].

The publication of TAS/TFRS does not mean that Uniform Accounting System has completed its function, but rather increases its importance in order to ensure their healthy implementation, comparability, and reliability [7].

With the application of TAS/TFRS, there are some changes in the format and classification of the financial statements in the Uniform Accounting System which have been compulsory since 1994, and the formal structure is changing. The importance of information is emphasized in the standards, and it is required to present important information in a truthful manner. TAS/TFRS also made significant changes in the valuation of balance sheet items and income statement items. These changes in measurement and valuation significantly affect the reported amounts of items in financial statements [8].

In the POAASA's decision dated 13.09.2018 and numbered (03/161), TAS are redefined. TAS cover TFRS mentioned above and FRS for LMEs and other standards determined by the agency. TAS of the decision for determining the scope of application is detailed below [4]:

- Institutions that require TFRS to be used in the preparation of their individual and consolidated financial statements: (A)—(i) Joint stock companies traded on the stock exchange; (ii) Investment institutions; (iii) Collective investment institutions; (iv) Portfolio management organizations; (v) Mortgage financing institutions; (vi) Housing finance and asset financing funds; (vii) Asset leasing companies; (viii) Central clearing houses; (ix) Central lending organizations; (x) Data storage organizations; and (xi) Capital market instruments that are not traded on the stock exchange, but which provide at least two of the three criteria from publicly traded companies in the capital market: (a) Total of 15 million and more Turkish Lira; (b) Annual net sales revenue of 20 million and more Turkish Liras; (c) Number of employees 50 and above. (B)—Businesses subject to the regulation and supervision of the BRSA in accordance with Banking Law—(i) Banks; (ii) Financial leasing companies; (iii) Factoring companies; (iv) Financing companies; (v) Asset management companies; (vi) Rating agencies; (vii) Financial holding companies; (viii) Companies holding financial holding companies; and (ix) Payment agencies. (C)—Insurance, reinsurance, and pension companies. (D)—Corporations, which are allowed to operate in Borsa Istanbul, joint stock companies.

- Subject to independent audit and the implementation of FRS for LMEs in the preparation of individual and consolidated financial statements of institutions, establishments, and enterprises other than the above; however, it will be able to implement the TFRS upon request of the institutions, organizations, and enterprises in question.

TFRS is the basis of the policy. It does not give businesses the chance to use the account name according to their needs. In the principle-based approach, only the account framework is determined, and the chart of accounts is not compulsory. Each business determines its own chart of accounts.

4. Financial Reporting Standards for Large- and Medium-Sized Enterprises (FRS for LMEs)

IFRS is divided into two as IFRS full set and SME sets in the world and in Turkey, although TFRS for SMEs that was in line with IFRS for SMEs was published in 2010 in the official gazette, but there has been no chance to implement it. Later, FRS for LMEs was issued by POAASA instead of the TFRS for SMEs. FRS for LMEs, entered into force on the date of publication, to be applied in the accounting periods beginning on or after 01.01.2018. It consists of 27 chapters and sets out the accounting principles for all accounting transactions that the companies may encounter in general.

The purpose of the IASB is to prepare a separate standard for SMEs as a set of accounting principles derived from IFRS, which will be used by small, simplified entities whose stocks are not listed on the stock exchange. IASB has defined IFRS for SMEs as a set of simple and simplified Financial Reporting Standards derived from IFRS that will be used by nonpublic entities [9].

FRS for LMEs was issued by replacing TFRS for SMEs with certain additions. This standard includes interest costs, fair value application, receivables and asset provisions, depreciation, severance pay, receivable and debt aging, inflation adjustment, and consolidation matters. It is essential that the fair value application and active registered fixed assets are valued. This issue has been added to FRS for LMEs while it was not included in TFRS for SMEs.

The purpose of FRS for LMEs is to provide financial statements that are in compliance with financial information needs. Four financial statements are required to be prepared in the FRS for LMEs [10]:

- Statement of financial position
- Statement of Profit or loss
- Statement of changes in equity
- Statement of cash flows

The objectives of the FRS for LMEs are defined as follows:

- Real
- Fits financial information requirements
- Providing the preparation of comparable financial statements

Since the FRS for LMEs has set a standard for the determination of commercial profit in Turkey and can be applied more easily than TFRS, it will contribute to a clearer understanding and proper implementation of the full set of standards. Thus, the main task of accounting, real activity result reporting, real situation detection, planning, and control will be available to produce information, and commercial purposes to use financial data will be become a habit [11].

FRS for LMEs is an important development in terms of accounting practices in Turkey. It is a financial reporting standard that enables the financial statements of the enterprises that are subject to independent audit and are not obliged to be accountable to the public and do not prefer to apply TFRS, to be understood by investors and lenders with their fair presentation, needed for financial information, and provides comparability of financial statements [12].

FRS for LMEs prepared by POAASA has been prepared in a simpler language than TAS/TFRS, avoiding the details that are not necessary for large- and medium-sized enterprises and addressing the basic principles of related subjects only [13].

The primary objective of the Accounting System Application General Communiqué issued based on the Tax Procedure Law is to show the financial position of the taxpayers against tax. FRS for LMEs is subject to independent audit and replaces the Accounting System Application General Communiqué and Supplementary Issues in entities not applying TFRS. The financial reporting framework that will be the basis for the preparation of the financial statements of the companies which are subject to independent audit and which will not be subject to TFRS in accordance with the Turkish Commercial Code; the base financial report must be FRS for LMEs.

In the POAASA's decision dated 13.09.2018 and numbered (03/161), FRS for LMEs are redefined. FRS for LMEs is a financial reporting framework that is subject to independent audit and is valid for financial statements presented to general assemblies of enterprises that do not apply the TFRS. It has been decided that may apply FRS for LMEs in the preparation of the individual and consolidated financial statements of the institutions, establishments, and enterprises other than those which are obliged to apply TFRS. However, it has been decided that such institutions, organizations, and enterprises may apply TFRS on demand [4].

5. Conclusion

Accounting shows the financial structure of the enterprise and the results of its economic activities. Accounting practices constitute the source of accounting practices including basic accounting concepts, accounting principles, and accounting policies. Accounting standards is a set of rules that are formed in order to enable the comparison of the information generated in the accounting information system as a result of the business activities with confidence in accordance with the reality without considering the distinction of events of the same nature.

The subject of accounting is the financial statements and reports to be prepared for public disclosure. Therefore, the concept of public disclosure is in a sense also the purpose of generally accepted accounting standards. In order to meet the increasing need for trust in an environment of global competition, the rapid increase of standardization movements has also emerged in the accounting field.

Works aiming for standardization of both national and international dimensions of accounting are implemented in Turkey. Uniform Accounting System which has been in use in Turkey for a long period of time is a topic of discussion. However, as a result of this change, the Turkish translations of IAS/IFRS are in use as TAS/TFRS. TAS and TFRS-applicable Financial Reporting Standards also continues to use the Uniform Accounting System which has been applied for a long time. In addition, the financial reporting standard has been enacted for large- and medium-sized enterprises. Despite the fact that Full Set TAS/TFRS is translated into Turkish, FRS for LMEs has been put into effect instead of IFRS for SMEs because it is a closer match. Companies that are subject to independent auditing and that do not apply TFRS are required to apply FRS for LMEs in the presentation of their consolidated financial statements.

International Accounting Standards have emerged as a result of the harmonization of financial reports prepared according to the standards determined by the legal or financial structures of different countries in order to enable users to compare the same criteria for information users in different countries. The consolidated financial statements of companies whose stock is traded in the stock exchange and that are subject to independent auditing must be prepared in accordance with TFRS.

Banks, insurance, reinsurance and pension companies, factoring companies, finance companies, financial leasing companies, and asset management companies are required to implement TFRS and have their own chart of accounts.

FRS for LMEs is applied by companies subject to independent audits but do not apply TFRS and provides comparable financial information. The purpose of FRS for LMEs is to provide financial statements that are in compliance with financial information needs. The companies that are not subject to independent audits are generally required to prepare their financial statements in accordance with Uniform Accounting System.

The application of the Accounting System Application General Communiqué and its annexed Uniform Chart of Accounts as prescribed by the Tax Procedure Law is still in progress. The combination of all these applications may cause confusion in some cases. In addition, it is necessary to review Uniform Chart of Accounts as soon as possible to define and reorganize the account codes that businesses need due to new applications.

Author details

Birsel Sabuncu
Pamukkale University, Denizli, Turkey

*Address all correspondence to: bsabuncu@pau.edu.tr

IntechOpen

© 2019 The Author(s). Licensee IntechOpen. This chapter is distributed under the terms of the Creative Commons Attribution License (http://creativecommons.org/licenses/by/3.0), which permits unrestricted use, distribution, and reproduction in any medium, provided the original work is properly cited.

References

[1] Seri No'lu Muhasebe Sistemi Uygulama Genel Tebliği. Available from: http://gib.gov.tr/node/88229 [Accessed: November 8, 2018]

[2] Available from: http://www.kgk.gov.tr/ContentAssignmentDetail/2530/Finansal-Raporlama-Standartlar%C4%B1na-Uygun-Hesap-Plan%C4%B1-Taslag%CC%86%C4%B1 [Accessed: January 02, 2019]

[3] Bankacılık Düzenleme ve Denetleme Kurumu, Tekdüzen Hesap Planı Tebliğ. Available from: http://www.resmigazete.gov.tr/eskiler/2007/01/20070126M1-1.htm [Accessed: November 10, 2018]

[4] Available from: http://www.kgk.gov.tr/Portalv2Uploads/files/Duyurular/v2/Kurul%20Kararlar%C4%B1/TMS%20Kurul%20Kararlar%C4%B1/RG-TMS%20Uygulama%20Kapsam%C4%B1na%20%C4%B0li%C5%9Fkin%20Kurul%20Karar%C4%B1.pdf [Accessed: December 10, 2018]

[5] Available from: http://www.kgk.gov.tr/Portalv2Uploads/files/Duyurular/v2/TMS/TMS%201%20Finansal%20Tablolar%C4%B1n%20Sunulu%C5%9Fu.pdf [Accessed: November 9, 2018]

[6] Sabuncu B. TMS/TFRS Kapsamında Türev Araçlar (Forward, Futures, Opsiyon, Swap). Ekin Kitabevi, Bursa: Muhasebe Uygulamaları; 2016. 140 p

[7] Akdoğan N, Sevilengül O. Türkiye muhasebe standartlarına uyum için tekdüzen hesap planınında yapılması gereken değişiklikler. Mali Çözüm Dergisi. 2007;**84**:30-70

[8] Gücenme GÜ, Arsoy AP, Ertan Y, Bora T. TMS/TFRS' ye dönüştürülen finansal tabloların denetim modelinin özellikleri. Muhasebe Finansman Dergisi. 2014;**64**:1-26

[9] IFRS. Available from: https://www.ifrs.org/issued-standards/ifrs-for-smes [Accessed: November 15, 2018]

[10] Büyük ve Orta Boy İşletmeler için Finansal Raporlama Standardı. Available from: http://www.resmigazete.gov.tr/eskiler/2017/07/20170729M1-1-1.htm [Accessed: December 3, 2018]

[11] Gençoğlu ÜG. Temel konularda BOBİ FRS ve TMS/TFRS karşılaştırması. Muhasebe Finansman Dergisi. 2017;**76**:1-24

[12] Öztürk C. Yerel finansal raporlama çerçevesi ve büyük ve orta boy işletmeler için finansal raporlama standardı: Bir literatür incelemesi (2016-2017). Journal of Business Research. 2018;**10**(1):763-781

[13] Ataman B, Cavlak H. Büyük ve Orta boy İşletmeler için finansal raporlama standardı (BOBİ FRS) ile tam set türkiye muhasebe ve türkiye finansal raporlama standartlarının (TMS/TFRS) karşılaştırılması. Finans Ekonomi ve Sosyal Araştırmalar Dergisi. 2017;**2**(3):153-168

Chapter 3

CSR Training and Financial Statement "Disclosure": The Case of Italy

Maria-Gabriella Baldarelli

Abstract

The objective of the chapter is to analyse the impact that CSR (corporate social responsibility) training had on non-financial information that is contained in financial statements following the EU directive 2014/95/EU and the Italian law: 254/2016. Literature review is based on institutional theory and social and environmental accounting as emancipatory tool to ameliorate the quality of life. The case analysed is the territory of Rimini, Italy. The case has been chosen because Rimini represents one of the most important concentrations within the hospitality and tourism industry in Europe. Another reason is that in the area of Rimini, 13 years ago, CSR training courses had been promoted for profit-making enterprises by the Chamber of Commerce along with public and private organisations. These training courses created a very important background for collocating the new EU directive regarding non-financial information that must necessarily be contained within the financial statements of enterprises. The objective of the chapter is to analyse the impact that the CSR training had on non-financial information that is contained in financial statements following the EU directive 2014/95/EU and Italian law: 254/2016.

Keywords: non-financial information, corporate social responsibility, financial statements, institutional theory, social and environmental accounting

1. Introduction

The idea behind the research project, which we will deal with in this study, derives from the combined results of two important processes which in the area of Rimini developed at different times, though they were both directed towards making the content of the financial statements and of the information the company supplies to the various stakeholders more complete.

Of these two processes, the former has had, and continues to have, the function of "fertilising" culture towards appreciation of environment and stakeholders and this has been realised by way of the PERCORSI project since 2008.

The latter process is more recent; indeed, it dates to 2014 and concerns the progressive attention of the EU to include non-financial information to the financial statements of companies quoted on the stock exchange. This, in Italy, culminated in law 254/2016, which makes inclusion of non-financial information in financial statements of companies quoted on the stock exchange mandatory.

The subject of this work is to verify, through an analysis of financial statements of companies that attended the training courses on CSR within the PERCORSI project, in order to be able to understand what the starting point is from which information can be extracted for investigations following the implementation of law 254/16.

We may define corporate social responsibility as the awareness by the company of the existence of an external environment that cannot be measured in purely economic terms, but which refers to the expectations that various stakeholders have. It also implies the attempt of the company to meet these expectations, in order to increase the consent or, at least, to reach the legitimacy that it needs to perform its objectives and to create strategic value [1].

Our work considers CSR training courses that are a useful framework towards implementing accounting disclosure and making social, environmental and sustainability reporting.

Therefore, the research question is thus: "How much did the training courses on CSR influence the area of Rimini in order to increase non-financial information contained in company financial statements?"

The research design is based upon use of institutionalist theory [2] and particularly upon part of the same which was declined in terms of sustainable development considering certain, more recent, thoughts on the role of accounting and accountability [3].

Literature review also points to social and environmental accounting as an emancipatory tool to ameliorating the quality of life. The case analysed is the Rimini area of Italy. The case had been chosen because Rimini has one of the most important hospitality and tourism industry concentrations in Europe.

Another reason is that in the Rimini territory, 13 years ago, CSR training courses had been stimulated for for-profit enterprises by the Chamber of Commerce along with public and private organisations.

These training courses created a really important background within which to collocate the new EU Directive about non-financial information, which must be compulsorily contained in the financial statements of enterprises.

The empirical section, instead, concerns use of "content analysis" which was applied in the first 6 months of 2018 in order to summarise it in the second half of the year.

The chapter includes the following sections: the first section contains the literature review, the second one concerns the Italy-Rimini background for social accounting and CSR training courses, the third one analyses empirical research results and the fourth one contains our conclusions.

2. Literature review regarding institutional theory and social accounting

In choosing the theoretical framework as the basis of empirical research, institutionalist theory helps us to understand up to what point we may consider the process that, put into operation by the norm, induces a uniformity of behaviour in organisations, and therefore, in companies as well. Indeed, such an institutionalist theory takes the phenomenon of "isomorphism" [2] into consideration. On opposing sides, however, are authors who, together with such a phenomenon, have analysed companies which tend towards behaviours that are similar just in certain aspects.

Focus of institutionalist literature is inertia as well as the tendency to conform to behaviours that spread throughout organisations in order to adhere to the requests made by the economic, political, and environmental system wherein they operate.

Indeed, those aspects that characterise the process of homogenisation of the organisations, from which institutionalisation derives, are called "mechanisms" by [4], whereas Scott [5] names them "structures". Such mechanisms can be grouped into three main areas, among which the first is labelled "coercive" and is relative to the pressures that prevalently come from politics, law and the market. They influence the behaviour that the organisations, therefore the companies as well, have to follow in order to survive. The second area regards those mechanisms defined as "normative", which have their origin in the strong influence of the professional networks, education and common values. The third area concerns the mechanisms defined as "mimetic". In this area, organisations emulate others that are considered best practices [4]. These mechanisms are respectively defined as regulative, normative and cognitive structures by Scott [5].

Catturi and Riccaboni [6], partially basing their theory on business administration [1, 7, 8], mainly focus on unwritten rules. The unwritten rules represent mechanisms/structures to push organisational change. From this, it emerges that for each company, the standard indicators represent a starting point but, at the same time, a limited part of performance evaluation. Companies' performance evaluation is also better monitored through qualitative social and institutional-type features [6].

Lai [9] traces the evolution of business theory basing it on the explanation supporting Zappa's theory [7]. According to Lai [9], institutionalism theory firstly affirms itself as opposed to the individualistic behaviour logics of social phenomena interpretation. Moreover, the principal aspect of the institutionalist theory is that there are norms and rules which allow, by way of the institution, the resolution of conflicts and the promotion of civil life. The third important aspect to underscore regards a dimension internal to institutionalist theory, which is relative to the governance of the company. Alongside this final one is the context external dimension wherein the norms, usages and coordination allow its development, since they are able to direct development of organisational layouts ([9], pp. 66–67). The point of view of institutionalism considers the company as a collection of upper subsystems, segments and sub-segments, which develop relational behaviours representing bonds of interdependence to such an extent as to require a direct approach [9].

Otherwise in the literature, "engagement research" has been put forward as a strong approach in developing theories in order to understand SEAR/SER (Social and Environmental Accounting Research/Social, Environmental Reporting) and to enhance organisational practices and performances [10, 11, 33] as well as in order to explore diverse issues, including change within organisations and mediation pathways [12–15].

Furthermore, Larrinaga-González [3, 16, 17] has adopted the institutionalist [4, 5] and neo-institutionalist theories [18] as a theoretical framework to point out the adjustment or, at least, to explain the procedures of SER (social and environmental reporting) and to understand the drivers of institutional modification. So they have also released up new directions in institutional theory [3, 19].

Considering the presence of numerous organisational fields around the issue of SR (sustainability reporting) as a possible clarification, the question, that Larrinaga-González [15] emphasises, is whether there is a sole global organisational field rather than different local organisational grounds for SR. Reflecting on the variance of SR practices, he pointed out the presence of locally based SR fields (i.e.: Environmental Management and Audit Scheme; European Commission, 2001; Triple bottom line; Global Reporting Initiative; ISO 14001, UN SDGs, etc.).

Especially, outlining prior research that had been conducted in sustainability management and reporting using a neo-institutional lens, Larrinaga-González [16] uses the neo-institutional viewpoint to construct an institutional description of the expansion of SR and to ascertain the consequences of the institutionalisation of SR.

He remarks that the observed studies of sustainability management and social reporting deliver some evidence for the institutionalisation of such practices. SR is the consequence of a combination of the mechanisms/structures that we described above. Larrinaga-González [16] hints that SDR (sustainability development reporting) is not yet entirely institutionalised and may be the outcome of additional general social/environmental awareness being institutionalised in some situations. Lastly, Larrinaga-González [16] faces the matter of modification and institutionalisation of SR and recognises some investigation designs. The former is about the initiating event that may change the institutional engagements and what elements play a role in changes to coercive/normative/cognitive mechanisms/structures. The latter is about the analysis of what interactions occur between competitive forces and institutional constructions in the progression of institutionalisation [19]. In the next section, we focus on the Italian peculiarities.

3. Framework of institutional factors and SEAR in Italy

In Italy, there are some important factors that sustain the need for social and environmental accounting and reporting to emancipate economic, social relationships and democracy.

Following Contrafatto and Rusconi's idea [20], social and environmental accounting (SEAR) theory in Italy had its origins in the 1970s [21], but until the 1990s, we did not find empirical social and environmental reporting excluding the Merloni-Battelle experience. The reasons of interest in this theory are due to the fact that financial accounting and reporting law was not so exhaustive so there was a wide-open space to cover with social and environmental accounting theory.

In the 1980s, there was a lot of publications concerning this subject [17, 22–25], but organisation culture was not yet ready to accept these issues, as we can read: "This was partly, at least, a result of mutual distrust between corporations, public authorities and trades unions" ([8], p. 4).

Despite difficulties, there are some important factors in favour of SEAR; among them firstly, there is a common platform of "Economia Aziendale" [20] where financial accountants and social and environmental accountant scholars can dialogue quite easily.

The concept of an organisation following institutional theory [9] opens a holistic view of it and involves social and environmental subjects too.

Secondly, involvement in the economic system of different organisations, such as the state with public organisations, for-profit organisations and private, not-for-profit ones, pushed transparency in order to face crisis and to go on globalisation challenges. So, this evolution made it more and more important to have one common platform of information for dialogue so all economic partners had been more oriented towards disclosing their behaviour as regards the stakeholder.

In the following years, SEAR developed into a theory [2, 19, 26–30], but above all in practice, due to a lot of European initiatives (Stakeholder forum and Green papers 2001 and subsequent).

We have to remember, among other things, two important projects regarding SEAR: the first one had its origins in 1988 and its name is SBG (social balance group) that stated and is stating some principles that form SEAR (http://www.gruppobilanciosociale.org/).

The second important initiative for Italy was CSR-SC, 2003, which saw the progressive spreading of the debate on corporate social responsibility, both because the Italian government placed it among the five priorities of its European semester and because the European Campaign promoted by the European Commission, the CSR-Europe and the Copenhagen Centre realised events and actions.

The EU Commission identifies some guiding trains of thought regarding socially responsible conduct like the codes of conduct, managerial rules and regulations, accounting, auditing and the drawing up of reports, labels, and socially responsible investments.

From the point of view of provisions and policies issued by the Government, the partial Italian response is also substantiated in the CSR (corporate social responsibility)—SC (social commitment) project promoted by the Italian Ministry for Labour and Social Policy (http://www.camcom.gov.it/cdc/).

Until 2006, SEAR implementation in Italy was voluntary, from this date on, there has been the Italian law of 24th March 2006, No. 155: "The discipline of social enterprise", which at article 10 requires the compulsory drawing up of a social report. It is the first time that the Italian Parliament deliberated on a compulsory rule concerning SEAR.

The legislative decree of 30th December 2016, No. 254 was issued following implementation of directive 2014/95/EU of the European Parliament and of the European Council of 22nd October 2014, which modified directive 2013/34/EU as far as the communication of non-financial information and information on diversity by certain companies and certain large-scale groups are concerned. The decree came into effect on 25th January 2017 and the measures therein contained apply to financial years starting from 1st January 2017. This decree represents an important novelty since, for the first time in Italian for-profit businesses, the need for non-financial accounting is ratified.

4. CSR training and activities in the Rimini region

We chose the Rimini case, because the influence on the environment especially of tourist entities (17,000) is great, as there is the main concentration of the hospitality industry in Europe.

In 2004, Rimini began a process to spread CSR (corporate social responsibility) in the territory and to create a new culture using networks (PERCORSI Project). In the

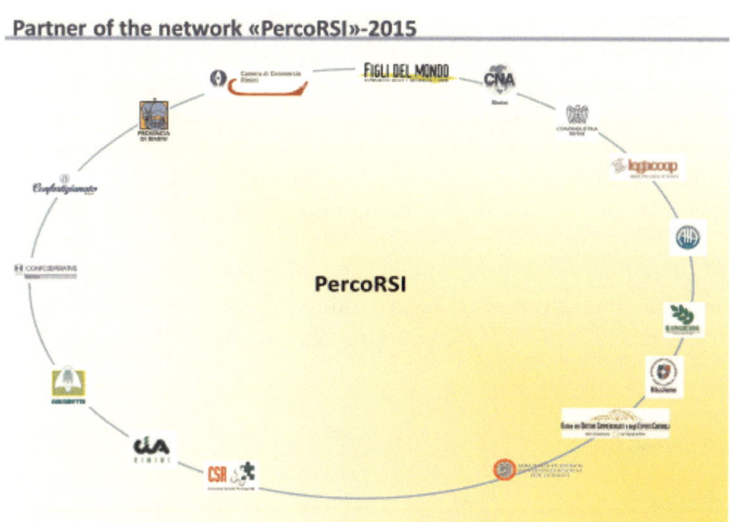

Table 1.
Parters of network percoRSI 2015.

Mechanisms ([4], p. 164)
Coercive • Since 2003, the Italian administration pushed Chambers of Commerce to spread social responsibility throughout the territory. • In 2006, the regulation has made it obligatory for Social entities to draw up a social report. • Directive 2014/95/EU regarding non-financial information had been originated. • Development of competing tourist alternatives has pushed Rimini to think of a socially responsible area. • Certain sector obligations relating to rules relative to workplace safety, climate change, environment emissions, etc. (for example: gas distribution sector and companies who manage waste). • Since 2016, law 254 has been issued concerning the obligation of non-financial information in the financial statements.
Normative • Development of a system of social and environmental accounting is the answer to a CSR culture, which is also based on essential reasons. • Organisation of training courses for companies had been started. • Involvement of professional associations: accountants and lawyers, as well as various business associations partly planned and coordinated by the Chamber of Commerce of the territory.
Mimetic: • Additional Italian areas have been involved by the local Chambers of Commerce in a process of applying social responsibility. • It is an imitative procedure founded on a series of choices of various agents, among whom the Chamber of Commerce, not-for profit organisations, etc. create a network of players all with the aim of developing a region that will become more and more socially responsible and sustainable.

Source: our elaboration.

Table 2.
Institutionalist in the Rimini area.

meantime, at the beginning, we carefully followed the whole "PERCORSI" project process. We had been impressed by the final objective: not only to realise SEAR but to modify entity culture into one of emancipation and social equality [10, 16].

In the case of Rimini, in a positive perspective of research, for-profit, non-profit entities, public administrations (municipality, Chamber of Commerce, and province), trade associations; Chartered Accountant Associations and other associations together with the University of Bologna formed narrowly joined networks to create a more socially responsible area [31], as you may see in the following **Table 1**.

The situation of the Rimini region using neo-institutional theory is represented in **Table 2**.

After the context presentation, we wonder if training activities have an impact on companies financial statements. In the following section, we will analyse the data of the empirical research, which help to provide an answer to such a question.

5. Research results[1]

The methodology avails of "content analysis" by way of searching eight keywords within consolidated financial statements. Especially, we focus the attention

[1] For the elaboration of data, we whole-heartedly thank Doctor Angelica Giammaria for the precious work done.

on the section of the balance sheet, where are the financial and non-financial information.

The elaborations were carried out using Excel sheets, which were later represented through graphics. Five enterprises were singled out: Pesaresi Giuseppe Ltd. (Limited), Petroltecnica Ltd., SCM Group Ltd., Focchi Ltd. and SGR Group Ltd [34].

The companies were chosen on the basis of the attendance of CSR courses activated within the PERCORSI project, as per above. Besides, they all belong to different sectors and are of different size.

This allows checking the different methods of impact upon the environment and society. Of these companies, only SGR draws, from 2011, the sustainability report, whereas the others throughout the period examined by the research project do not draw up any independent document where social, environmental or sustainability information is described.

For this reason, the case of SGR will be the last to be treated, in which we have included the sustainability report into the research as well. The choice does not make the data homogeneous but allows us to highlight the varying frequency due to a precise strategic choice of the company. Such a different approach allows us to highlight certain interesting considerations which we will later show.

The keywords have been chosen on the basis of relevance to the themes of work and are: social responsibility, sustainability, environment, development, safety, territory, stakeholder and innovation. An analysis of each financial statement from every company in the period 2008–2017 was carried out, since 2008 represents the beginning of the PERCORSI project and 2017 represents the year of change due to the quoted legislation. The consolidated financial statements come from the AIDA Database.

Below are the outcomes of the analyses, which are preceded by a brief outline which introduces each company.

The SCM Group S.p.A. was established in 1958 for the production of machine tools and is included in large-scale enterprises[2].

It is a world leader in technologies for the working of wood and advanced materials. At the end of 2016, employees numbered 1512; turnover was €331,006,785 and gross total invested capital was €356,554,493.

The parent-company is headquartered in Rimini, and has sister companies in: Europe, Asia, Australia and South America. At its overseas companies, employees numbered 3600 with turnover greater than € 700 million.

Companies in the group operate within the following sectors: construction, furniture-making, aerospace, shipping, automotive, and plastic material-making. **Graph 1** shows the summarised results of the conducted analyses whose values, on the vertical line, represent the frequency with which keywords are included in the financial statements.

From the graph, the words 'development' and 'safety' emerge in the main. The former is repeated 30 times and the latter 62.

This certainly means that the SCM Group S.p.A. in its activities surely pays particular attention to the development of new technologies for constructing industrial machines which are ever more advanced, but also to the training

[2] The European classification considers a SME a company with up to 250 employees. Above this quota, a company is considered a large-scale enterprise. There is also a size relating to turnover, that is a turnover above €10 million for a small-scale company and above €50 million for a medium-scale one. Legislative reference is recommendation 003/361/EC of the European Commission.

Graph 1.
Keyword trend of SCM group. Source: our elaboration.

development of its employees, wishing to improve the competences of human capital, fulcrum of success for this company.

'Safety' is repeated 11 times in the following 3 years, 2009, 2010 and 2011, as it is in 2014 and 2015. This frequency underlines the commitment of the company to guarantee safety in the workplace for its employees considering the dangerousness of the company's activities.

Instead, the terms 'social responsibility' and 'stakeholder' are not to be found, notwithstanding the company robustly deals with this issue, just as 'stakeholder' is never mentioned. Therefore, attention of the company, as research matters stand, is more directed towards the development of activities rather than to the disclosure and communication of the same in the financial statements.

As evidence of this, the SCM Group S.p.A. was the winner of the Sodalitas Social Award in 2008 in the category "Best Social Marketing campaign" with the project "International community, human capital, local community, suppliers and environment".

The company, for years, has started on a social responsibility path, training its own employees, collaborating with schools and universities and paying attention to environmental impact reduction during the single production phases.

Besides, thanks to the collaboration with the Figli del Mondo association and the Pope John XXIII Community, it has been possible to realise vocational training centres, teaching the technologies of working wood to the communities of more disadvantaged countries.

It provided economic support through the donation of machinery used in training centres for a refuge community for orphans of parents who died of AIDS in Ndola (Zambia) and in Foz de Iguaçu (Brazil). In 2011, the company went to the finals of Sodalitas Social Award in the category "The Jury's Special Award", an award given over to companies and associations which, via projects, favour youths' entry into the world of work (www.figlidelmondo.org).

The second company that we will look at is Pesaresi Giuseppe S.p.A.

The Pesaresi Giuseppe S.p.A. company was established in 1961, operates in the building industry and is a medium-sized company. Its main activities are in the construction of roads, motorways and airport runways and in the production of bituminous conglomerates and concrete. At the end of 2016, it numbered 60 employees, turnover of €15,001,266 with gross invested capital of €52,417,577. In **Graph 2**, we show the results of our analysis.

Among the more-used keywords, we can pick out: 'safety' and 'development'.

'Safety' shows an up-hill trend till 2015 where it is repeated up to 20 times.

Top management has always been strongly committed to spreading this culture throughout the whole of the company. This culture practised through a team of specialised figures and directed towards the guarantee of an efficient system of safety management which has an impact on, besides the workers, the community with which the company interacts as well.

The word 'development' demonstrates a rise beginning in 2014 till arriving at 2016 where it is repeated at least 18 times. Company development was possible, besides due to the entrepreneurial competences of the founding Pesaresi brothers and the capabilities of company staff members, thanks to the policies of protection of the employees and the environment.

The third most-used word in financial statements is 'environment', showing an almost continuous trend in its use till 2015, excepting 2012, and in the final 2 years there has been an increase, especially in 2016, repeated more than eight times. Company management tries to identify the processes that harm the environment in order to reduce them using appropriate interventions, investments as well as adequate competences.

Pesaresi Giuseppe S.p.A. is concerned with, besides, training and making their employees sensitive to environmental issues, for example through "la promozione della differenziazione nella raccolta e smaltimento dei rifiuti prodotti, sia internamente allo stabilimento che sui cantieri esterni." (www.pesaresi.it) (*the promotion of separate collection and disposal of the produced waste, both internally within the actual factory as well as on external building sites*). Only since year 2016, the two most characteristic words on the topic of CSR, viz. 'social responsibility' and 'sustainability' are used.

Following these observations, we may affirm that the company puts the concept of social responsibility towards people, the environment and the area wherein it operates at the core of its activities. This is to respond to the economic and social needs though it does not show any inclination towards the disclosure of such information in the financial statements.

Graph 2.
Keyword trend of Pesaresi. Source: our elaboration.

Petroltecnica S.p.A. was established in 1950 and is headquartered in Rimini, but has companies in other Italian cities: Bari, Milan, Cagliari and Priolo Gargallo—in the province of Syracuse in Sicily. The company operates in the sector of "Environmental Ecology" offering a wide range of services: demolition and decontamination of industrial work plants and refineries, recovering of uncontaminated sites, building site services, waste treatment and management, decommissioning of structures containing asbestos, and environment rapid rescue, therefore working in the safeguarding of the soil and waters. At the end of 2016, annual turnover was €59,561,224, gross invested capital equals €60,105,635 and employees numbered 246 among whom are operators, technicians, clerks and management. In **Graph 3**, below, we can see the research results.

The most-used keywords are 'development' and 'safety'. Regarding the word 'development', note a non-continuous use till 2010 though with a later upturn. Petroltecnica S.p.A., over the years, has reached a professional and technological development such that it has consolidated its presence throughout the nation and abroad, as is the case in Taiwan and Turkey.

Particularly, technological development is directed towards strengthening the quality of service offered and worker safety. Indeed, safety is a word we meet increasingly, even if non-continuously. One of the objectives of the company is surely to maintain high safety standards for the operators and for all those who come into contact with the activities of the company, "consapevole dell'importanza di garantire le migliori condizioni di salute e sicurezza negli ambienti di lavoro" (Petroltecnica S.p.A. Code of Ethics) (*aware of the importance of guaranteeing the best health and safety conditions in the workplace*). Visit www.petroltecnica.com.

The company periodically conducts training courses regarding themes connected to safety: for example, in 2016, 248 hours of training were provided for a total number of 984 participants. Another important word is 'environment', seeing as this company makes protection of the eco-system, one of its main aims, also going on to draw up the environmental report.

For this reason, Petroltecnica S.p.A. makes its employees aware of environmental issues while creating greater knowledge on the consequence of operations on the ecosystem.

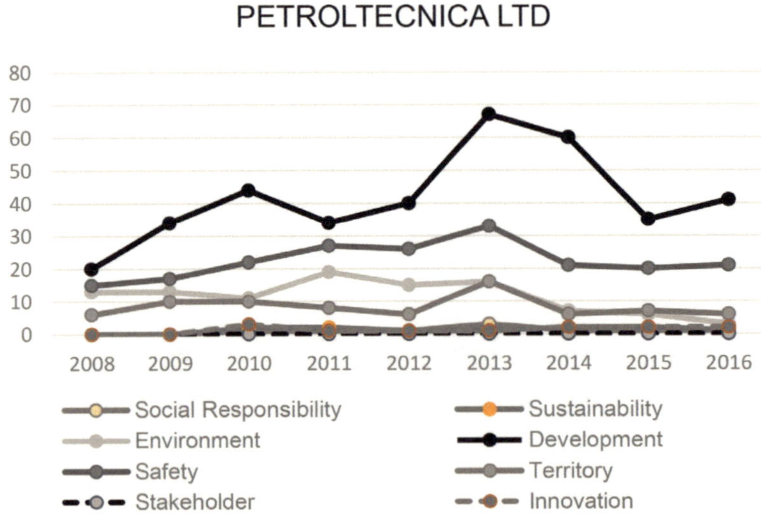

Graph 3.
Keywords regarding Petroltecnica. Source: our elaboration.

Thus, the company adopts socially responsible behaviours concerning various subjects, particularly as regards the environment, employees, collectivity and local community.

Let us analyse below Focchi S.p.A., established in 1914 and it operates in the sector of metal carpentry. In 1955, the company begins to produce aluminium windows and doors and 20 years later, concentrates on the production of "curtain walls", which is currently its main activity [www.focchi.it/il-gruppo].

The business has its headquarters in Rimini, London, Singapore and New York. Focchi S.p.A. is an "impresa leader nel settore degli involucri per edifici occupandosi direttamente sia della progettazione, della produzione e della posa in opera" (www.focchi.it/il-gruppo) (*world leader in the sector of building shells directly taking care of design, realisation and assembly*). At the end of 2016, the group had gross invested capital worth €105,412,000, employees numbered 163 and had a turnover of €56,361,000.

Graph 4 summarises the analysis conducted on consolidated financial statements.

Here too, as has already been seen, the keyword used the most is 'development', having been used the same number of times in 2008 and 2010 with 6 words, in 2011 and 2012 with 10 words and, finally, in 2013, 2015 and 2016 with 11 words. Focchi S.p.A. has a study and research office that is strongly committed "allo sviluppo ed alle prove di laboratorio, in modo da ottenere sempre le massime qualità prestazionali ed i prodotti più innovativi del settore." (www.focchi.it/il-gruppo) (*to development and laboratory testing, in such a way as to always obtain optimal quality performance as well as the most innovative products in the sector*).

The word 'safety' is often repeated in financial statements as well. Indeed, the technical staff periodically carry out safety checks in the workplace to get the certificate that allows them to access building sites. The word 'sustainability' is used at least once in the majority of the consolidated financial statements. Focchi S.p.A. chooses to make the concept of sustainability its own, building sustainable façades using innovative technologies. Lastly, the word 'environment' is constantly used in financial statements since it highlights the importance that Focchi S.p.A. places on environmental protection.

Research of sustainability goes hand-in-hand with developing new technologies which may permit the realisation of sustainable structures: since 2011, indeed, the word 'innovation' is starting to be used.

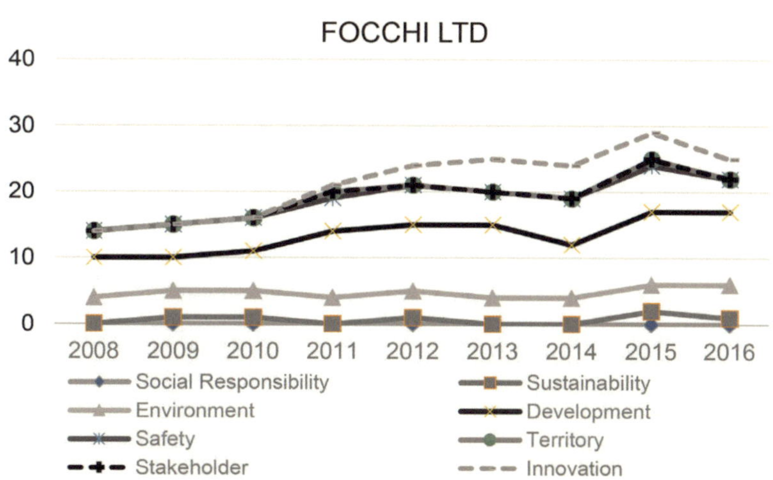

Graph 4.
Keyword trend of Focchi. Source: our elaboration.

In 2009, Focchi S.p.A. was finalist in the Sodalitas Social Awards in the category "Building the best community partnership programme". Management of the company financially sustained the Rimini association: *Cittadinanza Onlus*, "riconoscendone la validità delle iniziative e contribuendo al sostegno e allo sviluppo economico di progetti a favore dei disabili psichici." (www.figlidelmondo.org) (*recognising its initiatives' validity and contributing to sustaining and economically developing projects in favour of the psychologically disabled*). In that year, an international meeting was organised by that association in Rimini, where projects on these issues were discussed in order to implement them in developing countries.

Continuing the work, we will analyse the case of the SGR Group which, compared to those previously analysed, draws up a document separate from the financial statement, wherein it merges social and environmental information.

Indeed, the case of the SGR Group is different from the others because since 2011 it regularly draws up a sustainability report [32]. This allows us to make a comparison between keywords, in that the other companies we have previously treated do not draw up, according to specific standards, social, environmental, sustainability and integrated reports.

The SGR Group is a multi-utility and family-owned company founded in 1956, unlisted, based in Italy, made up of several companies, and with a long experience in gas distribution, over 50 years, during which time it has always been a one-family-owned enterprise, and this feature has been able to create a very strong connection with the territory wherein the enterprise is operating.

Its activities are carried out in Italy by SGR Rimini and in Bulgaria by the CityGas Company (Trakia region) [32].

In 2016, the company had 408 workers, a turnover of €290,042,198 and a total of invested capital of €411,820,001. In **Graph 5**, the research results are highlighted.

There being a sustainability report may be noted from the graph, which shows that the use of the word 'sustainability' has started since 2011 and has a growing trend. SGR adheres to the international strategy of the UN 2030 SDGs Agenda, ensuring "l'accesso a sistemi di energia affidabili, sostenibili e moderni, incentivando una crescita economica duratura e inclusiva [...]; ad assicurare salute e benessere per i propri dipendenti e fornire un'educazione di qualità, equa ed inclusiva per la comunità." (www.grupposgr.it/bilancio-di-sostenibilita) (*access to systems of reliable, sustainable and modern energies and incentivating lasting and inclusive economic growth [...], ensuring health and wellbeing for its employees and providing fair and inclusive quality education for the community*).

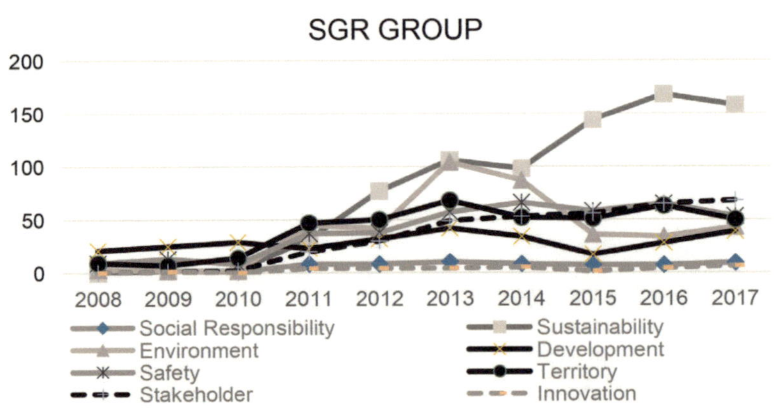

Graph 5.
Keyword trend of SGR. Source: our elaboration.

The second keyword we find is 'environment', which is fundamental in the strategic and development vision of the company.

Besides, there also feature the words 'stakeholder' and 'territory' which, often, are repeated to the same extent, starting from the sustainability report of 2014 till that of 2017.

SGR promotes innumerable initiatives throughout the territory and for the whole local community which may be of a cultural, sporting or social nature.

We may affirm that the SGR Group, differently from the others we have met, carries out aspects of social responsibility, and it is aware of disclosure, also with fitting documents containing such aspects.

6. Conclusions

In this work, we have wondered how training activities contributed to develop and to spread a culture that includes social responsibility in Rimini territory [31]. This relatively new culture may perhaps influence company management and financial reporting disclosure before and during the innovation in regulation, such as law 254/16.

In order to analyse the Rimini case, we adopted the institutionalist theory and, in particular, that part which may be expressed in terms of social, environmental and sustainable accounting [16]. This has created the theoretical bases, in order to be able to understand, among the various activities, training initiatives that have been started and consolidated over time.

Reading these activities has, on the one hand, validated the theory adopted which has isomorphism as its central point [4, 5], since, throughout the area analysed, the training activities of spreading the culture of CSR have been promoted by a process, the European one, that required alignment to them.

If, from one viewpoint, this might be true, then from another, the training activities in the Rimini area have pushed some of the companies, that we analysed, to become more and more aware of the importance of CSR and, especially, of the keywords through which to translate the same into concrete actions.

Analyses conducted on financial statements and consolidated financial statements demonstrate a positive trend in almost one of the key words used. This trend is one expression of interest of CSR. Even if the company, at the beginning, did not know anything about CSR, such a cultural process influenced the contents of the accounting increasing the attention to the key words, such as sustainability and/or safety and/or environment, etc. even before it became a semi-mandatory process by law.

Therefore, after our analysis, we can reply to the following research question: "How much did the training courses on CSR influence the area of Rimini in order to increase non-financial information contained in company financial statements?"

We formulate an answer, which considers that the training courses on CSR, which had been promoted in the Rimini area, had influence to increase the disclosure of non-financial information in financial statements and consolidated financial statements of some companies here examined. That trend is considered even if there is not an increase of the CSR considered as key word, because the word CSR may be implemented in practice in different ways. The conclusion of this work is partial, because a positive trend of the identified keywords is not always present in the cases analysed. This first result can be a starting point for further analysis through the analysis of other companies.

Another limitation of the work can be found in the initial hypotheses that we implicitly assumed and that is that the diffusion of the culture on CSR could lead to an influence on the disclosure of non-financial information. This hypothesis was completely verified in particular with the analysis of the SGR case [32] but should be further investigated for the other companies that we analysed.

Besides, it is a pity that the four companies analysed are not yet aware of the importance of drawing up, in an appropriate document, these interesting activities that they have developed over time both externally and internally. Just one has initiated a process of reporting which is mirrored in the high presence, in comparison with the other companies, of the keywords we have chosen for the analysis.

Reading this process with the theory presented here [4, 5] and extracting some inspiration for further research, we may affirm that this does not seem to us to be a purely isomorphic process, because the facets that emerge from this research and from the evolution of the CSR spreading process are multiple and they make this region more than directed towards standardisation, directed towards social innovation and creativity. This is demonstrated by the initiative, called, "Primo Miglio" (www.primomigliostartup.com), which promotes and sustains socially responsible initiatives for start-up enterprises. Of this, we will speak in further research works.

Author details

Maria-Gabriella Baldarelli[1,2]

1 Department of Management, University of Bologna, Italy

2 University Institute of Sophia, Florence, Italy

*Address all correspondence to: maria.baldarelli@unibo.it

IntechOpen

© 2019 The Author(s). Licensee IntechOpen. This chapter is distributed under the terms of the Creative Commons Attribution License (http://creativecommons.org/licenses/by/3.0), which permits unrestricted use, distribution, and reproduction in any medium, provided the original work is properly cited. (cc) BY

References

[1] Masini C. Lavoro e Risparmio. Torino: UTET; 1982

[2] Del Baldo M. Una Responsabilità Sociale per più Generazioni nel Gruppo FBL-Della Rovere Economia Aziendale On Line. Vol. 22007. pp. 55-72

[3] Bebbington J, Higgins C, Frame B. Initiating sustainable development reporting: Evidence from New Zealand. Accounting, Auditing & Accountability Journal. 2009;**22**(4):588-625. DOI: 10.1108/09513570910955452

[4] DiMaggio PJ, Powell WW. The iron cage revisited: Institutional isomorphism and collective rationality in organizational fields. American Sociological Review. 1983;**48**(1):147-160

[5] Scott WR. Institutions and Organizations. Thousand Oaks, CA: Sage; 1995

[6] Catturi G, Riccaboni A. Economia aziendale e teoria istituzionale: Affinità concettuali ed implicazioni operative per il controllo di gestione. Contabilità e Cultura Aziendale. 2001; (2):133-176

[7] Zappa G. La Nozione di Azienda Nell'economia Moderna. Milano: Giuffrè; 1954

[8] Ceccherelli A. Il Linguaggio dei Bilanci Formazione e Interpretazione dei Bilanci Commerciali. Firenze: Le Monnier; 1968

[9] Lai A. Paradigmi Interpretativi dell'impresa Contemporanea. Teorie Istituzionali e Logiche Contrattuali. Milano: Angeli; 2004

[10] Adams C, Larrinaga González C. Engaging with organizations in pursuit of improved sustainability accounting and performance. Accounting, Auditing & Accountability Journal. 2007;**20**(3): 333-355. DOI: 10.1108/09513570710748535

[11] Contrafatto M. The institutionalization of social and environmental reporting: An Italian narrative. Accounting, Organizations and Society. 2014;**39**:414-432. DOI: 10.1016/j.aos.2014.01.002

[12] Adams C. Internal organizational factors influencing social and ethical reporting: Beyond current theorizing. Accounting, Auditing & Accountability Journal. 2002;**15**(2):223-250

[13] Cooper C, Taylor P, Smith N, Catchpowle L. A discussion of the political potential of social accounting. Critical Perspectives on Accounting. 2005;**16**(7):51-74

[14] Parker LD. Social and environmental accountability research: A view from the commentary box. Accounting, Auditing & Accountability Journal. 2005;**18**(6):842-860

[15] Larrinaga González C. Paper Presented to the 7th Italian Conference on Social and Environmental Accounting Research, "Accounting, Accountability and Society"; 20-21 September 2018; Urbino, Italy; 2018

[16] Larrinaga González C. Sustainability reporting: Insights from neo-institutional theory. In: Unerman J, Bebbington J, O'Dwyer B, editors. Sustainability Accounting and Accountability. London: Routledge; 2007. pp. 150-167

[17] Bandettini A. Responsabilità Sociali Dell'azienda e Bilancio D'esercizio. Miscell. Bilancio D'esercizio e Amministrazione Delle Imprese. Studi in Onore di Pietro Onida. Milano: Giuffrè; 1981

[18] DiMaggio PJ, Powell WW. The New Institutionalism in Organizational Analysis. Chicago and London: University of Chicago Press; 1991

[19] Baldarelli MG, Del Baldo M, Nesheva-Kiosseva N. Environmental Accounting and Reporting. Theory and Practice, CSR, Sustainability, Ethics & Governance-Series. Springer International Publishing; 2017. ISBN: 978-3-319-50918-1. Available from: http://www.springer.com/us/book/9783319509167#otherversion=9783319509181. ISSN: 2196-7075

[20] Contrafatto M, Rusconi G. Social accounting in Italy: Origins and developments. Social and Environmental Accountability Journal. 2005;**25**(2):3-9

[21] Miolo Vitali P. Problemi Ecologici Nella Gestione delle Aziende. Milano: Giuffrè; 1978

[22] De Santis G, Ventrella AM. Il Bilancio Sociale dell'impresa. Milano: Angeli; 1980

[23] Matacena A. Impresa e Ambiente. Il Bilancio "Societale". Bologna: Clueb; 1984

[24] Vermiglio G. Il Bilancio Sociale nel Quadro Evolutivo del Sistema D'impresa. Messina: Grafo Editori; 1984

[25] Rusconi G. Il Bilancio Sociale Dell'impresa Problemi e Prospettive. Milano: Giuffrè; 1988

[26] Buscarini C. Un'ipotesi di Bilancio Sociale per le PMI. Torino: Giappichelli; 2005

[27] Molteni M. PMI: Quale Responsabilità Sociale? Economia & Management. Vol. 12004. pp. 111-125

[28] Bagnoli L. Quale Responsabilità Sociale per L'impresa? Milano: Angeli; 2004

[29] Rusconi G, Dorigatti M, editors. Impresa e Responsabilità Sociale. Milano: Angeli; 2006

[30] Arena P, editor. The Corporate Social Responsibility. Scientific Development and Implementation. Roma: ARACNE; 2006

[31] Baldarelli MG, editor. Verso-Valori e Responsabilità Sociale. Rimini: Bookstones; 2018. EAN: 9788898275694. Available from: https://www.ibs.it/verso-valori-responsabilita-sociale-ebook-vari/e/9788898275694

[32] Baldarelli M-G, Del Baldo M, Nesheva Kiosseva N. Implementing sustainability reporting (SR): (neo-) institutional theory insights in the analysis of the SGR Group, Italy and CityGas, Bulgaria. Journal of Modern Accounting and Auditing. 2014;**10**(11): 1067-1104. ISSN: 1548-6583

[33] Bebbington J, Russel S, Thomson I. Accounting and sustainable development: Reflections and propositions. Critical Perspectives on Accounting. 2017;**48**(10):21-34. DOI: 10.1016/j.cpa.2017.06.002

[34] Matacena A. The subject we are dealing with: MGA model of the enterprise. In: Baldarelli M-G, Del Baldo M, Nesheva-Kiosseva N, editors. Environmental Accounting and Reporting. Theory and Practice. Cham, Switzerland: Springer International Publishing; 2017. pp. 1-13

Section 3

Accounting and Financial Statements

Chapter 4

Nonlinear Effect of Financial Development and Foreign Direct Investment in Integration Economies Among ASEAN-5 Countries Following IFRS Adoption

Elya Nabila Abdul Bahri and Nor Hakimah Haji Mohd Nor

Abstract

This chapter examines the role of financial development on foreign direct investment (FDI) inflows in ASEAN-5 countries over the period of 1980–2017. The ASEAN-5 countries include Malaysia, Thailand, Indonesia, Singapore, and the Philippines. The panel cointegration of second generation is used in order to address the existence of economic integration among ASEAN-5 as proven in cross-sectional dependency test. The results from fully modified ordinary least square (FMOLS) and cross-sectional dependency autoregressive distributed lag (CS-ARDL) consistently shows that the financial development has a nonlinear relationship with FDI of U-shape, whereby the financial development will benefit the FDI after it beyond the threshold point at 70% of total GDP. Investors will make decision based on the financial status as shown in the financial accounting report, whereby the quality of financial accounting representing transparent information that leads on reducing asymmetric information between investor and the financial institutions in host countries. In addition, the causality analysis based on panel vector error correction model (VECM) confirms the presence of both long-run relationship and short-run dynamic among FDI, financial development, consumer price index, and real gross domestic product per capita.

Keywords: foreign direct investment, financial development, cross-sectional dependence, nonlinear, financial information quality

1. Introduction

Following the financial liberalization attempted, especially during the 1980s, foreign direct investment (FDI) has being as more important catalyst for accelerating the economic growth. The economic activities are integrated during the transition economies among developing countries as well as ASEAN-5 countries. The formation of ASEAN Economic Community (AEC) at the Ninth ASEAN Summit in

October 2003 represented an important milestone in ASEAN economic cooperation. It stimulates FDI inflows by reducing business costs associated with multinational activities in the ASEAN region that has always been a primary objective of the economic cooperation. Strengthening the financial sector with remaining stable economic condition may establish an attractive business environment for multinational firms to invest in the ASEAN-5 countries.

The trend of FDI inflows in ASEAN-5 countries fluctuate from year 1980 to 2017 as shown in **Figure 1**. There is high volatility in FDI inflows after the Asian financial crisis in 1997–1998 that leads FDI inflows dropped for Singapore, Malaysia and Indonesia, but increased for Philippines and Thailand. While Malaysia and Philippines drop in 2001 in another event of crisis, that is, bubble.com, but the FDI inflows to Singapore, Thailand and Indonesia are increased. The trend of FDI inflows among ASEAN-5 countries are however increases following AEC formation in 2003. The trend of FDI inflows dropped after the global financial crisis in 2007–2008, which affected ASEAN-5 countries, but its effects are delayed for Malaysia and Indonesia. Fluctuation of FDI inflows may reliance on the uncertainty of its enabler such as financial development in ASEAN-5 countries.

To materialize the benefit of FDI, thus, the role of financial development was found as an enabler of the FDI performance. There are five major functions of a financial system: producing information and allocating capital; monitoring firms and implementing corporate governance; meliorating risk; pooling the savings; and easing exchange [1]. These functions contribute to the stimulation of economic growth. Thus, the financial institutions and financial markets can exert a strong influence on economic development, where the incremental of economic growth enabling the FDI to perform better. Financial development is discovered as assistance to the FDI especially in technology transfer that needed more capital to funding the technology expenses.

It is important to consider the effect of financial development on FDI inflows are influenced by the quality of financial information. The investors would review the financial strength via the bank's financial report in the host countries. A quality of financial reporting relies on regulatory accounting standards through the International Financial Reporting Standard (IFRS). In general, ASEAN-5 countries already

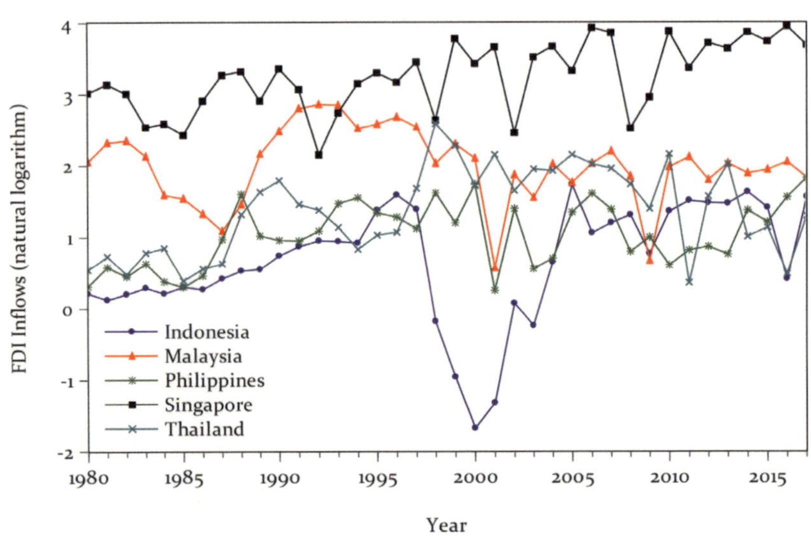

Figure 1.
FDI inflows (% of GDP) in ASEAN-5 countries from 1980 to 2017. Source of data: UNCTAD website (Accessed on 9 March 2019).

complied the regulation starting from 2005 onwards. The effect of regulatory quality is found as an incentive for quality of accounting information that benefited to all stakeholders. Thus, improving the regulatory quality of financial reporting enhances transparency that reducing the asymmetric information between investors and the financial institution in host countries.

2. Literature review

Early theories of the determinants of FDI were encompassed in eclectic approach [2]. The key requirements for FDI as follows: the firm must possess stable specific advantages; the firm must find it beneficial to utilize these advantages directly instead of selling or leasing them (so called as internalization advantages); and the firm must firm must find it profitable to combine these advantages with at least one factor input abroad (so that local production dominates exporting or locational advantages). These advantages include proximity to markets, specialized suppliers, evasion of protective barriers, and factor endowment advantages.

Financial development is found as one of the significant determinants of FDI [3, 4]. The financial markets are measured by the domestic credit provided by banks and domestic credit provided to the private sector as a percentage of GDP [5]. Domestic credit to the private sector refers to the financial resources provided to private sector through loans, purchases of non-equity securities, and trade credits and other receivable accounts. Meanwhile, the domestic credit provided by banks is nonguaranteed long-term commercial bank loans from private banks and other private financial institutions. The other measurement financial development is financial freedom as a catalyst for FDI inflows. Financial freedom is a measurement of banking security as well as independence from government control. The state ownership of banks and other financial institutions is seen as an inefficient burden, and political favoritism has no place in a free capital market [6]. Thus, the financial information quality also affected the investment efficiency because the investors need the information of financial health in the particular host countries [7].

Financial development, as better accounting and disclosure rules and better corporate governance, reduces the spread between domestic and foreign cost of capital [8]. It requires a sound financial reporting system that produces reliable and transparent accounting information for both domestic and foreign investors [9, 10]. Lack of financial reporting systems credibility is likely to have adverse effects on the ability of particular countries to attract foreign investments, because it retards the equity markets development [11]. In fact, the effect of financial development on economic growth, which enhances FDI inflows, is contingent the adoption of financial accounting quality by host countries [12].

The application of IFRS in host countries is considered a way to attract the FDI [13, 14]. IFRS leads to higher earnings quality and more foreign investment [15]. Furthermore, the short-run and long-run causality existed between IFRS adoption and FDI inflows [16]. The quality of financial reporting according to IFRS has potential to enhance transparency that reduces asymmetric information and cost for foreign investors [17]. IFRS adoption requires strong governance and internal controls within a bank to give confidence to the investors resulting in the quality financial information. The effect of regulatory quality is found as an incentive for quality of accounting information and compliance to the IFRS by firms. Hence, the financial statement act as organization's resume that indicate the strength of finance by banks in host countries. Quality of financial statement among bans in host countries will build good reputation that leads the confidence of decision maker from foreign firms. The information of financial development, that is, credit market, through the financial reporting can reduce the asymmetric information that attracts the FDI inflows into the host countries.

The development of strong financial market can increase an economy's ability to absorb and efficiently manage FDI inflows and take advantage of potential FDI benefits [18]. Although recent studies discover that financial development influences FDI performance to be realized, the long run relationship between the variables including FDI, financial development and macroeconomic variables need adequately addressed in the field of study. This study is therefore attempting to contribute to the existing literature in the dimensions of nonlinearity and cross-sectional dependency dimensions. The investigation on the effects of financial development on the FDI inflows employs both linear and quadratic models in the estimations. Incorporating the cross-sectional dependency due to economic integration, financial openness, economic freedom and spillover effects among ASEAN-5 countries over the period 1980–2017, the panel cointegration for second generation is used and the long-run coefficient estimated by considering the cross-sectional dependence in this study.

3. Econometric model and data description

The effect of the financial development on foreign direct investment is investigated by using ASEAN-5 countries panel data with the econometric model specified as follows:

$$FDI_{it} = \beta_0 + \beta_1 FinDev_{it} + \beta' X_{it} + \omega_i + \varepsilon_{it} \quad (1)$$

where, the *FDI* in Eq. (1) represents foreign direct investment, and *FinDev* indicating financial development. The domestic credit of private sector, liquid liabilities and private sector credit to deposit money are used as a proxy for financial development. Parameter X comprises with the vector of control variables: real gross domestic product per capita (*RGDPPC*) and consumer price index (*CPI*). The cross-sections are denoted by subscript i (i = 1, 2, ..., N), the time period by subscript t (t = 1, 2, ..., T), ω_i is the country fixed effect and ε is the stochastic random term.

Incorporating the nonlinear effect of financial development on FDI, Eq. (1) has been extended by using a quadratic specification as expressed in Eq. (2):

$$FDI_{it} = \alpha_0 + \alpha_1 FinDev_{it} + \alpha_2 FinDev_{it}^2 + \alpha' X_{it} + \theta_i + \mu_{it} \quad (2)$$

where $FinDev_{it}^2$ is a square term for financial development that indicates the nonlinearity of the relationship between financial development and FDI inflows. The focal parameters in the quadratic model in Eq. (2) are α_1 and α_2. If there exists a nonlinear relationship between financial development and FDI, an anti-Kuznets curve is anticipated since higher the financial development indicating the financial health that can assist the foreign entrepreneur and thus attracting the FDI inflows in ASEAN-5 countries. The anti-Kuznets curve is verified by significantly negative sign in parameter α_1 and significantly positive in α_2. The threshold level is computed by first order derivation ($\delta FDI/\delta FinDev$). Based on the quadratic model in Eq. (2), the turning point of financial development can be calculated as $-\alpha_1/2\alpha_2$.

This study uses FDI inflows as the percentage of GDP. Real GDP per capita in constant of US dollar (US$) is used to measure economic development. Domestic credit to private sector by banks as a percentage share of GDP (*DCPS*), liquid liabilities as a percentage of GDP (*LL*) and private credit to deposit money (*PCDM*) are used as proxies for financial development. Five countries have been selected among the ASEAN countries including Malaysia, Thailand, Indonesia, Singapore, and the Philippines. The study covers 38 years for the period of 1980–2017. The sources of the data are World Development Indicators, UNCTAD Database and Financial Structure Dataset.

The last few decades, the ASEAN-5 economies have witnessed an increasing economic freedom and financial integration implies a strong interdependence between these countries. To measure the existence of economic integration among ASEAN-5 countries, the cross-sectional dependency (CD) test is used for all variables [9]. The existence of cross-sectional dependency among ASEAN-5 countries are proven in **Table 1** indicated by the p-value of CD statistics which are lower than 0.01 for all variables that against the null hypotheses of cross-sectional independence among countries, CD $\sim$ N(0,1). Consumer price index is the highest absolute correlation among ASEAN-5 countries at 0.976, means the changes of price of one country closely affected price the other countries. Meanwhile, liquid liabilities are the highest absolute correlation in ASEAN-5 region among other proxies for financial development. This may involve the integrated economic process especially when the countries are neighbors. Furthermore, the cross-sectional dependence can arise for several reasons, such as spatial spillovers, financial contagion, and socioeconomic interactions [19].

Table 2 shows the descriptive statistics of the variables. Jarque-Bera for normality test shows that all variables are not normally distributed. The median for

Variable	CD test	Breusch-Pagan LM test	Absolute correlation
Foreign direct investment inflows	2.833***	17.103*	0.193
Domestic credit to private sector	10.776***	134.678***	0.553
Liquid liabilities	15.234***	239.532***	0.781
Private credit by deposit money	11.116***	146.326***	0.570
Real GDP per capita	17.571***	313.179***	0.901
Consumer Price Index	19.020***	361.854***	0.976

****Significant at 1% level.*
**Significant at 10% level.*

Table 1.
Result of cross-sectional dependency test for ASEAN-5 countries.

	FDI	DCPS	LL	PCDM	RGDPPC	CPI
Minimum	−2.583	9.681	10.400	6.490	1230.840	5.554
25% quantile	0.979	34.884	39.500	33.940	1880.193	46.614
Median	2.269	75.908	72.295	74.725	3571.915	71.617
75% quantile	5.058	106.363	105.290	102.210	8635.566	95.117
Maximum	26.084	166.504	136.63	163.210	55,235.500	142.182
St. deviation	5.841	40.852	35.270	40.304	13,085.87	32.874
Skewness	1.997	0.230	0.062	0.193	2.066	−0.248
Kurtosis	6.389	1.875	1.733	1.782	6.224	2.237
Jarque-Bera	217.175	11.701	12.827	12.924	217.441	6.566
Probability	0.000	0.003	0.002	0.002	0.000	0.038

Table 2.
Descriptive statistics of variables.

Variables	FDI	DCPS	LL	PCDM	RGDPPC	CPI
FDI	1.000					
DCPS	0.622***	1.000				
LL	0.620***	0.889***	1.000			
PCDM	0.651***	0.981***	0.901***	1.00		
RGDPPC	0.839***	0.665***	0.687***	0.700***	1.00	
CPI	0.489***	0.568***	0.764***	0.613***	0.551***	1.00

****Significant at 1% level.*

Table 3.
Correlations of variables.

DCPS, *LL* and *PCDM* is 76, 72 and 75%, respectively indicating the small differences between all proxies. The kurtosis of FDI is the highest at 6.389 indicating the fluctuation of FDI as compared to other variables. The correlation matrix presented in **Table 3** reveals that the variables are positive correlated among other with more than 0.5. The highest correlation between *DCPS* and *PCDM* is at 0.981 since both variables are the proxies of same variable of financial development.

Table 4 presents the mean of all variables of each country in the ASEAN-5 economies. As shown in this table, there is a considerable variation in RGDPPC across these countries, ranging from as low as US$2332.82 for Indonesia to as high as US$32,385.13 for Singapore. The mean of financial development demonstrates that Malaysia is consistently the highest for DCPS and LL, but Singapore is the highest in PCDM.

Each country in ASEAN-5 region has different starting date on complying IFRS in accordance with IFRS Foundation's Jurisdictional Profiles as shown in **Table 5.** Philippines is the earliest country adopting IFRS since 2005, while Malaysia is most recent among others. In addition, **Table 6** shows the different scores for each

Country	FDI	DCPS	LL	PCDM	RGDPPC	CPI
Malaysia	4.08	106.60	107.67	95.93	6757.70	77.30
Indonesia	0.86	32.42	31.07	25.53	2332.82	51.41
Thailand	2.18	102.20	83.29	97.29	3566.14	72.54
Singapore	14.20	95.26	95.80	99.92	32,385.13	84.84
Philippines	1.34	31.72	44.63	31.88	1801.00	59.97

Table 4.
Mean of variables over 1980–2017 for each country.

ASEAN-5 countries	Year of IFRS adoption
Malaysia	2012
Indonesia	Has not adopted (public commitment in support of moving towards IFRS from 2012)
Thailand	2011
Singapore	2010
Philippines	2005

Source: Yousefinejad et al. [16].

Table 5.
The IFRS adoption date of ASEAN countries.

Score	Characteristics of IFRS adoption	Malaysia	Indonesia	Thailand	Singapore	Philippines
1	Has the jurisdiction made a public commitment in support of moving towards a single set of high-quality global accounting standards?	1	1	1	1	1
2	Has the jurisdiction made a public commitment towards IFRS Standards as that single set of high-quality global accounting standards?	1	1	1	1	1
3	For domestic companies are IFRS Standards required or permitted?	1	0	0	1	1
4	Are IFRS Standards also required or permitted for more than the consolidated financial statements of companies whose securities trade in a public market?	1	0	0	1	1
5	Are all or some foreign companies whose securities trade in a public market either required or permitted to use IFRS Standards in their consolidated financial statements?	1	0	1	1	1
6	Are IFRS Standards incorporated into law or regulations?	1	0	0	0	1
7	Has the jurisdiction adopted the IFRS for SMEs Standard for at least some SMEs?	1	0	0	1	1
Total		7	2	3	6	7

Source: Yousefinejad et al. [16].

Table 6.
IFRS adoption scores.

country. The score was constructed based on IFRS Foundation's Jurisdictional Pro files. Malaysia and the Philippines achieved full scores based on the characteristics that has been constructed. Although Indonesia has not adopted IFRS, however, this country has a score of 2 because it has made a public commitment in support of moving towards a set of high-quality global accounting. Hence, ASEAN-5 countries have financial information quality following IFRS on presenting the financial position for each country. The quality of financial reporting and financial development soundness would lead better decision among foreign investors.

4. Methodology

The flow of methodology started with the panel unit root test with cross-sectional dependency to ensure the variables are integrated at first difference before proceed with cointegration test of panel cointegration second-generation. Since the existence of cross-sectional dependency among ASEAN-5 countries, hence, this study considers the using Westerlund's cointegration test as the second-generation of panel cointegration. In addition, the fully modifies ordinary least square

(FMOLS) and cross-sectional dependency autoregressive distributed lag (CS-ARDL) are used to estimate the long run coefficient in the specification. FMOLS estimator is used to overcome the endogeneity and heterogeneity problem. Meanwhile, the cross-sectional dependency needs to take into account in the estimation by using CS-ARDL. The causality test by using panel vector error correction model (VECM) is used to investigate the direction of causality among the variables.

4.1 Panel unit root test

The panel unit root test of second-generation in used in order to incorporating the cross-sectional dependency. For the case of ASEAN-5 countries, the common stochastic trends may occur due to global developments and strong relationships between economies. The heterogenous panel unit root test with cross-sectional dependence is employed for clarity [20]. The standard Dickey-Fuller (DF) or augmented Dickey-Fuller (ADF) regressions are augmented with the cross-section averages of lagged levels (x_{it-1}) and first-differences (Δx_{it-1}) of the individual series to eliminate cross-sectional dependence. Hence, the cross-sectional dependence ADF (CADF) test [20] expressed as follows:

$$\Delta x_{it} = \alpha_i + \rho x_{it-1} + v_{it} \tag{3}$$

where, $v_{it} = g_i \theta_t + \mu_{it}.\theta_t$ is a common factor and is white noise.

The CADF model is given by, without the autocorrelation of μ_{it} can be written as follows:

$$\Delta x_{it} = \alpha_i + \rho x_{it} + c_i \overline{x}_{t-1} + d_i \Delta \overline{x}_{t-1} + \varepsilon_{it} \tag{4}$$

The cross-sectionally augmented Im, Pesaran and Shin (IPS) or CIPS [20] is given by:

$$CIPS\ (N, T) = \frac{1}{N} \sum_{i=1}^{q} t_i(N, T) \tag{5}$$

where t_i indicates the statistics from each CADF model for each individual i of the panel and significance level based on the critical value (see [20]). If the p-value of CIPS statistics is lower than 0.05 indicating the null hypothesis of non-stationary of the variables is rejected.

4.2 Panel cointegration test

The second-generation panel cointegration test with cross-sectional dependence has four error-correction-based tests, which allows for large degree of heterogeneity in both long-run cointegration and short-run dynamics [21]. The presence of cointegration is tested by determining whether or not the existence of error-correction for individual panel of also the panel as a whole. Transforming Eq. (1) to the error-correction model can be expressed as follows:

$$\begin{aligned}\Delta \ln FDI_{it} = \mathrm{c}_{1j} + \varphi_{i1} \sum_{j=1}^{p} \Delta \ln FDI_{it-j} + \varphi_{i2} \sum_{j=0}^{p} \Delta \ln FinDev_{it-j} + \varphi'_i \sum_{j=0}^{p} \Delta X_{it-j} \\ + \theta_i [\beta_{1i}(\ln FDI_{it-1}) - \beta_{2i}(\ln FinDev_{it-1}) - \beta'_i(X_{it-1}) + \varepsilon_{it}\end{aligned} \tag{6}$$

where θ_i measures the speed of error-correction towards the long-run equilibrium, $FDI_{it} = -(\varphi_i/\theta_i) \times x_{it}$ for the series i. The null hypothesis, $H_0 : \theta_i = 0$ for all i

versus H_1: $\theta_i < 0$ for at least one i for the G_α and G_τ statistic tests. If H_0 is rejected, it means that cointegration exists for at least one of the cross-sectional units. Meanwhile, the P_α and P_τ test statistics pooled the information over all the cross-sectional units to test $H_0 : \theta_i = 0$ for all i versus $H_1 : \theta_i < 0$ for all i. The rejection of H_0 indicating the evidence of cointegration for the panel as a whole.

4.3 Long-run estimation

The long-run coefficient estimation is proceeded if the evidence of cointegration among the variables is proven. Fully modified ordinary least square (FMOLS) estimator is used to estimate the long-run coefficient for financial development and FDI inflows relationship. The long-run coefficients are further estimates by using cross-sectional dependency autoregressive distributed lag (CS-ARDL) in order to considering the element of cross-sectional dependency among ASEAN-5 countries for robustness.

4.3.1 Fully modified ordinary least square (FMOLS)

The FMOLS estimator allows for cross-sectional heterogeneity in the alternative hypothesis. The endogeneity and serial correlation problems are allowing in the FMOLS long-run coefficients estimation in order to obtain consistent and asymptotically unbiased estimates of the cointegrating vectors [22, 23]. The definition of in FMOLS estimator can be expressed as follows:

$$\hat{\beta} = N^{-1} \sum_{i=1}^{N} \left[\sum_{t=1}^{T} \left(X_{it} - \overline{X}_1 \right)^2 \right]^{-1} \left[\sum_{t=1}^{T} \left(X_{it} - \overline{X}_1 \right) Y_{it}^* - T\hat{\tau}_i \right] \quad (7)$$

where, $Y_{it}^* = \left(Y_{it} - \overline{Y} \right) - \frac{\hat{L}_{21i}}{\hat{L}_{22i}} \Delta X_{it}$, $\hat{\tau}_i \equiv \hat{\Gamma}_{21i} + \hat{\Omega}_{21i}^0 - \frac{\hat{L}_{21i}}{\hat{L}_{22i}} \left(\hat{\Gamma}_{22i} + \hat{\Omega}_{22i}^0 \right)$ and $\hat{L}_i$ is a lower triangular decomposition of $\hat{\Omega}_i$. The associated t-statistic is assumed to be normally distributed and given by:

$$t_{\hat{\beta}^*} = N^{-1/2} \sum_{i=1}^{N} t_{\hat{\beta}^*}, i; \text{where } t_{\hat{\beta}^*}, i = \left(\hat{\beta}_i^* - \beta_0 \right) \left[\hat{\Omega}_{11i}^{-1} \sum_{t=1}^{T} \left(X_{it} - \overline{X} \right)^2 \right]^{1/2} \quad (8)$$

The long-run relationship between financial development and FDI inflows is measured by the coefficient $(\hat{\beta})$ from FMOLS estimator.

4.3.2 Cross-sectional dependency autoregressive distributed lag (CS-ARDL)

Cross-sectional dependency autoregressive distributed lag (CS-ARDL) estimator is used for robustness that allows for cross-sectional dependency among ASEAN-5 countries in the alternative hypothesis. The dataset shows cross-sectional dependency existed for all variables (refer to **Table 1**), which might be due to integrational economies among neighbor countries in ASEAN-5. This element needs to consider in estimating the long-run coefficient by using CS-ARDL estimator [24]. The baseline model for generic ARDL (p,q) can be expressed as follows:

$$\overline{y}_{i,t} = \sum_{k=1}^{p} \varphi_{i,k} \overline{y}_{i,t-k} + \sum_{l=0}^{q} \beta'_{i,l} \overline{x}_{i,t-1} + u_{i,t} \quad (9)$$

while its cointegrating form would be:

$$y_{i,t} = \theta_i x_{i,t} + \alpha_i'(L)\Delta x_{it} + \widetilde{u}_{i,t} \tag{10}$$

In CS-ARDL, Eq. (9), the errors (u) is postulated a common unobserved factor structure for the errors. It can be written as:

$$u_{i,t} = \gamma'_i F_t + \varepsilon_{i,t} \tag{11}$$

CS-ARDL is an augmented model from generic ARDL (p,q) by averaging cross-sectional of the dependent and explanatory variables, as well as their lags, which are supposed to proxy for the unobserved common factors.

4.4 Panel vector error-correction model

The panel Granger causality in the framework of the panel VECM is employed to analyze the direction of the causal effect among FDI, financial development and the control variables, CPI and GDP per capita. The long-run model specified in Eq. (1) is estimated by using FMOLS to obtain the estimated residual, followed by Granger causality model estimation based on the error-correction model as follows:

$$\begin{aligned}\Delta \ln FDI_{it} = \alpha_{1i} &+ \sum_{k=1}^{m} \lambda_{11ik} \Delta \ln FDI_{i,t-k} + \sum_{k=1}^{m} \lambda_{12ik} \Delta \ln FinDev_{i,t-k} \\ &+ \sum_{k=1}^{m} \lambda_{13ik} \Delta \ln RGDPPC_{i,t-k} + \sum_{k=1}^{m} \lambda_{14ik} \Delta \ln CPI_{i,t-k} + \phi_{1i} EC_{i,t-1} + \mu_{1it}\end{aligned} \tag{12}$$

$$\begin{aligned}\Delta \ln FinDev_{it} = \alpha_{2i} &+ \sum_{k=1}^{m} \lambda_{21ik} \Delta \ln FinDev_{i,t-k} + \sum_{k=1}^{m} \lambda_{22ik} \Delta \ln FDI_{i,t-k} \\ &+ \sum_{k=1}^{m} \lambda_{23ik} \Delta \ln RGDPPC_{i,t-k} + \sum_{k=1}^{m} \lambda_{24ik} \Delta \ln CPI_{i,t-k} + \phi_{2i} EC_{i,t-1} + \mu_{2it}\end{aligned} \tag{13}$$

$$\begin{aligned}\Delta \ln RGDPPC_{it} = \alpha_{3i} &+ \sum_{k=1}^{m} \lambda_{31ik} \Delta \ln RGDPPC_{i,t-k} + \sum_{k=1}^{m} \lambda_{32ik} \Delta \ln FinDev_{i,t-k} \\ &+ \sum_{k=1}^{m} \lambda_{33ik} \Delta \ln FDI_{i,t-k} + \sum_{k=1}^{m} \lambda_{34ik} \Delta \ln CPI_{i,t-k} + \phi_{3i} EC_{i,t-1} + \mu_{3it}\end{aligned} \tag{14}$$

$$\begin{aligned}\Delta \ln CPI_{it} = \alpha_{4i} &+ \sum_{k=1}^{m} \lambda_{41ik} \Delta \ln CPI_{i,t-k} + \sum_{k=1}^{m} \lambda_{42ik} \Delta \ln FinDev_{i,t-k} \\ &+ \sum_{k=1}^{m} \lambda_{43ik} \Delta \ln RGDPPC_{i,t-k} + \sum_{k=1}^{m} \lambda_{44ik} \Delta \ln FDI_{i,t-k} + \phi_{4i} EC_{i,t-1} + \mu_{4it}\end{aligned} \tag{15}$$

where, EC is error-correction term comes from the FMOLS estimation, and m is the lag length. The short-run causality is determined by the statistical significance of the F-statistic associated with the corresponding right hand side variables. The presence or absence of long-run causality can be established by examining the significance of the t-statistic on the coefficient ϕ, in Eqs. (12)–(15).

5. Empirical findings

The result of the second-generation panel unit root for the option of with and without trend are presented in **Table 7**. The panel unit root of CIPS using one lag order due to parsimony principle. The null hypothesis of non-stationary is fail to rejected in level in lag 1 (q = 1) for both variables option without trend (Z_α) and with trend (Z_τ), but rejected in first difference indicating that all series are integrated of order one or $I(1)$ for the panel of ASEAN-5 countries.

Variable	Level				First difference			
	Z_α		Z_τ		Z_α		Z_τ	
	$q = 0$	$q = 1$	$q = 0$	$q = 1$	$q = 0$	$q = 1$	$q = 0$	$q = 1$
FDI	−4.0*	−1.5	−4.3*	−1.4	−10.0*	−8.3*	−9.7*	−7.5*
DCPS	−1.0	−1.0	0.6	0.5	−8.5*	−5.0*	−8.0*	−4.4*
LL	0.8	−0.8	0.9	−1.5	−4.6*	−5.1*	−3.6*	−4.3*
PCDM	−0.2	−2.0	1.9	−0.3	−3.7*	−4.0*	−3.7*	−3.4*
RGDPPC	2.2	1.0	4.4	3.1	−3.5*	−3.1*	−3.8*	−4.2*
CPI	1.0	−0.8	1.0	0.0	−5.5*	−3.7*	−4.7*	−3.2*

**Significant at 1% level.*

Table 7.
Second-generation panel unit root test of CIPS.

	Model 1: FinDev = DCPS		Model 2: FinDev = LL		Model 3: FinDev = PCDM	
	Statistics	*p*-value	Statistics	*p*-value	Statistics	*p*-value
G_τ	−2.975***	0.003	−3.054***	0.002	−3.398***	0.000
G_α	−8.162	0.450	−9.576	0.264	−8.249	0.438
P_τ	−5.701**	0.018	−5.240**	0.040	−6.038***	0.009
P_α	−6.939	0.172	−7.178	0.152	−7.051	0.162

****Significant at 1% level.*
***Significant at 5% level.*

Table 8.
Second-generation panel cointegration results.

Since all variables are integrated of order one, the panel cointegration test is employed to measure the existence of long-run relationship in Eq. (1). The results of second-generation of panel cointegration presented in **Table 8**. The panel cointegration test shows that P_α and P_τ test statistics reject the null hypothesis of no cointegration at 1, 5 and 10% significance level for both models using DCPS and PCDM, for the specification without trend. The P_α and P_τ test statistics have the highest power and are the most robust to cross-sectional correlation [21]. Thus, the evidence from the second-generation panel cointegration test supports the presence of a cointegrating relationship among FDI, financial development, price and economic development in ASEAN-5 countries.

Due to the existence of cointegration among variables in the region, the FMOLS estimator is used to estimate the long-run equilibrium. **Table 9** reports that Model 1–3 estimates the linear and nonlinear relationship between financial development and FDI in the long-run, by using DCPS, LL and PCDM as a proxy for financial development, respectively. Long-run covariance estimates pre-whitening with lag 1, where the automatic bandwidth selection is based on Newey-West fixed bandwidth and Bartlett kernel. In the linear specification, the relationship between financial development and FDI are not significant in all models. However, in contrast, there is significant of nonlinear relationship between financial development and FDI. The nonlinear relationship between these variables is anti-Kuznets or U-shape curve, where α_1 the coefficient of financial development and α_2 the financial development square coefficient (Eq. (2)) is negative and positive,

	Linear model			Quadratic model		
	Model 1	Model 2	Model 3	Model 1	Model 2	Model 3
FinDev	0.21	0.07	0.03	−7.98*	−3.78	−8.46*
$FinDev^2$	—	—	—	0.93*	0.34	0.93*
RGDPPC	2.38***	1.92***	2.44***	2.57***	1.84***	2.69***
CPI	−2.22***	−1.58***	−2.11***	−2.80***	−2.05***	−2.29***
Threshold value (% of GDP)				73.00	—	94.48

DCPS is used as a proxy in Model 1, while PCDM in Model 2.
****Significant at 1% level.*
**denotes significant at 10% level.*

Table 9.
Panel FMOLS long-run estimation (dependent variable: FDI).

respectively. The U-shape curve indicating that the financial development exceeded the threshold level, its incremental will attract more FDI inflows.

The result show that the negative effect of low level of financial development below the threshold level, in general at 70% of GDP that portrays the financial illness in host country. Initially, the negative relationship associated with the under-developed financial sector that may discourage the investor's decision to invest to the host country for those investors who are preferring on resource-seeking and market-seeking. Low level in credit market will reduce the purchasing power of parity among the citizens, and as a result, the innovative products produced by foreign firms may become unmarketable or over-supplied in the host country.

In the other hand, the high financial development reflects high financial strength that might attracts the inflows of FDI that related to assist them to set-up new business in host country and survival in their day-to-day business. When the level of financial development that above the 70% of GDP threshold point, it influences the positive impact on FDI inflows. Specifically, based on the quadratic specification of Model 1 in **Table 7**, the financial development threshold point is 73% of GDP $(-7.98/(0.93(2)))$, and Model 3 is 94.48% of GDP $(-8.46/(0.93(2)))$. The result showed that the DCPS surpassed the threshold point at median value to accelerate inflows of FDI, while PCDM should beyond the 75% quantile. The nonlinear relationship between financial development and FDI in this study is however differed from those of previous studies which examined the linear relationship (see [25–27]). The U-shape curve commensurate with the argument that well-developed financial market benefited FDI in host country [28].

The result from FMOLS is however not considering the cross-sectional dependency among ASEAN-5 countries. The long-run coefficients are further estimated by using CS-ARDL for robustness check as shown in **Table 10**. Similar with **Table 9** previously, the relationship between financial development on FDI inflows are not significant by using linear specification for all models in short-run and long-run estimations as shown in **Table 10**. The relationship between financial development and FDI inflows is however absence in the short-run estimation. But the relationship is existed in quadratic model specification, but result shows that only Model 2 has significant U-shape relationship between LL and FDI inflows in the long-run. The relationship is significant only in the long-run due to time lag effect in materializing the benefit of FDI inflows influenced by financial development. The U-shape relationship indicates by the negative coefficient for $\hat{\alpha}_1$, and positive coefficient for $\hat{\alpha}_2$ (refer to Eq. (2)), which both coefficients are significant at 1% level. It means, when the level of liquid liabilities as a percentage of GDP beyond the

	Linear model			Quadratic model		
	Model 1	Model 2	Model 3	Model 1	Model 2	Model 3
Short-run estimation						
FinDev	0.61	0.32	−0.11	10.36	−3.42	9.20
$FinDev^2$	—	—	—	−1.34	0.54	−1.27
RGDPPC	7.11***	5.39***	5.86***	10.02	5.73	4.89
CPI	2.69	2.85	2.98*	2.16	1.88	1.76
Long-run estimation						
FDI	−0.96***	−1.00***	−0.96***	−1.04***	−1.12***	−1.16***
FinDev	−0.05	−0.07	−0.53	1.33	−29.34***	−4.16
$FinDev^2$	—	—	—	−0.21	3.46***	0.38
RGDPPC	4.66**	2.37	3.93**	8.90	4.92	6.55
CPI	1.85	2.03	1.07	2.00	0.66	−0.73**
R^2	0.64	0.65	0.65	0.74	0.72	0.76
CD statistic	−3.91	−3.86	−3.60	−3.55	−3.91	−3.80
p-value	0.00	0.00	0.00	0.00	0.00	0.00
RMSE	0.37	0.36	0.36	0.33	0.34	0.31
Threshold value (% of GDP)				—	69.66	—

DCPS is used as a proxy in Model 1, while LL and PCDM in Model 2 and Model 3, respectively.
****denotes significant at 1% level.*
***denotes significant at 5% level.*

Table 10.
Cross-sectional dependence ARDL (CS-ARDL) estimation (dependent variable: FDI).

threshold point at median level, it will attract the FDI inflows to ASEAN-5 countries. Specifically, by considering the cross-sectional dependency among ASEAN-5 countries the liquid liabilities threshold point is 70% of GDP (−29.34/(3.46(2))) in long-run.

Although the more financial development can attract the FDI inflows, but the quality of financial reporting is important channel of its information for investors, since the ASEAN-5 countries are committed in complying IFRS rules. The information in financial reporting provided by firms and financial institution in host countries as a canal on presenting it financial position that leads better decision for foreign investors. The foreign investors may unable to make decision of lacking information on financial condition in host countries. A superior financial reporting system lowers the cost of capital and improves capital allocation efficiency [29]. The quality of financial reporting would lead to transparent and clear information that reducing asymmetric information between foreign investors and financial institutions in ASEAN-5 countries.

In further investigation, the causality between variables are tested by using Granger causality test based on VECM model as shown in **Table 11**. The lag length is based on Akaike information criterion. All models shown the negatively significant of error-correction term for the FDI equation, that suggesting the existence of long-run relationship when the FDI is dependent variable. Similarly, there are exists the long-run causality when financial development and CPI as a dependent variable for all models. The causality between financial development and FDI inflows is however occurred only in the long-run. As shown in **Table 11**, there is a unidirectional causal effect running from real GDP per capita to DCPS and PCDM, CPI to FDI, CPI

Dependent variable	Independent variables				ECT_{t-1}
	ΔFDI	*ΔFinDev*	*ΔRGDPPC*	*ΔCPI*	Coefficient (*t*-statistic)
	Wald F-statistics				
Model 1: FinDev = DCPS					
ΔFDI	—	2.106	0.935	2.201	−0.136*** (−4.292)
ΔFinDev	0.052	—	0.345	3.213	−0.016*** (−2.375)
ΔRGDPPC	2.263	24.223***	—	0.902	−0.001 (−0.55)
ΔCPI	17.821***	23.688***	4.793*	—	−0.012*** (−4.064)
Model 2: FinDev = LL					
ΔFDI	—	0.331	1.021	1.871	−0.081*** (−3.921)
ΔFinDev	4.542	—	0.726	0.996	−0.005* (−1.923)
ΔRGDPPC	2.929	3.101	—	1.827	−6.947 (−0.047)
ΔCPI	17.615***	1.287	3.332	—	−0.010*** (−5.375)
Model 3: FinDev = PCDM					
ΔFDI	—	2.210	1.029	2.224	−0.084*** (−3.588)
ΔFinDev	1.300	—	0.254	1.420	−0.011*** (−3.489)
ΔRGDPPC	2.000	44.048***	—	2.940	−0.001 (−0.342)
ΔCPI	14.846***	9.414***	4.079	—	−0.011*** (−5.106)

ECT is error-correction term.
****Significant at 1% level.*
**Significant at 10% level.*

Table 11.
Granger causality based on panel VECM estimation.

to DCPS and PCDM and CPI to real GDP per capita. The linkage is however broken between financial development-FDI nexus in short term. The error-correction terms presented in the last column of all models demonstrated that the burden of the short-run endogenous adjustment (in long-run trend) to bring the system back to its long-run equilibrium borne by the FDI, financial development and CPI equations.

6. Conclusion

The financial development, FDI, CPI and real GDP per capita is cointegrated by considering cross-sectional dependence. The relationship between the financial development and FDI in ASEAN-5 countries is found nonlinear of U-shape over the period from 1980 to 2017. Based on the findings from the quadratic model, financial development contributed towards promoting FDI after surpassed the threshold level at 70% of GDP or median score. It is important to understand how the financial development affects FDI. The results demonstrated that there exists cross-sectional dependence among ASEAN-5 countries which this supports the argument of geographic influence on financial integration as well as FDI inflows. Economic integration has a direct effect on internationalization by reducing transaction costs and partial information costs [30].

Financial integration among the ASEAN-5 countries strengthened financial development as well as ease transaction activities among the regional players [31]. The integration in the AEC Blueprint 2015 towards strengthening the financial institutions, enhance commitment in implementation and in monitoring and evaluation of finance. Since all ASEAN-5 countries are complying the financial accounting standards of IFRS (including Indonesia who committed to comply the IFRS), the quality of financial information is needed in order to monitor and evaluate the financial position. In addition, ASEAN seeks to achieve a well-integrated and smoothly functioning ASEAN region financial system, as well as the IFRS accounting standard, that characterized by more liberalized capital account regime and inter-linked capital markets. Strengthening financial integration as well as financial market infrastructure with the quality of financial reporting are therefore aimed at facilitating intra-ASEAN trade and investment by increasing the role of ASEAN indigenous banks. It also augmenting the integrated insurance and capital markets that leading to safe, cost-efficient and more connected regional economy. The attraction of FDI inflows is an important goal of the AEC and largely conditional to the success of the ASEAN-5 economies efforts.

Conflict of interest

This chapter has been declared as no conflict of interest.

Author details

Elya Nabila Abdul Bahri[1]* and Nor Hakimah Haji Mohd Nor[2]

1 University of Malaya, Kuala Lumpur, Malaysia

2 Kolej Universiti Islam Antarabangsa Selangor, Kajang, Malaysia

*Address all correspondence to: elyanabila@um.edu.my

IntechOpen

© 2019 The Author(s). Licensee IntechOpen. This chapter is distributed under the terms of the Creative Commons Attribution License (http://creativecommons.org/licenses/by/3.0), which permits unrestricted use, distribution, and reproduction in any medium, provided the original work is properly cited. (cc) BY

References

[1] Alfaro L, Chanda A, Kalemli-Ozcan S, Sayek S. FDI and economic growth: The role of local financial markets. Journal of International Economics. 2004;**64**(1):89-112. DOI: 10.1016/S0022-1996(03)00081-3

[2] Alfaro L, Kalemli-Ozcan S, Sayek S. FDI, productivity and financial development. The World Economy. 2009;**32**(1):111-135. DOI: 10.1111/j.1467-9701.2009.01159.x

[3] Bekana DM. Determinants of foreign direct investment in Ethiopia: Time series evidence from 1991-2013. The Journal of Developing Areas. 2016;**50**(1):141-155. DOI: 10.1353/jda.2016.0002

[4] Sawalha N, Mazouz A, Pellet P. Foreign direct investment inflows and determinants in selected developed, developing, and underdeveloped countries with limited resources. International Journal of Business and Economics Perspectives. 2013;**81**(1):76-92

[5] Sankaran C. An analysis of the determinants of FDI inflows: The case of the Dominican Republic. IUP Journal of Applied Economics. 2015;**14**(1):47-68

[6] Benáček V, Lenihan H, Andreosso-O'Callaghan B, Michalíková E, Kan D. Political risk, institutions and foreign direct investment: How do they relate in various European countries? The World Economy. 2014;**37**(5):625-653. DOI: 10.1111/twec.12112

[7] Rad SSE, Embong Z, Mohd-Saleh N, Jaffar R. Financial information quality and investment efficiency: Evidence from Malaysia. Asian Academy of Management Journal of Accounting and Finance. 2016;**12**(1):129-151

[8] Rajan R, Zingales L. Financial dependence and growth. American Economic Review. 1998;**88**(3):559-586

[9] Larson RK, York-Kenny S. An empirical analysis of International Accounting Standards, equity markets, and economic growth in developing countries. Journal of International Financial Management and Accounting. 1995;**6**(2):130-157

[10] Ndubizu GA. Accounting Standards and Economic Development: The Third World in Perspective. International Journal of Accounting. London: Prentice; 1984;**19**:181-196

[11] Saudagaran SM, Diga JD. Financial reporting in emerging capital markets: Characteristics and policy issues. Accounting Horizons. 1997;**11**(2):41-65

[12] Akisik O. Accounting regulation, financial development, and economic growth. Emerging Markets Finance and Trade. 2013;**9**(1):33-67

[13] Akpomi ME, Nnadi MA. The impact of International Financial Reporting Standards (IFRS) adoption on foreign direct investment (FDI): Evidence from Africa and implications for managers of education. Journal of Accounting and Financial Management. 2017;**3**(2):51-65

[14] Jinadu O, Ojeka SA, Ogundana OM. IFRS adoption and foreign direct investment: Evidence from Nigerian quoted firms. Mediterranean Journal of Social Sciences. 2016;**7**(2):99-105

[15] Barth ME, Landsman R, Lang M. International accounting standards and accounting quality. Journal of Accounting Research. 2005;**46**(3):467-498

[16] Yousefinejad M, Ahmad A, Salleh FM, Rahim RA. Causal relationship between International Financial Reporting Standard (IFRS) and foreign direct investment (FDI): A panel data analysis of ASEAN countries. Asian

Journal of Accounting and Governance. 2018;**10**:61-72. DOI: 10.17576/AJAG-2018-10-06

[17] Billet MT, Yu M. Asymmetric information, financial reporting, and open-market share repurchases. Journal of Financial and Quantitative Analysis. 2016;**51**(4):1165-1192

[18] Blonigen BA, Piger J. Determinants of foreign direct investment. Canadian Journal of Economics. 2014;**47**(3): 775-812. DOI: 10.1111/caje.12091

[19] Pesaran MH. General diagnostic tests for cross section dependence in panels. Cambridge Working Papers in Economics. 2004. 0435. DOI: 10.17863/CAM.5113

[20] Pesaran MH. A simple panel unit root test in the presence of cross-section dependence. Journal of Applied Econometrics. 2007;**22**(2):265-312. DOI: 10.1002/jae.951

[21] Westerlund J. Testing for error correction in panel data. Oxford Bulletin of Economics and Statistics. 2007;**69**(6): 709-748. DOI: 10.1111/j.1468-0084.2007.00477.x

[22] Pedroni P. Fully modified OLS for heterogenous cointegrated panels. Adv. Econom. 2000;**15**:93-130

[23] Pedroni P. Purchasing power parity tests in cointegrated panels. The Reviews of Economics and Statistics. 2001;**83**(4):727-731

[24] Chudik A, Mohaddes K, Pesaran M, Raissi M. Long-run effects in large heterogenous panel data models with cross-sectionally correlated errors. In: Hill R, Gonzalez-Rivera G, Lee TH, editors. Advances in Econometrics: Essays in Honour of Aman Ullah. Vol. 36. Emerald Publishing; 2016. pp. 85-135

[25] di Giovanni J. Whar drives capital flows? The case of cross-border M&A activity and financial deepening. Journal of International Economics. 2005;**65**:127-149. DOI: 10.1016/j.jinteco.2003.11.007

[26] Stein E, Daude C. Longitude matters: Time zones and the location of foreign direct investment. Journal of International Economics. 2007;**71**: 96-122. DOI: 10.1016/j.jinteco.2006.01.003

[27] Mohamed SE, Sidiropoulos MG. Another look at the determinants of foreign direct investment in MENA countries: An empirical investigation. Journal of Economic Development. 2010;**35**(2):75-95

[28] Levine R. Finance and growth: Theory and evidence. In: Handbook of Economic Growth. Netherlands: Elsevier; Vol. 1. 2005. pp. 865-934

[29] Bushman R, Smith A. Financial accounting information and corporate governance. Journal of Accounting and Economics. 2001;**31**:237-333

[30] Guerin SS. The role of geography in financial integration: Foreign direct investment. LLEE working paper no. 36. Rome: LUISS Lab of European Economics: 2005

[31] Bahri ENA, Nor AHSM, Nor NHHM. The role of financial development on foreign direct investment in ASEAN-5 countries: Panel cointegration with cross-sectional dependency analysis. Asian Academy of Management Journal of Accounting and Finance. 2018;**14**(1):1-23. DOI: 10.21315.aamjaf2018.14.1.1

Chapter 5

The Roles of Accounting Valuations and Earnings Management in the Survivorship of Technology Firms during the Global Financial Crisis

Oliver Neoh and Matthias Nnadi

Abstract

This study examines the survivorship of technology firms listed on the NASDAQ market during the immediate and post-2008 global financial crisis period. Underpinned by contingency theory, this study demonstrates the varying roles of accounting valuation and earnings management metrics in the technology industry. Findings in this chapter show during the global financial crisis periods, technology firms have greater survivorships when they are undervalued, and possess a lesser degree of discretionary earnings (DA). The DA factor is a double-edged sword for technology firms since it has positive and negative effects on the returns and survivorships, respectively. The research and development (R&D) variable remains a positive component for both returns and survivorships of these firms.

Keywords: accounting valuation, earnings management, technology, prediction, performance, survivorship

1. Introduction

The technology sector has rapidly gained importance in the past two decades. The NASDAQ Composite, which measures all the stocks on the tech-heavy NASDAQ market, surpassed its dotcom high in 2015.[1] Valuations are also close to the general market today, in sharp contrast to the extreme valuation gap seen in 2000.[2] Firms within the technology sector possess distinct aspects relative to non-technology firms, examples being: innovation and technical advancement leading to high economic growth, opaque information causing moral hazard, longer investment horizon required for technology innovations, greater uncertainties in cash

---

[1] The transformation of the industry over the past 17 years has been significant. Technology companies are expected to generate more than 25% of the S&P 500's overall earnings in the fourth quarter of this year, compared with 15% at the index's peak in 2000. At that time, tech stocks accounted for more than a third of the benchmark S&P 500. Today that figure has fallen closer to 23%.

[2] Tech stocks trade at 19.4 times 2017 full-year earnings, while the S&P 500 is at 19 times. As per Bloomberg data, in the first quarter of 2000, S&P 500 technology groups traded at 73 times earnings.

IntechOpen

flows and growth potentials. It is further asserted that greater litigation risk arises from information asymmetry [1]. Therefore, our goal is to analyse the reasons behind the rapid performance growth and change in survivorship in technology sector during of financial turmoil and its initial recovery of our 2008 global financial crisis. We evaluate the impact of accounting valuation and earnings management indicators considering special aspects of technology firms and our lesson from the technology bubble in 2000.

Currently there exists a gap in the academic discourse comparing the state of the technology sector during the crisis of 2008. Our study fills the present gap by re-evaluating the sector in the aftermath of the most recent crisis and reflects on the findings drawn from preceding crisis of 2000. The current study seeks to evaluate whether the same findings drawn from the earlier crisis literature, referring to [2], are still relevant for capital providers and participants in the post-crisis period. Accordingly, our study sets a considerable contribution to current studies on the effects of valuation measures in the technology sector on the financial crisis, and whether these effects vary over the various periods of the crisis.

Secondly, previous studies limit their scope to analysing performance of stock returns [3, 4]. This chapter however focuses on examining both future performance and the survivorship of firms within the technology-sector. This is linked to our base understanding that accounting valuation and earnings management variables play a much greater role in determining future survivorship within this particular sector that has developed with large strides since the late twentieth century. Moreover, given the uniqueness of the technology sector, such as the larger existence of intangible assets, we anticipate that our evaluation encompassing linked variables such as discretionary accruals (DA) and research and development (R&D) expenditure will be a valuable contribution to the current technology sector literature. We examine how these effects are contingent on different time horizons, short and long run, and morality (ETHICS) of these technology firms' behaviours.

This chapter contributes to ongoing discussion in relation to the determining the effects of accounting valuation and earnings management variables in the performance and survivorship of technology firms during the financial crisis. The three core objectives of the study linked particularly towards future performance and survivorship are:

I. to determine the level of explanatory power of valuation and accounting metrics in determining the short- to long-term performance of securities within the technology sector throughout the most recent global financial crisis (GFC) period;

II. to examine whether valuation and accounting metrics hold predictive power in explaining the future survivorship or failure of technology firms in the immediate and post-crisis period; and

III. to investigate how the technology firms' moral hazard (i.e. opportunistic behaviour) affect their future returns and survivorships during and after the GFC period.

We find that our accounting valuation and earnings management techniques including R&D, market capitalization (MktCap), earnings to price ratio (EP), book to market ratio (BM), and discretionary accruals (DA) have stronger positive effects on longer term performance of the technology firms. In other words, undervaluation, larger firm size, and more discretionary earnings and R&D lead to higher returns which increase more with longer terms. On the other hand, undervaluation, large firm size (MktCap), more R&D but with less discretionary earnings (DA) increase the survivorships of these firms. We confirm that practising earnings management introduces a double-edged sword for the technology firms since DA has positive and negative effects on the returns and survivorships, respectively particularly during the global financial crisis. R&D mostly increases both the returns and survivorships although the significance may vary. Furthermore, the moral hazard (ETHICS) within the technology firms tends to reduce their future returns while not significantly affecting their survivorships.

1.1 Development of the research hypothesis

The underlying research question for this chapter considers whether in times of financial crisis, capital providers, including investors, equity providers, and debt holders could use accounting valuation, and earnings management variables in predicting the future performance and survivorship/failure of technology-firms. To investigate the research question, we consider the contingency theory as a fitting framework as it suggests that an optimal course is contingent on the circumstance or environment confronting the phenomenon being examined. This is because our study seeks to provide an explanation to phenomena that are dependent upon times of crisis.

Keystone works utilising this contingency approach explore a diverse range of subjects, such as the usefulness of virtual learning environments in education amongst different sectors or groups [5–7], or the optimal conditions in organisations where certain IT implementations will be beneficial [8]. The same framework is applied to our investigation as we seek to identify which accounting valuation, and earnings management techniques are useful in explaining future performance and probability of failure contingent on times of crisis and periodic outlooks. Accordingly, we propose two hypotheses directly relevant to the issues for equity-holders on future performance, and creditors on future probability of failure:

H_1: *Accounting* valuation *and earnings management metrics will be significant at different times of crises in explaining short- to long-term performance, and where they are significant they should follow the conventional relationship these measures have with future returns.*

H_2: *Accounting* valuation, *earnings management, and past performance variables will be significant at different times of crises in explaining future firm failure, and when they do, they adhere to conventional relationships.*

For equity providers and investors of stock markets, this issue of being able to predict the performance of the firm is highly relevant as these groups typically seek to invest in businesses that are set for future success. If the investment is successful, they are rewarded with stock price returns. For debt holders, it is important that they are able to determine the probability of future failure and to assess whether the recipient of the loan would be able to pay-back or risk defaulting on their loan. For accurate prediction of these outcomes, a more robust risk-assessment of lending agreements is essential.

The rest of the chapter is structured as follows: Section 2 is the literature review, Section 3 is the data and research design, Section 4 is the results and discussion of findings and Section 5 is the conclusion, implication of study and limitations.

2. Review of literature

Arguably, since the late twentieth century the technology sector has increased its presence within society in every aspect. Its continued rise is now imminent and a robust understanding of the sector is well in demand. Alongside the sector's rise, a myriad of listed technology companies have also grown in presence and influence in capital markets since its origins. Throughout its history, these high-technology investments have at times showcased behaviours of high-growth characteristics [9]. To understand the issues that the chapter aims to address, we provide the four core areas covered in the existing literature, consisting of the technology sector, financial crises, financial measures, and earnings management.

2.1 The technology sector

Technology stocks have increased immensely in both presence and importance to global financial markets. For instance, from 1990 to 2000, technology firms have emerged from barely being existent to occupying six of the ten largest firms in the US, in terms of market capitalisation [10]. The National Association of Securities Dealers Automated Quotations (NASDAQ) index which is heavily weighted by technology stocks has grown over 29 times to reach a total market capitalisation of $8.4tn.[3] Moreover, the largest stocks in today's financial markets are concentrated to the largest of the technology pioneer companies, namely Apple, Alphabet, and Microsoft. This is testament to the prominence of technology in today's world. Technology firms arguably have adopted increasingly important roles in today's capital market [11].

Nevertheless, technology stocks have faced significant challenges. Particularly, in consequence of the technology bubble and crash of late-1999 to Spring 2000, the world experienced a monumental downturn in the technology sector that sent ripples of distress throughout international financial markets. The next crisis following this was the most recent sub-mortgage crash in 2008, which similarly sent devastating effects to global markets. As such, the periodic context that coincided with the growth of the technology sector was significantly more volatile than its preceding bull-run of the 1950s to 80s.

The explanations that have arisen behind the crisis in 2000 is critical for our study, and its objective of examining the technology sector's reaction to the most recent global financial crisis (GFC) that erupted in 2008. The causes of this credit crunch and its huge impact is explored by many scholars in the field, refer to [3, 12, 13]. The crisis is a prime foundation behind this study which explores the varying dynamics of the financial crises. Close to 8 years after, this presents an opportune moment to look into the consequent shaping of the technology sector and investigate the determinants behind the performance and survivorship in this post-crisis period.

2.2 Financial crises

The cause of financial crisis is a recurring subject in economic history. Referring to the earlier 2000 technology-bubble, much discussion in the academic landscape has evolved in this field where its origins are commonly pointed to the over-optimistic behaviours expressed in the market towards the rise of Internet stocks. For example, some scholars advocate that the absence of traditional valuation metrics proved to be evidence for collective investor irrationality during the initial crash of the tech-sector [4]. As a result of this downturn, there has been a vast array of studies in the financial literature which has examined causes and impacts of the bubble. Two key areas of study have observed technology stocks in: (1) the valuation space, and how traditional metrics may not be as applicable to tech-stocks; and (2) the investment perspective looking at the various co-movements of technology

[3] Data from Bloomberg Terminal.

stocks in comparison to other sectors as well as the behaviour of different investment clienteles.

To exemplify the scale of destruction that crises like the GFC produces, there has been examination of nine crisis episodes, including the GFC and the technology bubble, and their implications on equity markets and the incidence of contagion [3]. They identify the GFC as a global crisis having contagion effects in all the channels tested. Tests on the technology bubble show that it possess normal interdependencies as a result of the downturn. Our research will therefore seek to further evaluate the effects of the GFC specifically towards the technological sector.

2.3 Accounting and valuation metrics

There has been ongoing discussion on the predictive power that accounting and valuation metrics serve in explaining future performance and survivorship of listed technology firms [2, 4, 14]. Due to the nature of tech-firms' wide-ranging business operations, they tend not to have the same drivers of business growth or failure as other sector firms. Within technology remit, the general nature of operations may be heavily service-oriented when referring to the likes of software providers, or more product-driven such as those companies specialising in computer hardware. A lot of these companies may be travelling through their early growth stages, meaning investments and capital expenditure do not necessarily translate to earnings until much later on in their life-cycle.

Furthermore, investment behaviour towards the sector has a track-record of being over-optimistic, as argued in the sector's early development and in the emergence of the 2000 technology-bubble [14]. The chapter recognises this thinking as a reflection based on the origins of the past tech-bubble, however our chapter seeks to enhance understanding in the contemporary links and similarities that the market expressed in the most recent crisis of 2008, and the key implications for equity holders and creditors active within the sector.

2.4 Earnings management

There has also been a wide-ranging scope of literature examining the phenomenon of companies practising earnings management. There has been explanations that earnings management occurs when managers use judgement in financial reporting and in structuring transactions to alter financial reports to either mislead some stakeholders about the underlying economic performance of the company or to influence contractual outcomes that depend on reported accounting practices [15]. The academic discussion on this topic explores the possibility that earnings management may imply on the immediate and future performance of tech companies during financial crisis. There is further evidence showing that firms are likely to manipulate earnings in order to report small profits instead of losses during periods of financial distress [16].

Earnings management generally refers to the management of items such as increasing accruals, as well as cutting discretionary expenditure, such as advertising, selling, general and administrative expenses, and research and development (R&D). This is particularly relevant when looking at how companies manage earnings in preparation for reporting season and meeting analyst expectations. Certain findings indicate high-tech firms are more likely to use discretionary accruals to reward CEOs to meet their earnings expectations compared to low-tech firms [17]. Moreover, some studies make conclusions that firms who show higher levels of earnings management to beat forecasts typically outperform firms who do not engage in this practise as much [18].

3. Data and research design

To explain the sample selection of technology-firms from the NASDAQ market index, we firstly used the COMPUSTAT database to generate a list of firms that were active on the NASDAQ market as of January 2007. The rationale behind selecting the NASDAQ market is that its constituent members are heavily weighted towards the technology sector, a decision similarly applied in reference to [4]. As we identify the pinnacle of the GFC to be the latter half of 2008, we generate a shortlist which comprises of firms which were active as of January 2007 to ensure that each firm within our final sample have been active for at least a year. We also ensure that all firms within our final sample were active as of third quarter of 2008.

After generating the shortlist of firms, we further break down our sample into defined Industry Classification Benchmark (ICB) codes within the NASDAQ index to arrive at a total of 350 firms. These represent the chapter's scope in evaluating technology firms. The earliest starting point of our data is selected as the year-ending 2007 so that we may calculate figures which require a year-on-year change, such as the calculation of yearly sales growth. In other words, if the financial data involves a year-on-year change, the earliest data point used to calculate this change figure will be the 2007 value. All financial and valuation data are generated from the financial year end of 2008, which is December. This is the same for 2009 and 2010.

In addition, we collect the overall firms' ethical behaviour scores (ETHICS) in US from The Global Competitiveness Report (2008–2010). This variable measures the effect of ethical behaviour of the technology firms to their returns and bankruptcies in our further analysis. The descriptions of the accounting valuation and earnings management metrics used in the study are detailed in **Table 1**.

In terms of their data extraction, we use yearly data where it refers to an accounting reported figure (Sales_g, EBITDA, RD, and calculation of DA) and quarterly data for valuation metrics (EP, BM, PS and MktCap) from the Bloomberg Terminal ranging from 2007 to 2015. Where the variables are extracted on an annual basis, we use the base year figures 2008, 2009, and 2010. Where the variables are quarterly, we use the Q4 data from the base years, 2008, 2009, 2010 to form the independent variable pool as this aligns with the financial year-end figures of our accounting data.

All accounting items required to conduct the DA procedures are extracted using yearly data from the Bloomberg database for each year, 2008–2010. The 2007 figures are also generated considering that some items require a calculation to measure the change. The *non-discretionary accruals using the Jones Model* is express as:

$$NDA_t = a_1\left(\frac{1}{A_{t-1}}\right) + a_2(\Delta REV_t) + a_3(PPE_t) \quad (1)$$

The total accruals using the Jones Model is express as:

$$TA_t = a_1\left(\frac{1}{A_{t-1}}\right) + a_2(\Delta REV_t) + a_3(PPE_t) + \vartheta_1 \quad (2)$$

where ΔREV_t = revenues in year t less revenues in year $t-1$ scaled by total assets at $t-1$; PPE_t = gross property plant and equipment in year t scaled by total assets at $t-1$; A_{t-1} = total assets at $t-1$; $a_1 a_2 a_3$ = firm-specific parameters; TA_t = total accruals scaled by lagged total assets; income before extraordinary items—CF from continuing operations.

Type	Variables	Description
Valuation	Earnings to price ratio (EP)	Also known as the earnings yield, the EP ratio is the inverse of the PE ratio measuring the earnings per share of a company divided by the share price of the company. Generally, the higher an EP ratio gets, the company can either be seen by the market with low confidence in its earnings, or is currently undervalued
	Book to market ratio (BM)	The book to market ratio is a ratio that calculates the book value of a firm relative to its market value. This ratio is also used in valuation practices to identify undervalued, typically over 1.00, or overvalued securities, if less than 1.00
	Price to sales ratio (PS)	Another ratio that divides a company's stock price by its revenues. It is used to indicate the value attached to each dollar of a company's sales or revenues. Lower values tend to suggest undervaluation while higher values may indicate overvalued stocks or those with higher confidence assigned by the market
Accounting	Market capitalisation (MktCap)	Market capitalisation is derived by calculating the company's shares outstanding multiplied by the current market price. It is used as a close proxy for size of the firm, and typically larger firms tend to survive crises
	Sales growth ($Sales_g$)	Sales growth in our study is defined as the growth in sales revenue year on year in Q4 of a particular year. Those with higher growth are assumed to enjoy greater return prospects and lower likelihood of failure
	Earnings before interest, taxes, depreciation, and amortisation (EBITDA)	An accounting item that acts as a strong indicator of a company's financial performance. It is useful in displaying the earning potential of a business as well as a gauge to analyse profitability that eliminates the effects of financing and accounting decisions
	R&D expenditure (R&D)	Refers to expenses linked to the research and development of a company's goods or services. Importantly, the technology sector is amongst the highest users of RD and the level may be viewed by market participants as either an expense or investment initiative, a key part to our study
	3m, 6m, 1y, 3y, 5y	These are the firm's 3-month, 6-month, 1-year, 3-year, and 5-year performance measured by stock returns. These explain the firm's performance from the short to long term
	Failed_6m, Failed_1y, Failed_3y, Failed_5y	These are the dummy variables assigned 1 if the firm fails and 0 if otherwise in 6-month, 1-year, 3-year, and 5-year
Earnings management	Discretionary accruals (DA)	Along with non-discretionary accruals, they make up the total accruals practiced by the company. Discretionary accrual is associated with management choices whereas non-discretionary accruals originate from business conditions. DA is a good proxy for earnings quality and our study employs the Jones and Modified Jones Model for its calculation
Qualitative data	ETHICS	The scores of firm's ethical behaviour in US collected from The Global Competitiveness Report (2008–2010)

The table shows the description of the variables we use. It shows the description for our valuation (EP, BM, PS), accounting (MktCap, $Sales_g$, EBITDA, R&D, (3m, 6m, 1y, 3y, 5y) returns, (Failed_6m, Failed_1y, Failed_3y, Failed_5y) failures, DA, and ETHICS).

Table 1.
Variable description.

We also compute the non-discretionary accruals using the Modified Jones Model which is express as:

$$NDA_t = a_1\left(\frac{1}{A_{t-1}}\right) + a_2(\Delta REV_t - \Delta REC_t) + a_3(PPE_t) \tag{3}$$

where ΔREC_t = net receivables in year t less net receivables in year $t-1$ scaled by total assets at $t-1$; NDA_t = non-discretionary accruals scaled by total assets of prior year.

Once we have calculated the discretionary accruals value for each firm in their MJDA and JDA forms, we then measure each firm's value relative to the sample median in percentage terms similar to the methodology adopted by past studies, refer to [18]. Consequently, a positive value refers to a firm who has greater discretionary accruals for that particular year relative to its peer group. The descriptive statistics of our overall data is shown in the following **Table 2**.

3.1 Research design

The core methodologies used for our research include the linear regression and logistic regression models. In our linear regression section investigating H_1, we conduct OLS regressions for each base year measuring 3-month, 6-month, 1-year, 3-year, and 5-year performance as the dependant variable. In addressing H_2,

	3m	6m	1y	3y	5y	EP	BM	PS	MktCap	$Sales_g$
Average	7.95	21.59	35.16	59.73	80.14	0.06	0.74	2.28	2.66	0.08
Median	7.06	11.53	20.84	33.86	37.94	0.05	0.55	1.69	2.65	0.01
Min	−59.69	−100.00	−100.00	−100.00	−100.00	0.00	0.02	0.05	0.27	−4.21
Max	103.87	314.04	1358.18	1089.90	2880.31	0.84	11.01	27.98	5.47	9.07
Std.	20.44	48.84	85.65	148.61	231.81	0.07	0.79	2.21	0.86	0.49
10th per	−16.59	−18.92	−34.44	−100.00	−100.00	0.01	0.23	0.43	1.57	−0.06
90th per	32.29	70.69	117.39	214.18	290.98	0.12	1.30	4.44	3.74	0.28
N	866	866	866	866	866	866	866	866	866	866

	EBITDA	Failed_6m	Failed_1y	Failed_3y	Failed_5y	MJDA	JDA	R&D	ETHICS
Average	1.16	0.01	0.04	0.15	0.26	0.43	0.39	0.10	5.36
Median	0.00	0	0	0	0	0	0	0.09	5.40
Min	−29.32	0	0	0	0	−16.37	−15.71	0.00	5.20
Max	350.67	1	1	1	1	35.87	34.33	0.54	5.50
Std.	13.00	0.11	0.19	0.35	0.44	3.82	3.70	0.07	0.12
10th per	−0.05	0	0	0	0	−2.41	−2.39	0.02	5.20
90th per	0.59	0	0	1	1	3.26	3.10	0.18	5.50
N	866	866	866	866	866	866	866	866	866

The table shows the descriptive statistics of the data set we used. We show the average, median, minimum (min), maximum (max), standard deviation (Std.), 10th percentile (10th per), 90th percentile (90th per), and the total number of sample (N) for our overall data. Our overall data include 3m, 6m, 1y, 3y, 5y, EP, BM, PS, MktCap, $Sales_g$, EBITDA, Failed_6m, Failed_1y, Failed_3y, Failed_5y, MJDA, JDA, R&D, and ETHICS from 2008 to 2010.

Table 2.
Descriptive statistics.

we conduct logistics regression using 3-models for each base year measuring 1-year failure, 3-year failure, and 5-year failure. One year is chosen as the earliest measure of failure as any earlier, such as 6-months, would consist of an insignificant number of firms that have failed. Logistic regression results earlier than 1-year, in our case, are limited in their validity.

Our linear regression model for determining future performance is express as:

$$r_{a_t} = a + b_1EP + b_2BM + b_3PS + b_4MktCap + b_5EBITDA + b_6Sales_g + b_7MJDA + b_8RD \tag{4}$$

And our logistic regression model for determining future failure is express as:

$$Failed\ firmt = a + b_1EP + b_2BM + b_3PS + b_4Sales_g + b_5MJDA + b_6RD + b_7MktCap + b_8EBITDA + b_91y... + b_{11}5y \tag{5}$$

In Eq. (4), the methodology for the linear regression involves a similar framework used in reference to [4], examining the market value of securities in the tech sector during the tech-bubble of 2000. In our case, the dependent variable is selected as the percentage return over the prescribed period (3m, 6m, 1 y, 3 y, and 5 y) from the end of the selected base-years (2008, 2009, and 2010). The independent variables include our valuation, accounting, and earnings management variables of each underlying security in the sample, as described in **Table 1**. A positive significant coefficient would imply a positive relationship between the indicator and future returns.

Our linear regression model for determining future performance using our overall data including ETHICS is express as:

$$r_{a_t} = a + b_1EP + b_2BM + b_3PS + b_4MktCap + b_5EBITDA + b_6Sales_g + b_7MJDA + b_8RD + b_9ETHICS \tag{6}$$

Furthermore, our logistic regression model for determining future failure using our overall data including ETHICS is express as:

$$Failed\ firmt = a + b_1EP + b_2BM + b_3PS + b_4Sales_g + b_5MJDA + b_6RD + b_7MktCap + b_8EBITDA + b_9ETHICS + b_{10}6m... + b_{13}5y \tag{7}$$

Then the corresponding null and alternative hypotheses for linear (Eq. (6)) and logistic regressions (Eq. (7)) are coefficients equal zero and non-zero, respectively, as before.

4. Results and discussion of findings

4.1 Future performance

Table 3 is the result of the multivariate analyses using the Jones Model and shows that the R&D is significant[4], and follows the traditional relationship with

[4] In this paper, we mostly discuss the results that show significantly strong results (i.e., p-values ≤ 0.05). Weakly significant results (i.e., $0.05 < p$-values < 0.10) are sometimes aligned with the strongly significant results, but we do not actively discuss these.

	3m	6m	1y	3y	5y	EP	BM	PS	MktCap	$Sales_g$	EBITDA	Failed_6m	Failed_1y	Failed_3y	Failed_5y
3m	1^{***}														
6m	0.3^{***}	1^{***}													
1y	0.21^{***}	0.67^{***}	1^{***}												
3y	0.14^{***}	0.42^{***}	0.49^{***}	1^{***}											
5y	0.03	0.26^{***}	0.3^{***}	0.62^{***}	1^{***}										
EP	-0.08^{**}	0.19^{***}	0.23^{***}	0.15^{***}	0.13^{***}	1^{***}									
BM	-0.09^{***}	0.29^{***}	0.28^{***}	0.18^{***}	0.11^{***}	0.42^{***}	1^{***}								
PS	0.02	-0.15^{***}	-0.11^{***}	-0.09^{***}	-0.08^{**}	-0.27^{***}	-0.37^{***}	1^{***}							
MktCap	0.05	-0.14^{***}	-0.14^{***}	-0.09^{**}	-0.03	-0.19^{***}	-0.43^{***}	0.48^{***}	1^{***}						
$Sales_g$	0.02	0.01	-0.03	-0.02	-0.05	0.06^{*}	-0.01	-0.03	-0.09^{***}	1^{***}					
EBITDA	-0.04	-0.05	-0.04	-0.02	0	-0.01	0	-0.04	-0.06^{*}	0.01	1^{***}				
Failed_6m	0.05	-0.28^{***}	-0.17^{***}	-0.12^{***}	-0.09^{**}	-0.05	-0.02	0.01	-0.02	-0.1^{***}	0.06^{*}	1^{***}			
Failed_1y	0	-0.12^{***}	-0.32^{***}	-0.22^{***}	-0.16^{***}	-0.05	-0.04	-0.04	-0.02	-0.04	0.03	0.51^{***}	1^{***}		
Failed_3y	-0.05	-0.08^{**}	-0.17^{***}	-0.45^{***}	-0.32^{***}	-0.07^{*}	0.01	-0.09	-0.12^{***}	-0.02	0.01	0.24^{***}	0.49^{***}	1^{***}	
Failed_5y	-0.04	-0.03	-0.12^{***}	-0.32^{***}	-0.46^{***}	-0.06^{*}	0.07^{**}	-0.08^{**}	-0.18^{***}	0.04	-0.01	0.17^{***}	0.34^{***}	0.71^{***}	1^{***}

The table shows the correlation matrix of our data set we used. It uses a Pearson correlation calculation.
**Significance at 10% level.*
***Significance at 5% level.*
****Significance at 1% level.*

Table 3.
Correlation Table

both survivorship and future performance of tech-firms. Across two of the three years in the study, namely 2008 and 2009, the longer the returns are, the stronger the positive effects from BM, MktCap, MJDA, and R&D become. R&D acts a good predictor of future performance with significant coefficients of 155.73 in 6 months, 265 in 1 year and 335.09 in 3 years respectively in 2008 model. In the 2009 model, the R&D results remain significant at 56.41 and 286.02 respectively.

Sales growth is negatively significant to the medium-term returns (1y) in 2010 with a coefficient of −16.35 (**Table 3**). This suggests that firms who reported increased growth in year-on-year sales did not necessarily enjoy corresponding positive returns in medium-term. This finding corroborates with past literature suggesting there is a significant association between three-day market returns and Internet firm revenue announcements [19]. This finding and the PS results shed valuable insights into the characterisation of tech stock returns in the crisis periods. Our discretionary accruals (MJDA) variable also exhibits significant linkage to the explanation of increased short- to medium-term returns from 2008. As shown in **Table 3** the variable possesses positive coefficients of 1.23 and 4.25 in 2008s 6-month and 3-year outlook respectively which indicates a positive linkage to future returns. Therefore, discretionary accruals is a positive element for the returns which amplifies with longer returns.

As a whole, the accounting variables such as BM, MktCap, MJDA, and R&D have stronger positive effects on the longer-term performances in 2008 and 2009. In general, the large firm size (MktCap), undervaluation (EP and BM), more discretionary accruals and R&D are positive drivers for the returns of the technology firms. The negative effects from sales growth occur in 2010 sometime after the crisis. The overvaluation measured by PS seemed to positively affect the long-term returns but this effect fades out as time goes by.

4.2 Future survivorship

For future survivorship, **Table 4** shows that the R&D variable plays a significant role in a firm's 1-year outlook over the three crises years with negative coefficients of −88.16, −95.25, and −64 across 2008–2010. This reveals key points about the practice of R&D expenditure in the technology sector and why firms who practice higher levels tend to experience positive future returns and a higher likelihood of survivorship.

Furthermore, our findings suggest that firms with higher MktCap tend to survive and are more likely to outperform in the longer-term. With negative coefficients ranging from −2.17 and −0.48 in 2008, and −0.86 and −0.91 in 2010, the explanatory power for future failure is in line with the argument that undercapitalisation is a core reason to every technology business failure evaluated [20]. The R&D is significantly negative which highlights the importance of R&D expenditure in the technology sector as an investment rather than an asset. Support for this expenditure figure in providing a 'truer measure of a company's value because this spending often turns out to be money in an investor's pocket in the future' is also advocated in reference to [21]. This implies that expenditure in the technology industry leads to improved efficiency, increased sales, and ultimately increasing company value.

The level of discretionary accruals (MJDA) practiced by firms is also a good predictor of future failure in the medium- to long-term in 2008 and 2010. In predicting future failure in **Table 4**, the coefficients suggest that higher levels of discretionary accruals translate to greater likelihood of failure with positive coefficients of 0.18 in 2008; 0.36 and 0.27 respectively in 2010. Importantly, this conforms closely with our expectation that practising earnings management introduces

$$MODEL: r_{a_{t+p}} = a + b_1EP + b_2BM + b_3PS + b_4MktCap + b_5EBITDA + b_6Sales_g + b_7MJDA + b_8R\&D$$

	Years								
	2008				2009		2010		
	3m	6m	1y	3y	3m	1y	3m	1y	5y
Const	−4.45 (−0.76)	4.57 (0.31)	2.99 (0.11)	22.78 (0.51)	9.43 (1.56)	2.61 (0.13)	**14.33**** **(2.24)**	−1.6 (−0.13)	**−76.04*** **(−1.82)**
EP	−9.82 (−0.67)	47.8 (1.29)	**252.49***** **(3.85)**	**220.46*** **(1.95)**	−28.34 (−1.26)	−117.87 (−1.52)	−44.66 (−1.43)	14.96 (0.24)	23.24 (0.11)
BM	−0.59 (−0.42)	**12.85***** **(3.66)**	**19.83***** **(3.18)**	**23.41**** **(2.18)**	**6.37*** **(1.89)**	**22.56*** **(1.93)**	−0.61 (−0.13)	−5.36 (−0.57)	−2.03 (−0.07)
PS	0.44 (0.4)	0.54 (0.19)	1.77 (0.36)	**19.93**** **(2.35)**	0.42 (0.69)	**5.15**** **(2.46)**	−0.54 (−1.12)	0.63 (0.67)	−5.09 (−1.63)
MktCap	**4.72**** **(2.5)**	3.53 (0.74)	−1.31 (−0.16)	−5.51 (−0.38)	**−2.84*** **(−1.69)**	−5.57 (−0.96)	−0.8 (−0.47)	−2.96 (−0.89)	**33.94***** **(3.08)**
EBITDA	−0.27 (−1.1)	−0.63 (−1.01)	−0.72 (−0.64)	−2.6 (−1.35)	−0.07 (−0.46)	−0.41 (−0.8)	−0.04 (−0.75)	−0.08 (−0.79)	0.22 (0.64)
$Sales_g$	2.2 (1.08)	−3.84 (−0.75)	−6.07 (−0.67)	−0.1 (−0.01)	1.2 (0.4)	0.92 (0.09)	**6.23*** **(1.66)**	**−16.35**** **(−2.21)**	−16.96 (−0.69)
MJDA	0.32 (1.34)	**1.23**** **(2.07)**	0.93 (0.88)	**4.25**** **(2.34)**	−0.25 (−0.52)	0.9 (0.55)	−1.08 (−1.27)	2 (1.19)	−2.84 (−0.51)
R&D	−5.25 (−0.29)	**155.73***** **(3.36)**	**265***** **(3.22)**	**335.09**** **(2.37)**	**56.41***** **(3.15)**	**286.02***** **(4.61)**	4.55 (0.29)	30.75 (0.98)	127.85 (1.24)
F-test	**2.55*****	**4.81*****	**6.27*****	**3.46*****	**2.47****	**3.80*****	1.36	1.16	**1.96***
Adjusted R^2	0.03	0.08	0.11	0.05	0.05	0.09	0.01	0.00	0.03
N	350	350	350	350	241	241	275	275	275

The table shows the valuation and accounting variables' effect on the technology firms' returns from 2008 to 2010. The dependent variables are 3 months (3m), 6 months (6m), 1 year (1y), and 3 years (3y) returns of the technology firms. Then we use the earnings to price ratio (EP), book to market ratio (BM), price to sales ratio (PS), market capitalization (MktCap), EBITDA, sales growth ($Sales_g$), modified Jones for discretionary accruals (MJDA), and research and development (R&D) as our independent variables. The values in parentheses are the t-values to the corresponding coefficients.

**Significance at 10% level.*

***Significance at 5% level.*

****Significance at 1% level.*

Table 4.

Multivariate analysis of periodic returns of NASDAQ technology firms—using modified Jones model for discretionary accruals (MJDA).

a double-edged sword. On one hand, firms practising greater degrees of earnings management tend to enjoy greater returns up to a certain point as seen in our analysis on future returns, however at the same time high levels of discretionary accruals damages their earnings quality and heightens risk of potential failure.

4.3 Ethical behaviour of the firms

Furthermore, the earnings managements can be related with the opportunistic behaviour of the firm. We investigate this issue by analysing the effect of ethical behaviour of the firms using ETHICS variable on their future returns and future survivorship. Since the ETHICS is an annual value identical within a year while different across the years, we include this in our overall sample including all periods

$Model: Failed\ firmt = a + b_1EP + b_2BM + b_3PS + b_4MktCap + b_5EBITDA + b_6Sales_g + b_7MJDA + b_8R\&D + b_9 1y \ldots + b_{11} 5y$

	Years							
	2008		2009			2010		
	1y	5y	1y	3y	5y	1y	3y	5y
Const	**5.54* (1.66)**	0.77 (1.1)	3.29 (0.96)	1.24 (0.91)	−0.34 (−0.3)	1.12 (0.46)	1.35 (0.94)	1.77 (1.57)
EP	−16.07 (−1.01)	**−5.86** (−2.37)**	−7.98 (−0.56)	**−13.27* (−1.73)**	−4.81 (−1.07)	−4.52 (−0.47)	**−12.41** (−2.08)**	0.41 (0.1)
BM	−1.66 (−1.32)	0.19 (1.09)	−1.17 (−0.84)	0.18 (0.31)	0.47 (0.85)	−1.14 (−0.74)	−0.36 (−0.37)	−0.45 (−0.58)
PS	0.4 (0.78)	−0.21 (−1.18)	−0.28 (−0.65)	−0.27 (−1.35)	0.03 (0.27)	−0.34 (−1.29)	−0.06 (−0.53)	0.01 (0.12)
MktCap	**−2.17* (−1.81)**	**−0.48* (−1.89)**	−0.77 (−0.84)	−0.53 (−1.38)	−0.43 (−1.3)	−0.02 (−0.03)	**−0.86** (−2.29)**	**−0.91*** (−3.07)**
EBITDA	0.08 (1.44)	−0.01 (−0.58)	−0.05 (−0.52)	−0.01 (−0.26)	−0.01 (−0.4)	−0.09 (−0.29)	−0.04 (−0.31)	−0.01 (−0.36)
$Sales_g$	**−4.3* (−1.85)**	0.15 (0.78)	−1.43 (−1.23)	0.51 (0.99)	0.01 (0.02)	−1.06 (−0.65)	−0.05 (−0.1)	−0.82 (−1.63)
MJDA	**0.18* (1.85)**	0.03 (1.14)	−0.06 (−0.32)	0.14 (1.38)	−0.04 (−0.45)	0.19 (0.97)	**0.36*** (2.78)**	**0.27* (1.67)**
R&D	**−88.16*** (−3.04)**	0.4 (0.18)	**−95.25*** (−3.09)**	−3.4 (−0.84)	−2.49 (−0.78)	**−64*** (−3.92)**	−2.4 (−0.91)	−0.01 (0)
3m	0.05 (1.57)	−0.01 (−0.78)	0.02 (0.68)	**0.02* (1.92)**	**0.03*** (2.72)**	0.03 (1.51)	0 (−0.45)	0 (0.29)
6m	0 (−0.27)	0 (1.35)	**−0.04** (−2.02)**	0.01 (1.15)	0 (0.33)	−0.01 (−0.95)	0.01 (0.65)	0 (0.45)
1y		0 (0.28)		**−0.03*** (−4.2)**	0 (−0.35)		**−0.03*** (−4.52)**	0 (0)
3y		**−0.01*** (−5.85)**			**−0.02*** (−5.61)**			**−0.02*** (−5.26)**
χ^2	**38.78*****	**78.62*****	**60.26*****	**52.54*****	**77.59*****	**57.14*****	**49.95*****	**78.30*****
Pseudo R^2	0.57	0.20	0.63	0.26	0.29	0.49	0.21	0.24
N	350	350	241	241	241	275	275	275

The table shows the valuation and accounting variables' effect on the technology firms' returns from 2008 to 2010. The dependent variables are dummy variables indicating 1 if the technology firm fails in 1 year (1y), and 3 years (3y), and 5 year (5y) and 0 otherwise. Then we use the earnings to price ratio (EP), book to market ratio (BM), price to sales ratio (PS), market capitalization (MktCap), EBITDA, sales growth ($Sales_g$), modified Jones for discretionary accruals (MJDA), and research and development (R&D) as our independent variables. The values in parentheses are the t-values to the corresponding coefficients.

**Significance at 10% level.*
***Significance at 5% level.*
****Significance at 1% level.*

Table 5.
Logistic analysis of failed NASDAQ firms using modified Jones model for discretionary accruals (MJDA).

$$MODEL: r_{a_{t+p}} = a + b_1EP + b_2BM + b_3PS + b_4MktCap + b_5EBITDA + b_6Sales_g + b_7MJDA + b_8R\&D + b_9ETHICS$$

	3m	6m	1y	3y	5y
Const	−5.69 (−0.18)	**−147.8**** **(−2.05)**	**−971.81***** **(−7.98)**	−263.65 (−1.16)	**−1137.59***** **(−3.18)**
EP	−17.02 (−1.58)	**49.28**** **(2.03)**	**174.69***** **(4.26)**	**179.28**** **(2.35)**	**327.49***** **(2.72)**
BM	−1.63 (−1.5)	**15.35***** **(6.26)**	**23.33***** **(5.64)**	**25.51***** **(3.31)**	**21.78*** **(1.79)**
PS	−0.27 (−0.73)	−0.58 (−0.7)	1.58 (1.12)	−0.07 (−0.03)	−3.83 (−0.92)
MktCap	0.55 (0.55)	−0.71 (−0.32)	−2.08 (−0.55)	−4.01 (−0.57)	13.44 (1.21)
EBITDA	−0.06 (−1.07)	−0.16 (−1.29)	−0.16 (−0.77)	−0.2 (−0.52)	0.12 (0.21)
$Sales_g$	1.18 (0.83)	0.52 (0.16)	−2.97 (−0.55)	−7.71 (−0.76)	−20.39 (−1.28)
MJDA	0.2 (1.07)	**1.3***** **(3.1)**	**1.42**** **(2.01)**	**4.21***** **(3.19)**	0.32 (0.15)
R&D	8.64 (0.84)	**90.11***** **(3.9)**	**205.73***** **(5.27)**	**231.66***** **(3.19)**	**200.81*** **(1.75)**
ETHICS	2.64 (0.45)	**27.78**** **(2.09)**	**179.05***** **(7.97)**	52.32 (1.25)	**211.71***** **(3.2)**
F-test	1.49	**13.43*****	**21.88*****	**6.69*****	**4.01*****
Adjusted R^2	0.01	0.11	0.18	0.06	0.03
N	866	866	866	866	866

The table shows the valuation and accounting variables' effect on the technology firms' returns from 2008 to 2010. The dependent variables are 3 months (3m), 6 months (6m), 1 year (1y), 3 years (3y), and 5 years (5y) returns of the technology firms. Then we use the earnings to price ratio (EP), book to market ratio (BM), price to sales ratio (PS), market capitalization (MktCap), EBITDA, sales growth ($Sales_g$), modified Jones for discretionary accruals (MJDA), research and development (R&D), and firm's ethical behaviour score (ETHICS) as our independent variables. The values in parentheses are the t-values to the corresponding coefficients.

**Significance at 10% level.*

***Significance at 5% level.*

****Significance at 1% level.*

Table 6.

Full multivariate analysis of periodic returns of NASDAQ technology firms—using modified Jones model for discretionary accruals (MJDA).

(2008, 2009, and 2010) and run linear and logistic regressions as in our Sections 4.2 and 4.3. The results for the future returns and future survivorships are shown in **Tables 5** and **6**, respectively.

In **Table 5**, we do find highly similar relationship of EP, BM, MJDA and R&D with the future returns as in our **Table 3**. However, using the overall sample in **Table 5** shows more significant effects of these variables on the future returns. Then we find that the ETHICS has positive effect on the future returns in technology firms. In other words, the ethical behaviour of the technology firms tend to increase the future returns. Thus, the opportunistic behaviour of the technology firms is likely to decrease their future returns.

The similar relationship of EP, BM, PS, MktCap, Salesg, MJDA, and R&D on future survivorships are also found between **Tables 4** and **6** while the latter one using overall data tend to show more significant relationships. However, in this case, we do not find highly significant effect of ETHICS on the future survivorships of the technology firms. Therefore, the opportunistic behaviour of the technology firms are more likely to reduce their future returns while leaving their future survivorships not significantly affected.

4.4 Robustness check

To provide a robustness check on the results obtained, we have compare the output generated when running our returns and survivorship analyses using the Modified Jones Discretionary Accruals (MJDA) method. As shown in **Table 7**, there is no difference between the two results.

From **Table 7**, we can deduce that our results in evaluating predictors of future performance are robust in being replicated across the MJDA and JDA procedures. Each regression model provides the same significance and similar explanatory power through their R-squared. All variables across the two methods possess the same signage of coefficients and remain within the statistical significance zone. In testing the robustness of our results in finding predictors of the survivorship of tech-firms, we find similar result using the JDA as presented in **Table 8**. All models remain significant just as in the MJDA results and the variables hold the same meaning within the outputs. These confirm that our results are consistent with the initial findings. We further perform our robustness check on the results from **Tables 7** and **8** where we used our overall data including ETHICS.

$Model: Failed\ firmt = a + b_1EP + b_2BM + b_3PS + b_4MktCap + b_5EBITDA$
$+ b_6Sales_g + b_7MJDA + b_8R\&D + b_9ETHICS + b_{10}6m ... + b_{13}5y$

	6m	**1y**	**3y**	**5y**
Const	−4.75 (−0.33)	9.48 (1.02)	6.25 (1.37)	**7.22* (1.94)**
EP	**−26.86** (−2.33)**	**−12.62** (−2.17)**	**−7.12*** (−2.81)**	**−5.29*** (−3.14)**
BM	−0.54 (−0.57)	**−1.22* (−1.91)**	−0.27 (−1.4)	0.06 (0.54)
PS	0.01 (0.06)	−0.21 (−1.41)	−0.16** (−2.15)	−0.02 (−0.42)
MktCap	0.06 (0.11)	−0.44 (−1.25)	**−0.49*** (−3.08)**	**−0.61*** (−4.73)**
EBITDA	0.01 (0.72)	0 (0.08)	0 (−0.41)	−0.01 (−0.64)
$Sales_g$	**−2.32*** (−3.42)**	−0.97 (−1.33)	−0.12 (−0.61)	0.08 (0.58)
MJDA	**0.13* (1.77)**	**0.09* (1.71)**	0.03 (1.08)	0.02 (1.1)
R&D	**−53.67*** (−3.8)**	**−73.42*** (−6.29)**	**−5.84*** (−3.49)**	−1.23 (−1.03)
ETHICS	0.69 (0.26)	−1.3 (−0.76)	−0.99 (−1.17)	**−1.18* (−1.71)**
χ^2	**44.69*****	**138.67*****	**44.01*****	**48.06*****
Pseudo R^2	0.38	0.48	0.06	0.05
N	866	866	866	866

The table shows the valuation and accounting variables' effect on the technology firms' returns from 2008 to 2010. The dependent variables are dummy variables indicating 1 if the technology firm fails in 6 months (6m), 1 year (1y), and 3 years (3y), and 5 year (5y) and 0 otherwise. Then we use the earnings to price ratio (EP), book to market ratio (BM), price to sales ratio (PS), market capitalization (MktCap), EBITDA, sales growth ($Sales_g$), Modified Jones for discretionary accruals (MJDA), research and development (R&D), and firm's ethical behaviour score (ETHICS) as our independent variables. The values in parentheses are the t-values to the corresponding coefficients.
**Significance at 10% level.*
***Significance at 5% level.*
****Significance at 1% level.*

Table 7.
Full logistic analysis of failed NASDAQ firms using modified Jones model for discretionary accruals (MJDA).

$MODEL: r_{a_{t+p}} = a + b_1EP + b_2BM + b_3PS + b_4MktCap + b_5EBITDA + b_6Sales_g + b_7JDA + b_8R\&D.$									
	Years								
	2008				**2009**		**2010**		
	3m	**6m**	**1y**	**3y**	**3m**	**1y**	**3m**	**1y**	**5y**
Const	−4.46 (−0.76)	4.44 (0.3)	2.35 (0.09)	22.85 (0.51)	9.43 (1.56)	2.63 (0.13)	**14.36**** **(2.24)**	−1.73 (−0.14)	**−76*** **(−1.82)**
EP	−9.76 (−0.67)	48.08 (1.3)	**253.12***** **(3.86)**	**221.03*** **(1.96)**	−28.35 (−1.27)	−117.9 (−1.52)	−44.95 (−1.44)	15.39 (0.25)	22.41 (0.11)
BM	−0.59 (−0.42)	**12.86***** **(3.66)**	**19.89***** **(3.19)**	**23.38**** **(2.18)**	**6.37*** **(1.89)**	**22.56*** **(1.93)**	−0.6 (−0.13)	−5.35 (−0.57)	−1.98 (−0.06)
PS	0.44 (0.4)	0.53 (0.19)	1.74 (0.35)	**19.95**** **(2.36)**	0.42 (0.69)	**5.15**** **(2.46)**	−0.54 (−1.12)	0.63 (0.66)	−5.09 (−1.63)
MktCap	**4.74**** **(2.51)**	3.6 (0.76)	−1.08 (−0.13)	−5.44 (−0.38)	**−2.84*** **(−1.69)**	−5.58 (−0.96)	−0.8 (−0.47)	−2.94 (−0.88)	**33.93***** **(3.07)**
EBITDA	−0.27 (−1.1)	−0.63 (−1.01)	−0.71 (−0.64)	−2.61 (−1.36)	−0.07 (−0.46)	−0.41 (−0.8)	−0.04 (−0.76)	−0.08 (−0.79)	0.22 (0.64)
$Sales_g$	2.23 (1.1)	−3.71 (−0.73)	−5.91 (−0.65)	0.31 (0.02)	1.2 (0.4)	0.93 (0.09)	6.17 (1.64)	**−16.22**** **(−2.19)**	−17.08 (−0.7)
JDA	0.33 (1.34)	**1.27**** **(2.05)**	0.86 (0.78)	**4.46**** **(2.36)**	−0.24 (−0.52)	0.9 (0.56)	−1.07 (−1.27)	2.08 (1.26)	−2.77 (−0.51)
R&D	−5.25 (−0.29)	**155.82***** **(3.36)**	**266.04***** **(3.23)**	**334.49**** **(2.36)**	**56.41***** **(3.15)**	**286.01***** **(4.61)**	4.51 (0.29)	30.96 (0.99)	127.8 (1.24)
F-test	**2.55****	**4.80*****	**6.25*****	**3.48*****	**2.47****	**3.80*****	1.36	1.18	**1.95***
Adjusted R^2	0.03	0.08	0.11	0.05	0.05	0.09	0.01	0.01	0.03
N	350	350	350	350	241	241	275	275	275

The table shows the valuation and accounting variables' effect on the technology firms' returns from 2008 to 2010. The dependent variables are 3 months (3m), 6 months (6m), 1 year (1y), and 3 years (3y) returns of the technology firms. Then we use the earnings to price ratio (EP), book to market ratio (BM), price to sales ratio (PS), market capitalization (MktCap), EBITDA, sales growth ($Sales_g$), Jones for discretionary accruals (JDA), and research and development (R&D) as our independent variables. The values in parentheses are the t-values *to the corresponding coefficients.*
**Significance at 10% level.*
***Significance at 5% level.*
****Significance at 1% level.*

Table 8.
Multivariate analysis on periodic returns of NASDAQ technology firms—using Jones model for discretionary accruals (JDA).

5. Conclusion and implication of study

We find that, during the global financial crisis periods, the technology firms have larger returns with undervaluation, larger firm size, and more discretionary earnings and R&D which increases more with longer terms. On the other hand, these firms have greater survivorships when they are undervalued, larger in size, have more R&D but with less discretionary earnings (DA). DA is a double-edged sword for the technology firms since it has positive and negative effects on the returns and survivorships, respectively. R&D is a positive component for both returns and survivorships of these firms. The moral hazard (ETHICS) tend to reduce the returns of these firms but do not have significant effect on their survivorships.

The key implications for investors, equity holders and creditors derived from our results and analyses are threefold. Firstly, the alignment of our results and underlying expectation that certain variables should demonstrate significant explanatory power at various times of crisis illustrates the strong relevance of contingency theory in evaluating the phenomenon of tech firms during periods of financial turmoil. Secondly, through our analysis of results, we reiterate that traditional relationships of accounting valuation and earnings management measures may not always hold especially during crisis periods. Thirdly, from a general perspective, the results indicate that the variables employed in this study demonstrate greater predictive power in determining the phenomena of future tech-firm failure than performance.

Author details

Oliver Neoh and Matthias Nnadi*
School of Management, Cranfield University, Bedfordshire, UK

*Address all correspondence to: matthias.nnadi@cranfield.ac.uk

IntechOpen

© 2019 The Author(s). Licensee IntechOpen. This chapter is distributed under the terms of the Creative Commons Attribution License (http://creativecommons.org/licenses/by/3.0), which permits unrestricted use, distribution, and reproduction in any medium, provided the original work is properly cited.

References

[1] Francis BB, Hasan I, Zhou M. Strategic conservative earnings management of technology firms: Evidence from the IPO market. Financial Markets, Institutions and Instruments. 2012;**21**(5):261-293

[2] Demeres E, Lev B. A rude awakening: Internet shakeout in 2000. Review of Accounting Studies. 2001;**6**(2–3): 331-359

[3] Fry-Mckibbin R, Hsiao CYL, Tang C. Contagion and global financial crises: Lessons from nine crisis episodes. Open Economies Review. 2014;**25**:521-570

[4] Clark E, Zenaidi A, Baccar S. Market valuation of technology stocks before and after the crash. International Journal of Business. 2007; **12**(2):210-217

[5] Goodnow WE. The contingency theory of education. International Journal of Lifelong Education. 1982; **1**(4):341-352

[6] Gavira RL, Omoteso K. Perceptions of the usefulness of virtual learning environments in accounting education: A comparative evaluation of undergraduate accounting students in Spain and England. Accounting Education: An International Journal. 2013;**22**(5):445-466

[7] Nicolaou AI. A contingency model of perceived effectiveness in accounting information systems: Organizational coordination and control effects. International Journal of Accounting Information Systems. 2000;**1**:91-105

[8] Khazanchi D. Information technology (IT) appropriateness: The contingency theory of 'fit' and IT implementation in small and medium enterprises. Journal of Computer Information Systems. 2005;**45**(3):88-95

[9] Smith JA, Cordina R. The role of accounting in high-technology investments. The British Accounting Review. 2014;**46**:309-322

[10] Damodaran A. The Dark Side of Valuation: Valuing Young, Distressed and Complex Businesses. 2nd ed. New Jersey: Pearson Education; 2010

[11] Ritter JR, Welch I. A review of ipo activity, pricing, and allocations. Journal of Finance. 2002;**57**(4):1795-1827

[12] Carmassi J, Gros D, Micossi S. The global financial crisis: Causes and cures. Journal of Common Market Studies. 2009;**47**(5):977-996

[13] McKibbin WJ, Stoeckel A. The global financial crisis: Causes and consequences. Asian Economic Papers. 2010;**9**(1):54-86

[14] Ofek E, Richardson M. DotCom mania: The rise and fall of internet stock prices. The Journal of Finance. 2003; **58**(3):1113-1137

[15] Healy PM, Wahlen JM. A review of the earnings management literature and its implications for standard setting. Accounting Horizons. 1999;**13**(4): 365-383

[16] Leuz C, Nanda D, Wysocki P. Earnings management and investor protection: An international comparison. Journal of Financial Economics. 2003;**69**(3):505-528

[17] Kwon SS, Yin QJ, Han J. The effect of differential accounting conservatism on the "over-valuation" of high-tech firms relative to low-tech firms. Review of Quantitative Finance and Accounting. 2006;**27**(2):143-173

[18] Bhoraj S, Hribar P, Picconi M, McInnis J. Making sense of cents: An

examination of firms that marginally miss or beat analyst forecasts. Journal of Finance. 2009;**64**(5):2361-2388

[19] Davis AK. The value relevance of revenue for internet firms: Does reporting grossed-up or barter revenue make a difference. Journal of Accounting Research. 2002;**40**(2): 445-477

[20] Bruno AV, Leidecker JK, Harder JW. Why Firms Fail. Business Horizons. 1987;**30**(2):50-58

[21] Gollotto JC, Kim S. Market valuation of dot com companies; R&D versus hype. Managerial Finance. 2003;**29**(11): 61-72

Section 4

Banking

Chapter 6

Banks Financial State Analysis and Bankruptcy Risk Forecasting with Application of Fuzzy Neural Networks

Yuriy Zaychenko, Michael Zgurovsky and Galib Hamidov

Abstract

The problem of banks bankruptcy risk forecasting under uncertainty is considered. For its solution, the application of computational intelligence methods fuzzy neural networks ANFIS and TSK and inductive modeling method FGMDH was suggested and explored. Experimental investigations were carried out and estimation of the efficiency of the suggested methods was performed at the problems of bankruptcy risk forecasting for Ukrainian and leading European banks. The efficiency comparison with classic statistical methods such as ARMA, logit, and probit models was fulfilled. The comparative experiments with rating system CAMELS and matrix method were carried out. In general, the comparative analysis had shown that fuzzy forecasting methods and techniques give better results than conventional crisp methods for forecasting bankruptcy risk. On the whole, the conclusions of experiments with European banks completely confirmed the conclusions of experiments with Ukrainian banks. But at the same time, the crisp methods are more simple in implementation and demand less time for their adjustment. The set of informative bank financial factors for bankruptcy risk forecasting was determined and estimated.

Keywords: banks bankruptcy risk, forecasting, FNN, FGMDH, ARMA, logit, probit model, rating system CAMELS

1. Introduction

The problems of banks financial state analysis and bankruptcy risk forecasting are of great importance. The opportune discovery of coming bankruptcy allows top bank managers to make urgent decisions for preventing the bankruptcy. Nowadays, there are a lot of methods and techniques of banks state analysis and determination of bank rating—WEB Money, CAMEL [1], Moody's S&P, etc. But their common drawback is that all of them work with complete and reliable data and cannot give correct results in case of incomplete and unreliable input data. This is especially actual for the Ukrainian banking system where bank managers often provide the incorrect reports about bank financial state to obtain new credits and loans.

Therefore, it is very important to create new methods for banks bankruptcy risk forecasting under uncertainty. The main goal of present investigation is to consider

IntechOpen

and estimate novel methods of bank financial state analysis and bankruptcy risk forecasting under uncertainty and compare with classical methods. The implementation and assessment of the efficiency of the suggested methods are performed at the problems of bankruptcy risk forecasting for Ukrainian and European banks.

2. Bankruptcy risk forecasting of Ukrainian banks

2.1 Problem statement

As it is well known, the year 2008 was the crucial year for the bank system of Ukraine. If the first three quarters were periods of fast growth and expansion, the last quarter became the period of collapse in the financial sphere. A lot of Ukrainian banks faced the danger of coming default.

For this research, the quarterly accountancy bank reports used were obtained from National bank of Ukraine site. For analysis, the financial indices of 170 Ukrainian banks were taken up to the date January 01, 2008 and July 01, 2009, that is, about two years before crises and just before the start of crises [2].

The important problem that occurred before the start of the investigations is which financial indices are to be used for better forecasting of possible bankruptcy. Thus, another goal of this exploration was to detect the most relevant financial indicators for obtaining maximal accuracy of forecasting.

For analysis, the following indicators of banks accountancy were considered:

assets, capital, financial means, and their equivalents; and

physical person's entities, juridical person's entities, liabilities, and net incomes (losses).

The collected indicators were used for analysis by fuzzy neural networks as well as classic statistical methods. As output data of models for Ukrainian banks were two values:

1, if the significant worsening of bank financial state is not expected in the nearest future

−1, if the bank bankruptcy is expected in the nearest future.

2.2 FNN TSK model and hybrid training algorithm

For forecasting of banks bankruptcy risk, the application of fuzzy neural networks (FNN) ANFIS and TSK was suggested [3]. The application of FNN is determined by following reasons:

the capability to work with incomplete and unreliable information under uncertainty; and

the capability to use expert information in the form of fuzzy inference rules.

Let us consider the mathematical model and training algorithm of a fuzzy neural network TSK (Takagi, Sugeno, Kang'a), which is generalization of the neural network ANFIS. The rule base of FNN TSK with M rules and N variables can be written as follows [3]:

$$R_1 : if\ x_1 \in A_1^{(1)}, x_2 \in A_2^{(1)}, \ldots, x_n \in A_n^{(1)}\ \text{then}\ y_1 = p_{10} + \sum_{j=1}^{N} p_{1j} x_j;$$

$$R_M : if\ x_1 \in A_1^{(M)}, x_2 \in A_2^{(M)}, \ldots, x_n \in A_n^{(M)}\ \text{then}\ y_M = p_{M0} + \sum_{j=1}^{N} p_{Mj} x_j,$$

where $A_i^{(k)}$ is the value of linguistic variable x_i for the rule R_k with membership function (MF) of the form

$$\mu_A^{(k)}(x_i) = \frac{1}{1 + \left(\frac{x_i - c_i^{(k)}}{\sigma_i^{(k)}}\right)^{2b_i^{(k)}}} \tag{1}$$

$$i = \overline{1, N}; k = \overline{1, M}.$$

At the intersection of the TSK network rule conditions, R_k MF is defined as a product

$$\mu_A^{(k)}(x) = \prod_{j=1}^{N} \left[\frac{1}{1 + \left(\frac{x_j - c_j^{(k)}}{\sigma_j^{(k)}}\right)^{2b_j^{(k)}}} \right]. \tag{2}$$

With M inference rules, the general output of FNN TSK is determined by the following formula:

$$y(x) = \frac{\sum_{k=1}^{M} w_k y_k(x)}{\sum_{k=1}^{M} w_k}, \tag{3}$$

where $y_k(x) = p_{k0} + \sum_{j=1}^{N} p_{kj} x_j$. The weights in this expression are interpreted as the degrees of fulfillment of rule antecedents (conditions): $w_k = \mu_A^{(k)}(x)$, which are given by (2).

The fuzzy neural network TSK, which implements the output in accordance with (3), represents a multilayer network whose structure is shown in **Figure 1**.

This network has five layers with the following functions:

1. The first layer performs fuzzification separately for each variable $x_i, i = 1, 2, \ldots, N$, defining for each rule the k value MF $\mu_A^{(k)}(x_i)$ in accordance with the fuzzification function, which is described, for example, by Gaussian or bell-wise function. This is a parametric layer with parameters $c_j^{(k)}, \sigma_j^{(k)}, b_j^{(k)}$, which are subject to adjustment in the learning process.

2. The second layer performs the aggregation of individual variables x_i, determining the resulting degree of membership $w_k = \mu_A^{(k)}(x)$ for the vector x. This is not a parametric layer.

3. The third layer is a function generator TSK, wherein the output values are calculated $y_k(x) = p_{k0} + \sum_{j=1}^{N} p_{kj} x_j$. At this layer, also functions formed in the previous layer $y_k(x)$ on w_k are multiplied. This is a parametric layer, wherein the adaptation of linear parameters (weight) p_{k0}, p_{kj} for $j = \overline{1, N}, k = \overline{1, M}$, is carried out determining the rules output functions.

4. The fourth layer consists of two summing neurons, one of which calculates the weighted sum of the signals $y_k(x)$, and the second one calculates the sum of the weights $\sum_{k=1}^{M} w_k$.

5. The fifth layer is composed of a single output neuron. In it, weight normalizing is performed and the output signal determined in accordance with the expression:

$$y(x) = \frac{f_1}{f_2} = \frac{\sum_{k=1}^{M} w_k y_k(x)}{\sum_{k=1}^{M} w_k} \tag{4}$$

This is also nonparametric layer.

From this description follows that TSK fuzzy network contains only two parametric layers: first and third, the parameters of which are determined in the training process. Parameters of the first layer $\left(c_j^{(k)}, \sigma_j^{(k)}, b_j^{(k)}\right)$, we call nonlinear, and the parameters of the third layer $\left\{p_{kj}\right\}$—linear weights. The general expression for the functional dependence (4) for the network TSK is defined as follows:

$$y(x) = \frac{1}{\sum_{k=1}^{M}\prod_{j=1}^{M}\mu_A^{(k)}(x_j)} \sum_{k=1}^{M}\left(p_{k0} + \sum_{j=1}^{N} p_{kj} x_j\right)\prod_{j=1}^{N}\mu_A^{(k)}(x_j)$$

If we assume that at any given time moment, the nonlinear parameters are fixed, then the function $y(x)$ would be linear with respect to the variable x_j.

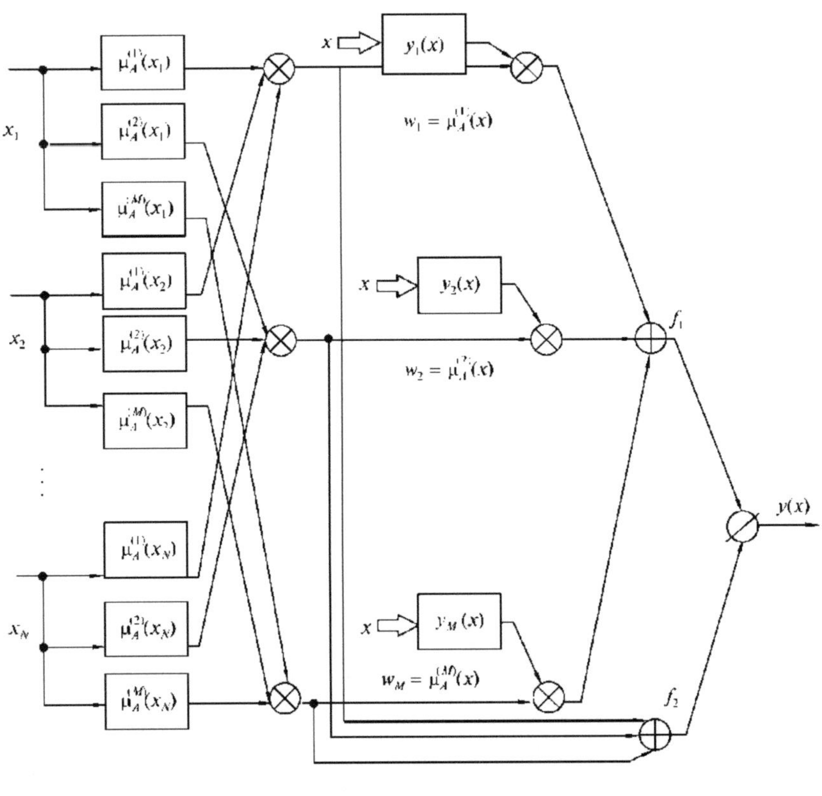

Figure 1.
The structure of TSK fuzzy neural network.

In the presence of N input variables, each rule R_k formulates $(N+1)$ variable $p_j^{(k)}$ of linear dependence $y_k(x)$. If M inference rules are present, then $M(N+1)$ linear network parameters are obtained. In turn, each MF uses three parameters (c, σ, b), which are subject to adjustment. With M inference rules, three MN nonlinear parameters are obtained. In total, this gives $M(4N+1)$ linear and nonlinear parameters that must be determined in the learning process. This is a very large value. In order to reduce the number of parameters for adaptation, we operate with fewer number of MF. In particular, it can be assumed that some of the parameters of one function MF $\mu_A^{(k)}(x_j)$ are fixed, e.g., $\sigma_j^{(k)}$ and $b_j^{(k)}$.

2.2.1 Hybrid learning algorithm for fuzzy neural networks

Considering a hybrid learning algorithm which is used for FNN TSK, all parameters can be divided into two groups. The first group includes linear parameters p_{kj} of the third layer, and the second group includes nonlinear parameters (MF) of the first layer. Adaptation occurs in two stages.

In the first stage after fixing the individual parameters of the membership function by solving a system of linear equations, linear parameters of polynomial p_{kj} are calculated. With the known values of MF dependence, input-output can be represented as a linear form with respect to the parameters p_{kj}:

$$y_k(x) = \sum_{k=1}^{M} w'_k \left(p_{k0} + \sum_{j=1}^{N} p_{kj} x_j. \right) \tag{6}$$

where

$$w'_k = \frac{\prod_{j=1}^{N} \mu_A^{(k)}(x_j)}{\sum_{r=1}^{M} \prod_{j=1}^{N} \mu_A^{(r)}(x_j)}, k = \overline{1, M}. \tag{7}$$

With the dimension L of training sample $\left(x^{(l)}, d^{(l)}\right)$, $(l = 1, 2, \ldots, L)$ and replacement of the network output by expected value $d^{(l)}$, we get a system of L linear equations of the form

$$\begin{bmatrix} w'_{11} & w'_{11}x_1^{(1)} \ldots & w'_{11}x_N^{(1)} \ldots. & w'_{1M} & w'_{1M}x_1^{(1)} \ldots & w'_{1M}x_N^{(1)} \\ w'_{21} & w'_{21}x_1^{(2)} \ldots & w'_{21}x_N^{(2)} \ldots & w'_{2M} & w'_{2M}x_1^{(2)} \ldots & w'_{2M}x_N^{(2)} \\ \ldots & \ldots & \ldots & \ldots & \ldots & \ldots \\ w'_{L1} & w'_{L1}x_1^{(L)} & w'_{L}x_N^{(L)} & w'_{LM} & w'_{LM}x_1^{(L)} & w'_{LM}x_N^{(L)} \end{bmatrix} \times \begin{bmatrix} p_{10} \\ p_{11} \\ \ldots \\ p_{1N} \\ \ldots \\ p_{M0} \\ p_{M1} \\ \ldots \\ p_{MN} \end{bmatrix} = \begin{bmatrix} d^{(1)} \\ d^{(2)} \\ \ldots \\ d^{(L)} \end{bmatrix} \tag{8}$$

where $w'_{\ell i}$ means normalized weight of the i-th rule at presentation of ℓ-th input vector x^{ℓ}. This expression can be written in matrix form:

$$Ap = d.$$

Matrix A dimension is equal to $L(N+1)M$. By thus, a number of rows L usually is much greater than a number of columns $(N+1)M$. The solution of this equations system may be obtained by conventional methods as well as using pseudoinverse matrix A at one step:

$$p = A^{+}d,$$

where A^{+} is a pseudoinverse matrix for matrix A.

In the second stage, after fixing the values of linear parameters p_{kj}, the actual output signals $y^{(\ell)}$, $\ell = 1, 2, \ldots, L$ are determined using a linear equations system:

$$y^{(L)} = Ap. \tag{9}$$

Then, the error vector $\varepsilon = (y - d)$ and the criterion E are calculated:

$$E = \frac{1}{2}\sum_{\ell=1}^{L}\left(y\left(x^{(\ell)}\right) - d^{(\ell)}\right)^2. \tag{10}$$

The error signals are sent through the network backward according to the method of back propagation until the first layer, at which gradient vector components of the objective function with respect to parameters $\left(c_j^{(k)}, \sigma_j^{(k)}, b_j^{(k)}\right)$ are calculated.

After calculating the gradient vector, a step of gradient descent method is made. The corresponding formulas (for the simplest method of the steepest descent) are the following:

$$c_j^{(k)}(n+1) = c_j^{(k)}(n) - \eta_c \frac{\partial E(n)}{\partial c_j^{(k)}} \tag{11}$$

$$\sigma_j^{(k)}(n+1) = \sigma_j^{(k)}(n) - \eta_\sigma \frac{\partial E(n)}{\partial \sigma_j^{(k)}} \tag{12}$$

$$b_j^{(k)}(n+1) = b_j^{(k)}(n) - \eta_b \frac{\partial E(n)}{\partial b_j^{(k)}} \tag{13}$$

where n is a number of iterations.

After verifying the nonlinear parameters, the process of adaptation of linear parameters TSK (first phase) restarts and nonlinear parameters are further adapted (second stage). This cycle continues until all the parameters will be stabilized.

Formulas (11)–(13) require the calculation of the gradient of the objective function with respect to the parameters of the MF. The final form of these formulas depends on the type of MF. For example, if using the generalized bell-wise functions:

$$\mu_A(x) = \frac{1}{1 + \left(\frac{x-c}{\sigma}\right)^{2b}}$$

the corresponding formulas for gradient of the objective function for one pair of data (x, d) take the form [3]:

$$\frac{\partial E}{\partial c_j^{(k)}} = (y(x) - d) \sum_{r=1}^{M} \left(p_{r0} + \sum_{j=1}^{N} p_{rj} x_j \right) \cdot \frac{\partial w'_r}{\partial c_j^{(k)}} \tag{14}$$

$$\frac{\partial E}{\partial \sigma_j^{(k)}} = (y(x) - d) \sum_{r=1}^{M} \left(p_{r0} + \sum_{j=1}^{N} p_{rj} x_j \right) \cdot \frac{\partial w'_r}{\partial \sigma_j^{(k)}} \tag{15}$$

$$\frac{\partial E}{\partial b_j^{(k)}} = (y(x) - d) \sum_{r=1}^{M} \left(p_{r0} + \sum_{j=1}^{N} p_{rj} x_j \right) \cdot \frac{\partial w'_r}{\partial b_j^{(k)}} \tag{16}$$

In the practice of the hybrid learning method implementation, the dominant factor in adaptation is considered to be the first stage in which weights p_{kj} are determined using pseudoinverse in one step. To balance its impact, the second stage should be repeated many times in each cycle.

It is worth to note that the described hybrid algorithm is one of the most effective ways of training fuzzy neural networks. Its principal feature is the division of the process into two stages separated in time. Since the computational complexity of each nonlinear optimization algorithm depends nonlinearly on the number of parameters subject to optimization, the reduction in the dimensions of optimization significantly reduces the total amount of calculations and increases the speed of convergence of the algorithm. Due to this, hybrid algorithm is one of the most efficient in comparison with conventional gradient-based methods.

3. The application of fuzzy neural networks for financial state forecasting

A special software kit was developed for FNN ANFIS and TSK application in bankruptcy risk forecasting problems. As input data, the financial indicators of Ukrainian banks in financial accountant reports were used in the period of 2008–2009 [2] . As the output values were used +1, for bank nonbankrupt and −1, for bank bankrupt. In the investigations, various financial indicators were analyzed, and different number of rules for FNN and the analysis of data collection period influence on forecasting accuracy were performed.

The results of experimental investigations of FNN application for bankruptcy risk forecasting are presented below.

In the first series of experiments, input data at the period of January 2008 were used (that is for two years before possible bankruptcy) and possible banks bankruptcy was forecasted at the beginning of 2010.

Experiment No. 1:

Training sample—120 Ukrainian banks, test sample—50 banks.

Number of rules = 5.

Input data—financial indices (taken from bank accountant reports):

assets, capital, cash (liquid assets), households deposits, liabilities.

The results of application of FNN TSK are presented in **Table 1**.

Results	
Total amount of errors	5
% of errors	10
First type of errors	0
Second type of errors	5

Table 1.
Results of FNN TSK forecasting.

The similar experiments were carried out with FNN ANFIS.

Experiment No. 2:

The goal of the next experiment was to find out the dependence of rule number on predicting accuracy. Input data—the same financial indices as in experiment 1.

The results of application of FNN TSK are presented in **Table 2**.

Results	
Total amount of errors	6
% of errors	12
First type of errors	1
Second type of errors	5

Table 2.
Results of FNN TSK forecasting.

The similar experiments were carried out with FNN ANFIS.

Experiment No. 3:

The comparative analysis of forecasting results versus the number of rules is presented in **Table 3** [4].

Network/number of rules	Total number of errors	% of errors	Number of first type errors	Number of second type errors
ANFIS 5	6	12	0	6
ANFIS 10	7	14	1	6
TSK 5	5	10	0	5
TSK 10	6	12	1	5

Table 3.
Comparative analysis of FNN ANFIS and TSK in dependence on rules number.

Comparing the results in **Table 3**, one may conclude FNN TSK has better accuracy than FNN ANFIS.

The goal of the next experiments was to explore the influence of training and test samples size on accuracy of forecasting.

Experiment No. 4:

Training sample—120 Ukrainian banks, test sample—50 banks, and number of rules = 10.

Input data—financial indicators:

assets, entity, cash (liquid assets), household deposits, and liabilities.

Results	
Total number of errors	7
% of errors	10
First type of errors	1
Second type of errors	6

Table 4.
Results of FNN TSK forecasting.

The results for FNN TSK are presented in **Table 4**.
The similar experiments were carried out with FNN ANFIS.
After analysis of the experimental results the following conclusions were made:

FNN TSK ensures the higher accuracy of risk forecasting than FNN ANFIS;

the variation of the number of rules in the training and test samples makes slight influence on the accuracy of forecasting; and

the goal of the next series of experiments was to determine the optimal input data (financial indicators) for bankruptcy risk forecasting. The period of input data was January 2008.

Experiment No. 5:

Number of banks and rules were the same as in previous experiment 4.

Input data—financial indicators (taken from banks financial accountant reports):

profit of current year, net percentage income, net commission income; and

net expense on reserves and net bank profit/losses.

The results of FNN TSK application are presented in **Table 5**.

Experiment No. 6:

Number of banks and rules were the same as in the previous experiment 5.

Results	
Total number of errors	13
% of errors	19
First type of errors	6
Second type of errors	7

Table 5.
Results of FNN TSK forecasting.

Input data—the financial indicators (taken from banks financial accountant reports):

general reliability factor (own capital/assets);

instant liquidity factor (liquid assets/liabilities);

cross coefficient (total liabilities/working assets);

general liquidity coefficient (liquid assets + defended capital + capitals in reserve fund/total liabilities); and

coefficient of profit fund capitalization (own capital/charter fund).

The results for FNN TSK are presented in **Table 6**.

It is worth to note that these financial indicators are also used as input data in Kromonov's method of banks bankruptcy [5–7], whose results are presented below.

Results	
Total number of errors	7
% of errors	10
First type of errors	1
Second type of errors	6

Table 6.
Results of FNN TSK forecasting.

Experiment No. 7:

Training sample—120 Ukrainian banks and test sample—70 banks.

Number of rules = 5.

Input data—following financial indicators (other than in experiments 5 and 6):

ROE—return on entity (financial results/entity);

ROA—return on assets (financial results/assets);

CIN—incomes-expenses ratio (income/expense);

NIM—net percentage margin; and

NI—net income.

The results of application of FNN TSK for forecasting with these input indicators are presented in **Table 7**.

It should be noted that these indicators are used as input in the method of Euro Money [1].

Results:	
Total number of errors	12
% of errors	17
First type of errors	5
Second type of errors	7

Table 7.
Results of FNN TSK forecasting.

Experiment No. 8:

Training sample—120 Ukrainian banks and test sample—70 banks.

Number of rules = 5.

Input data—financial indicators (banks financial accountant reports):

general reliability factor (own capital/assets);

instant liquidity factor (liquid assets/liabilities);

cross coefficient (total liabilities/working assets);

general liquidity coefficient (liquid assets + defended capital + capitals in reserve fund/total liabilities);

coefficient of profit fund capitalization (own capital/charter fund); and

coefficient entity security (secured entity/own entity).

The results of FNN TSK application with these financial indicators are presented in **Table 8**.

The comparative analysis of forecasting results using different sets of financial indicators are presented in **Table 9**.

Next experiment was aimed on finding the influence of data collection period on the forecasting results. It was suggested to consider two periods: January of 2008 (about 1.5 year before the crisis) and July of 2009 (just before the start of crisis).

Experiment No. 9:

Training sample—120 Ukrainian banks and test sample—70 banks.

Number of rules = 10.

Results	
Total amount of errors	8
% of errors	13
First type of errors	1
Second type of errors	7

Table 8.
Results of FNN TSK forecasting.

Experiment	**Total number of errors**	**% of errors**	**First type of errors**	**Second type of errors**
Experiment 5	13	19	6	7
Experiment 6	7	10	1	6
Experiment 7	12	17	5	7
Experiment 8	8	13	1	7

Table 9.
The dependence of forecasting accuracy on sets of input financial indices.

Experiment/number rules	Total number of errors	First type of errors	Second type of errors	Total % of errors
January 1, 2008 5 rules	7	0	7	10
July 1, 2009 5 rules	5	0	5	7
July 1, 2009 10 rules	7	3	4	10

Table 10.
Accuracy of forecasting in dependence on data collection period.

Input data—financial indices, the same as in experiment 8.

In **Table 10**, the comparative results of forecasting versus period of input data are presented.

3.1 The application of fuzzy GMDH for financial state forecasting

In the process of investigations, fuzzy group method of data handling (FGMDH) was also suggested for financial state of Ukrainian banks forecasting [3]. GMDH is the inductive modeling method that enables to construct a model automatically by experimental data [3]. As input data, the same indices were used as in the experiments with FNN TSK.

In **Table 11**, the forecasting accuracy of FGMDH is presented in dependence on input data collection period.

If we compare the results of FGMDH with the results of FNN TSK, one can see that FNN TSK gives better results for short-term risk forecasting (one year before possible bankruptcy) while FGMDH has better accuracy using older input data and so it has advantages in long-term forecasting (2 or more years).

3.2 The generalized analysis of crisp and fuzzy forecasting methods

In the concluding experiments, the comparative analysis of application of all the considered methods was carried out. The following methods were considered [4]:

fuzzy neural network ANFIS;

fuzzy neural network TSK; and

crisp forecasting methods: Kromonov's method and Byelorussian bank association method.

Input data period	Total error number	% of errors	First type of errors	Second type of errors
2004	10	14	3	7
2005	9	13	3	6
2006	8	11.4	3	5
2007	7	10	2	5
2008	6	8.5	1	5
2009	6	8.5	2	4

Table 11.
Comparative results of forecasting using method FGMDH in dependence on period of input data collection.

Method/period	Total amount of errors	% of errors	First type of errors	Second type of errors
ANFIS	7	10	1	6
TSK	5	7	0	5
GMDH	6	8.5	1	5
Kromonov's method	10	15	5	5
BBA method	10	15	2	8

Table 12.
Comparative results analysis of various forecasting methods.

As input data, the financial indices of Ukrainian banks on July 2007 year were used. The results of application of all methods for bankruptcy risk analysis are presented in **Table 12**.

3.3 Application of rating system CAMEL for assessment of financial state of Ukrainian banks

The most widely used approach of banks financial state analysis and bankruptcy risk forecasting is based on the application of rating systems. The determination of bank rating is one of the methods that enables to obtain complex financial assessment of bank financial state and compare them. There are various private and official banks rating systems. The most known of them are systems developed by world leaders in this sphere-rating companies Fitch, Standard & Poor's, Moody's, etc. Officially recognized banks rating system that is widely used in the world is system CAMELS. It's American rating system was developed and implemented by Federal reserve System (FRS) and Federal Deposit Insurance Corporation (FDIC) in 1978 [1].

Supervision over banks activity based on risk estimation by system CAMELS lies in determination of general bank state using the common criteria that defines all aspects and spheres of bank activity. This system is also widely used in Ukraine by National Bank of Ukraine (NBU) according to developed "Statement of order of rating estimates determination by rating system 'CAMELS'."

Rating system CAMELS allows NBU to estimate general financial state and stability of banking system of Ukraine. Such assessment enables to obtain information for priority determination in banking supervision activity and necessary materials and financial resources for performing adequate control over banking system.

At the same time, system CAMELS envisages the detail supervision and analysis of bank state. Such analysis may be performed only while complex inspecting checking of bank activity, which enables to determine how the top managers analyze and control bank risks.

The base of rating system, CAMELS, is risk assessment and determination of rating estimates by each component of the system: capital adequacy, assets quality, management, liquidity, and sensitivity.

Due to rating system, each bank obtain digital rating by all six components, and integral (complex) rating estimate is determined on the base of rating estimates of all components. Components of rating system are estimated by 5 balls scale in which estimate 1 is the highest, and estimate 5 is the lowest one. Integral estimate is also determined by 5 balls scale. Banks that obtained integral rating estimate 1 or 2 are considered reliable by all the factors capable to overcome economic depression and its management believed to be qualified.

Banks that got integral estimate 3 have substantial drawbacks, which may lead to serious problems with liquidity and solvency if these drawbacks won't be corrected in proper time. In this case, bank's supervision system should give recommendations to managers how to overcome existing problems.

Banks that got rating estimate 4 or 5 have serious problems, which demand strict supervision and special urgent actions to prevent possible bankruptcy (see **Table 13**).

3.3.1 Comparative experimental investigations of efficiency of bankruptcy risk forecasting by systems CAMEL and FNN TSK

For bankruptcy risk forecasting in banking sphere of Ukraine, a special data set was collected consisting of 160 Ukrainian banks in the period 2012–2014 . It was divided into training and test subsamples in ratio 70/30 for FNN TSK, i.e., training samples consisted of 110 banks and test samples of 50 banks. The experiments were carried out, and the following results were obtained for FNN TSK (in average, 20 experiments were performed for each year and rules number), which are presented in **Table 2**. The data were collected in the year indicated in the first column, and the forecasting was made for next year, e.g., 2012-5—means bankruptcy risk forecasting in 2013 with use of 5 rules in FNN TSK by data of 2012. Two types of experiments were carried out with fixed parameters of membership functions (MF) and with training MF parameters. In **Table 14**, forecasting results for FNN TSK with adaptation of parameters are presented and in **Table 15** with fixed parameters values.

In **Table 16**, forecasting results for FNN TSK with triangular MF are presented, while in **Table 17** with trapezoidal MF.

	Bank integral rating				
	"1"	"2"	"3"	"4"	"5"
Bank financial state	Bank is stable, reliable, and has skilled management		Bank has substantial drawbacks, which may lead to serious problems in future.	Bank faces very serious problems, which may lead to bankruptcy.	
Control from banks supervision service			Bank supervision service should give clear instructions to overcome existing problems.	Banks need urgent actions to prevent possible bankruptcy.	
Application of special actions			Proper influence actions are performed over bank due to demands of existing regulation laws of NBU.		

Table 13.
Comparative results analysis of various forecasting methods.

Year and number of rules	General number of errors	% of errors	Number of first type errors	Number of second type errors
2012—5	6	12	0	6
2013—5	9	18	0	9
2014—5	8	16	1	7
2012—10	7	14	2	5
2013—10	5	10	0	5
2014—10	10	20	4	6

Table 14.
Forecasting results for FNN TSK with FM parameters' adaptation.

Year and number of rules	General number of errors	% of errors	Number of first type errors	Number of second type errors
2012—5	8	16	1	7
2013—5	8	16	0	8
2014—5	9	18	1	8
2012—10	9	18	3	6
2013—10	7	14	1	6
2014—10	11	22	4	7

Table 15.
Forecasting results for FNN TSK with fixed parameters.

Year and number of rules	General number of errors	% of errors	Number of first type errors	Number of second type errors
2012—5	9	18	1	8
2013—5	7	14	1	6
2014—5	9	18	0	9
2012—10	11	22	4	7
2013—10	10	18	2	8
2014—10	13	26	4	9

Table 16.
Forecasting results for FNN TSK with triangular MF.

Year and number of rules	General number of errors	% of errors	Number of first type errors	Number of second type errors
2012—5	7	14	1	8
2013—5	8	16	1	6
2014—5	5	10	0	9
2012—10	9	18	4	7
2013—10	12	24	2	8
2014—10	11	22	4	9

Table 17.
Forecasting results for FNN TSK with trapezoidal MF.

The application of well-known matrix method by Nedosekin [11, 12] with level (threshold) of cut 0.7 gave the following results presented in **Table 18.**

Results obtained by rating system CAMEL are presented in **Table 19** (threshold of 4).

Year and number of rules	General number of errors	% of errors	Number of first type errors	Number of second type errors
2012—0,7	14	28	8	6
2013—0,7	11	22	3	8
2014—0,7	16	32	9	7

Table 18.
Forecasting results of matrix method by Nedosekin.

Year and number of rules	General number of errors	% of errors	Number of first type errors	Number of second type errors
2012—4	12	24	7	5
2013—4	14	28	5	9
2014—4	9	18	5	4

Table 19.
Forecasting results of rating system CAMEL.

In **Figure 2**, the probability of error for different forecasting methods and various MF is presented.

In **Figure 3**, dependence of error probability versus number of rules in FNN TSK is presented.

Analyzing the performed experiments, the following conclusions may be made.

1. The experimental investigations of efficiency of different forecasting methods FNN TSK, matrix method of Nedosekin, and rating system CAMELS were carried out for the problem of bankruptcy risk forecasting of Ukrainian banks.

2. Results obtained by FNN TSK are the best (min error%). Mean forecasting accuracy by TSK with Gaussian MF is equal to 85%, with trapezoidal MF mean accuracy is 82%, triangular MF gives 79%, matrix method of Nedosekin −70%, while standard rating system CAMELS has 75% accuracy.

3. With the increase of rules, number error probability first decreases, then attains minimum and then begins to raise.

3.4 General conclusions of investigations for Ukrainian banks

Various methods for Ukrainian banks financial state forecasting were considered and analyzed. The following methods were considered [3, 4]: fuzzy neural network

Figure 2.
Bankruptcy risk forecasting results for different methods.

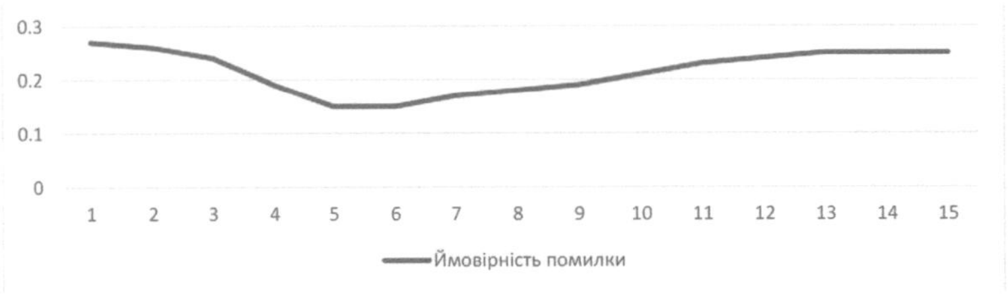

Figure 3.
Error probability for different rules number in FNN.

ANFIS, fuzzy neural network TSK, Kromonov's method, Byelorussian bank association method, rating system CAMELS, and matrix method (Nedosekin).

As the input data, the financial indices of Ukrainian banks were considered.

While experiments with the adequate financial indicators were detected using which the best forecasting results for Ukrainian banks were obtained:

general reliability factor (own capital/assets);

instant liquidity factor (liquid assets/liabilities);

cross coefficient (total liabilities/working assets);

general liquidity coefficient (liquid assets + defended capital + capitals in reserve fund/total liabilities); and

coefficient of profit fund capitalization.

1. It was established that FNN TSK gives much more accurate results than FNN ANFIS. With increase of rules, number error probability first decreases, then attains minimum and then begins to raise.

2. The fuzzy GMDH gives better results using older data that is, more preferable for long-term forecasting (two or more years).

3. The comparison of FNN TSK with standard rating system CAMELS has shown that TSK enables to obtain more accurate bankruptcy risk forecasting.

4. In general, the comparative analysis had shown that fuzzy forecasting methods and techniques give better results than the conventional crisp and rating methods for forecasting bankruptcy risk. But at the same time, the crisp methods are more simple in implementation and demand less time for their adjustment.

4. Comparative analysis of bankruptcy risk forecasting methods for European banks under uncertainty

4.1 Introduction

The results of successful application of fuzzy methods for bankruptcy risk forecasting of Ukrainian banks under uncertainty stimulated the further investigations of these methods application for financial state analysis of European leading banks.

The main goal of this exploration was to investigate novel methods of European banks bankruptcy risk forecasting, which may work under uncertainty with incomplete and unreliable data.

Besides, the other goal of this investigation was to determine which factors (indicators) are to be used in forecasting models to obtain results close to real data. Therefore, we used a set of financial indicators (factors) of European banks according to the International accountant standard IFRS. The annual financial indicators of about 300 European banks were collected in 2004–2008, preceding the start of crisis of bank system in Europe in 2009. The data source is the information system Bloomberg [8]. The resulting sample included the reports only from the largest European banks as system Bloomberg contains the financial reports only from such banks. For correct utilization, input data were normalized in interval [0,1].

4.2 Application of fuzzy neural networks for European banks bankruptcy risk forecasting

The period for which the data were collected was 2004–2008. The possible bankruptcy was analyzed in 2009. The indicators of 165 banks were considered among which more than 20 banks displayed the worsening of the financial state in that year. Fuzzy neural networks and Fuzzy Group Method of Data Handling (FGMDH) were used for bank financial state forecasting.

In accordance with the above stated goal, the investigations were carried out for detecting the most informative indicators (factors) for financial state analysis and bankruptcy forecasting. Taking into account incompleteness and unreliability of input data, FNN ANFIS and TSK were suggested for bankruptcy risk forecasting.

After performing a number of experiments, the data set of financial indicators was found using which FNN made the best forecast. These indicators are the following:

debt/assets = (short-term debt + long-term debt)/total assets;

loans to deposits ratio;

net interest margin (NIM) = net interest income/earning assets;

return on equity (ROE) = net income/stockholder equity;

return on assets (ROA) = net income/assets equity;

cost/income = operating expenses/operating income; and

equity/assets = total equity/total assets.

A series of experiments was carried out for determining the influence of the number of rules and period of data collection on forecasting results.

In the first series of experiments, FNN TSK was used for forecasting.

Experiment Nos. 1–5:

Training sample = 115 banks of Europe, testing sample = 50 banks, and number of rules = 5.

Input data period = 2004 (experiment 1), 2005 (experiment 2), 2006 (experiment 3), 2007 (experiment 4), and 2007 (experiment 5).

The total results of application FNN TSK for different rules number and data collection period are presented in **Table 20**.

Furthermore, the similar experiments were performed with FNN ANFIS, while the period of data collection varied since 2004–2007. The corresponding results for FNN ANFIS are presented in **Table 21** showing the influence of data collection period on forecasting accuracy.

After analysis of these results, the *following conclusions* were made:

1. FNN TSK has better forecasting accuracy than FNN ANFIS;

2. the best input variables (indicators) for European banks bankruptcy risk forecasting are the following:

Experiment/number of rules	Total errors number	% of errors	Number of first type errors	Number of second type errors
2004—5	8	16	0	8
2005—5	7	14	0	7
2006—5	5	10	0	5
2007—5	1	2	0	1
2004—10	8	16	0	8
2005—10	8	16	1	7
2006—10	11	22	7	4
2007—10	4	8	0	4

Table 20.
Forecasting results for FNN TSK versus number of rules and data period.

Experiment/number of rules	Total errors number	%% of errors	Number of first type errors	Number of second type errors
2004—5	8	16%	0	8
2005—5	8	16%	1	7
2006—5	8	16%	4	4
2007—5	4	8%	0	4

Table 21.
Forecasting results for FNN ANFIS versus number of rules and data period.

debt/assets = (short-term debt + long-term debt)/total assets;

loans to deposits;

net interest margin (NIM) = net interest income/earning assets;

return on equity (ROE) = net income/stockholder equity;

return on assets (ROA) = net income/assets equity;

cost/income = operating expenses/operating income; and

equity/assets = total equity/total assets.

Input data collection period (forecasting interval) makes influence on forecasting results.

4.3 The application of fuzzy GMDH for bank financial state forecasting

In next experiments, Fuzzy Group Method of Data Handling (FGMDH) was applied for European banks financial state forecasting. Fuzzy GMDH enables to construct forecasting models using experimental data automatically without expert [3]. The additional advantage of FGMDH is possibility to work with the fuzzy information.

As the input data in these experiments, the same indicators as in experiments with FNN TSK were used. In **Table 22**, forecasting results are presented in dependence on input data period collection for FGMDH

Input data period	Total number of errors	% of errors	Number of first type errors	Number of second type errors
2004	7	14	0	7
2005	6	12	1	5
2006	4	8	1	3
2007	2	4	0	2

Table 22.
Comparative analysis of forecasting results for FGMDH.

Method (period)	Total number of errors	% of errors	Number of first type errors	Number of second type errors
ANFIS (1 year)	4	8	0	4
TSK (1 year)	1	2	0	1
FGMDH (1 year)	2	4	0	2
ANFIS (2 years)	8	16	4	4
TSK (2 years)	5	10	0	5
FGMDH (2 years)	4	8	1	3

Table 23.
Forecasting results of different fuzzy methods.

If to compare the results of FGMDH with the results of FNN TSK, one can see that neural network has better accuracy at the short forecasting interval (1 year), while fuzzy GMDH has better accuracy at the greater intervals (2 or more years). This conclusion coincides with similar conclusion for Ukrainian banks.

In **Table 23,** the comparative results of application of different methods for bankruptcy risk forecasting are presented

4.4 Application of linear regression and probabilistic models

4.4.1 Regression models

For estimation of fuzzy methods' efficiency at the problem of bankruptcy risk forecasting the comparison with crisp method, the regression analysis of linear models was performed. As input data, the same indicators were used, which were found optimal for FNN. Additionally, the index net financial result was also included in the input set. This index makes great impact on forecasting results. Thus, input data in these experiments were eight financial indicators of 256 European banks according to their reports:

debt/assets—X1;

loans/deposits—X2;

net interest margin—X3;

ROE (return on equity)—X4;

ROA (return on assets)—X5;

cost/income—X6;

equity/assets—X7; and

net financial result—X8.

The input data were normalized before the application. The experiments were carried out with full regression ARMA model, which used eight variables and shortened models with six and four variables.

Each obtained model was checked on testing sample consisting of 50 banks. The comparative forecasting results for all ARMA models are presented in **Table 24**.

As one may see in **Table 24**, the application of all types of linear regression models gives the same error of 18%, which is much worse than application of fuzzy neural networks.

4.4.2 Logit models

Furthermore, the experiments were performed using logit models for bankruptcy forecasting [9, 10]. The training sample consisted of 165 banks and the testing sample of 50 banks.

The first one was constructed, linear logit model, using all the input variables. It has the following form (estimating and forecasting equations):

$$I_Y = C(1) + C(2) * X_1 + C(3) * X_2 + C(4) * X_3 + C(5) * X_4 + C(6) * X_5 + C(7) * X_6 + C(8) * X_7 + C(9) * X_8$$

$$Y = 1 - @CLOGISTIC(-(C(1) + C(2) * X_1 + C(3) * X_2 + C(4) * X_3 + C(5) * X_4 + C(6) * X_5 + C(7) * X_6 + C(8) * X_7 + C(9) * X_8))$$

The next constructed model was a linear probabilistic logit model with six independent variables. The final table including the forecasting results of all the logit models is presented below (**Table 25**)

Input data	Testing sample	First type of errors	Second type of errors	Total number of errors	% of errors
All variables (eight)	50	5	4	9	18
Six variables	50	5	4	9	18
Four variables	50	5	4	9	18

Table 24.
Comparative analysis of ARMA models.

Input data	Testing sample	First type of errors	Second type of errors	Total number of errors	% of errors
All variables (eight)	50	6	2	8	16
Six variables	50	6	2	8	16

Table 25.
Comparative analysis of logit models.

4.4.3 Probit models

The next experiments were carried out with probit models [9, 10]. The first constructed model was the linear probit model based on 206 banks using all the input variables. It has the following form:

$$
\begin{aligned}
I_Y &= C(1) + C(2) * X_1 + C(3) * X_2 + C(4) * X_3 + C(5) * X_4 \\
&\quad + C(6) * X_5 + C(7) * X_6 + C(8) * X_7 + C(9) * X_8 \\
Y &= 1 - @CLOGISTIC(-(C(1) + C(2) * X_1 + C(3) * X_2 \\
&\quad + C(4) * X_3 + C(5) * X_4 + C(6) * X_5 + C(7) * X_6 + C(8) * X_7 + C(9) * X_8))
\end{aligned}
$$

As the experiments had shown that the inputs net interest margin (X_3) and net financial result (X_8) very weakly influence on the forecasting accuracy, they were excluded in the next experiments. The next probit model included six variables.

Furthermore, in this model insignificant variables debt/assets (X_1) and loans/deposits (X_2) were excluded, and as a result, linear probit model with four variables was obtained.

Each of the constructed probit models was checked on the test sample of 50 banks. The results of application of all probit models are presented in **Table 26**.

As one may see from **Table 26**, the application of all the probit models gives relative error 14–18%, which is much worse than results obtained by fuzzy neural networks. It is worth to mention the decrease of model forecasting quality after exclusion of insignificant variables.

4.5 Concluding experiments

In the final series of experiments, investigations and detailed analysis of various methods for forecasting bankruptcy risk were performed. The following methods were investigated: FNN ANFIS, FNN TSK, FGMDH, regression models, logit models, and probit models.

Period of input data was 2007 (1 year before possible bankruptcy).

Comparative analysis of all the forecasting methods is presented in **Table 27**.

As one may see from this table, fuzzy methods and models show much better results than crisp methods: ARMA, logit models, and probit models. When forecasting by one year prior to current date, fuzzy neural network TSK shows better results than FGMDH. But when forecasting for longer intervals (several years), FGMDH is the best method.

In a whole, the conclusions of experiments with European banks completely confirmed the conclusions of experiments with Ukrainian banks.

Input data	Testing sample	First type of errors	Second type of errors	Total number of errors	% of errors
All variables (eight)	50	5	2	7	14
Six variables	50	5	2	7	14
Four variables	50	6	3	9	18

Table 26.
Forecasting results of probit models.

Method	Total number of errors	% of errors	First type of errors	Second type of errors
ANFIS	4	8	0	4
TSK	1	2	0	1
FGMDH	2	4	0	2
ARMA	9	18	4	5
Logit	8	16	2	6
Probit	7	14	2	5

Table 27.
Comparative analysis of methods for banks bankruptcy forecasting.

5. Conclusions

The problem of banks bankruptcy risk forecasting under uncertainty was considered.

For its solution, the application of novel methods of computational intelligence, fuzzy neural networks ANFIS and TSK and fuzzy GMDH, was suggested.

1. The experimental investigation of FNN TSK, ANFIS, and GMDH application in the problem of bankruptcy risk forecasting was carried out for Ukrainian and European banks.

2. The comparison of forecasting efficiency of FNN TSK and ANFIS with Fuzzy GMDH and conventional statistical methods ARMA, logit, and probit models was performed.

3. The experimental results show that FNN and FGMDH have much better accuracy than statistical methods. When forecasting by one year prior to current date, fuzzy neural network TSK shows better results than FGMDH. But when forecasting for longer intervals (several years), FGMDH is the best method.

4. While in experimental investigations, the best sets of financial indicators for bankruptcy forecasting were found for Ukrainian and European banks as well.

Author details

Yuriy Zaychenko*, Michael Zgurovsky and Galib Hamidov
Igor Sikorsky Kyiv Polytechnic Institute, Kiev, Ukraine

*Address all correspondence to: zaychenkoyuri@ukr.net

IntechOpen

© 2019 The Author(s). Licensee IntechOpen. This chapter is distributed under the terms of the Creative Commons Attribution License (http://creativecommons.org/licenses/by/3.0), which permits unrestricted use, distribution, and reproduction in any medium, provided the original work is properly cited.

References

[1] Camel Rating System. El access https://www.investopedia.com/terms/c/camelrating.asp

[2] Quarterly Report of Ukrainian Banks at NBU Site [Electronic Resource]/ National Bank of Ukraine; 2009. 4 p. (ukr). Available from: http://www.bank.gov.ua/Bank_supervision/Finance_b/2009/01.10.2009/fin_state.htm

[3] Michael Z, Yuriy Z. Fundamentals of Computational Intelligence: System approach. Studies in Computational Intelligence 652. AG, Switzerland: Springer International Publ. 308 p

[4] Ghamish ONAA, Yuriy Z, Voitenko O. Analysis of financial state and banks bankruptcy risk forecasting. System Research and Information Technologies. 2015:2. (Rus)

[5] Ghamish ONAA, Yuriy Z. Comparative analysis of methods of banks bankruptcy risk forecasting under uncertainty. International Journal of Engineering and Innovative Technology (IJEIT). 2015;**4**:5

[6] Meshcheryakov AA. Rating Estimate of Commercial Bank. Ukrainian Academy of Custom Service. 2010 (Ukr.) Available from: http://www.nbuv.gov.ua/portal/Soc_Gum/Ekpr/2009_25/mescheryakov.htm

[7] Shulik V. Method of Banks Credit-rating of the Largest Ukrainian Credit Agency. Credit-rating Academy. 2011. 6 p. Available from: http://www.credit-rating.ua/ru/analytics/analytical-articles/

[8] Bloomberg Inc. Available from: https://Bloomberg.com

[9] De Jong P, Heller GZ. Generalized Linear Models for Insurance Data. New York: Cambridge University Press; 2008. 197 p

[10] Hosmer DW, Lemeshow S. Applied Logistic Regression. New York: John Wiley & Sons, Inc.; 2000. 377 p

[11] Nedosekin AO. System of portfolio optimization of Siemens Services. Banking Technologies. 2003. p. 5 (Rus). See also: http: //www.finansy.ru/publ/fin/004.htm

[12] Nedosekin AO. The applications of fuzzy sets theory to problems of finance control. Section//Audit and Financial Analysis. 2000. p. 2 (Rus). Available from: http://www.cfin.ru/press/afa/2000-2/08-3.html

Chapter 7

The Colombian Banking Sector: Analysis from Relative Efficiency

Gloria Rodriguez-Lozano

Abstract

The banking sector is that sector of the modern economy that is primarily called upon to play the important role of intermediation between the surplus agents and the deficit agents. Based on this fact, this research presents and analyzes the behavior of banks in Colombia since 2002 and up to 2016 (15 years) through the application of data envelopment analysis, a nonparametric methodology of advanced linear programming, which generates a single efficiency indicator for each unit studied in each period, optimizing multiple resources (inputs) and multiple products (outputs). One aspect of the results shows that for the year 2014, 71% of the banks were efficient, this being the highest result within the period studied.

Keywords: Colombian banking sector, banking efficiency, data envelopment analysis, efficiency measurement, relative efficiency

1. Introduction

In the development of any economy, the banking sector plays a key role, since it is primarily that sector that advances the task of intermediation between the so-called surplus agents (they do not spend the total of their monetary resources, the savers) and the so-called deficit agents (require additional monetary resources); this is one of the reasons why banks exist in all countries. Additionally, because they transmit to the population in general, monetary and credit policies issued by central banks and/or governments in such a way that they become one of the key sectors in every economy since an important part of savings, investment and financing goes through their intermediation. Otherwise, banks play a preponderant role in determining living standards within modern economies, so much so that [1] banks have the ability to stimulate and collect the savings of a society and distribute them among companies and sectors that need capital as an input for their economic activities.

Banks are important within any financial system; for example, in the United States in mid-2017, there were 7,836 member banks of the Federal Deposit Insurance Corporation. According to [2], within the European Union as of November 2016, there were 127 larger banking groups and according to [3], in Latin America, there are 23 banks in Chile; according to [4], in Mexico, there were 47 institutions of multiple banking as of December 2016 and in Colombia 25 banks as of December 2016.

According to [5], within the Colombian financial system, banks are part of the credit facilities supervised by the Superintendencia Financiera de Colombia (SFC), along with financial corporations, traditional financing companies, leasing specialized finance companies, and financial cooperatives. According to [6], the credit institutions are financial institutions whose main function is to capture legal currency resources from the public, either in sight deposits (savings accounts and current accounts) or in term deposits (CDT and CDAT'S), to place them again through loans, discounts, advances, or other active credit operations. According to [7], being watched by the SFC means that there is an institution that authorizes and monitors the activity carried out by entities that receive monies from the public, where you save, invest your capital, and have a loan, insurance, or your pension. According to Decree 663 of April 2, 1993, published in the Official Gazette No. 40, 820, Organic Statute of the Financial System [8], in its Article 2 it is made explicit that the main function of banking establishments is the capture of resources in bank current account, as well as the collection of other sight or term deposits, with the primary objective of making active credit operations.

Given the importance of the banking sector worldwide, this research analyzes the behavior of the Colombian banking sector during the last 15 years, using the nonparametric methodology data envelopment analysis (DEA) to generate relative efficiency indicators for each of the banks and for every year throughout the study period; questions are answered: are there efficient banks throughout the period? What are the average efficiency levels of the sector for each year? In what year or years were there better results of relative efficiency?

2. Colombian banking sector

In 1923 and through Law 25, the Banco de la República was created, the second central bank created in the region, a year before that had been created in Peru. According to [9], between the years 1922 and 1950, 17 central banks were created in the region under the guidance of Edwin Walter Kemmerer. Along with the approval of Law 25 of 1923, which established the organic status of the Banco de la República, Law 45 was approved, which regulated private banks and established the Banking Superintendency, seeking to allow the stable functioning of the Colombian financial sector [10]. In the nineties, the transition to a universal banking system began, with an emphasis on commercial banks. In [11], a scheme of subsidiaries and matrices was standardized, the structure of the sector and its services was redefined, and the entry and exit of markets were liberalized.

The financial crises of the twentieth century resulted in the closure of entities and processes of internal mergers and acquisitions, which consolidated the system in the early twenty-first century transforming the financial sector. According to [12], the period from 2002 to 2009 was characterized by an environment of economic recovery, by the high flows of foreign capital and by the increase in the services provided by the banks. During these years, the transformations led to a reduction in the number of financial institutions, which went from more than forty banks between mortgages and commercials, in the mid-1990s, to less than twenty in 2009. The same is true for [13], those between 1995 and 2009, the financial sector had a consolidation process.

In such a way that the competition for the market resumed, always with two big banks at the head, Bancolombia and Grupo Aval. But Banco Davivienda bought Bancafé and it became the third bank; Granahorrar was bought by BBVA, positioning itself as the fourth bank in the country. For the year 2008, 72% of the Colombian banking market was distributed as follows, in importance: Aval, Bancolombia, Davivienda, and BBVA.

For the last few years, and in global terms, it is observed that the Colombian financial sector is monopolized by a few financial groups, which through their business conglomerates offer an extensive portfolio of banking services, securities administration, insurance, investment fund management, layoffs, and pensions, among others. According to [14], for 2014, only 10 of the 33 most representative companies in the sector accounted for 68% of total investments: Bancolombia, Porvenir, Banco de Bogotá, Davivienda, Protección, BBVA, Banco de Occidente, Corpbanca, Banco Agrario, and Colpatria; of these 10, eight are banks. Regarding income, 61% of the total was concentrated in ten entities: 7 banks, 2 insurers, and one administrator of pension and severance funds: Bancolombia, Banco Agrario, Banco de Bogotá, Banco de Occidente, Davivienda, Protección, BBVA and Banco Corpbanca, insurer Suramericana, insurer Positiva, and Porvenir administrator. In relation to profits only 8 concentrated 66% of the total profits of the sector; among these are Bancolombia, Banco Agrario, Banco de Bogota, Banco de Occidente, Davivienda, Banco Colpatria, Banco Popular, and Suramericana Vida.

For the year 2016, in Colombia, there are four important national financial groups with international operations: Grupo Aval, Grupo Bancolombia, Grupo Bolívar, and Grupo Colpatria.

According to [15], the Aval Group is one of the most important financial groups in Colombia; even on its website, it calls itself "Colombia's largest financial group". Specifically, it is the largest financial conglomerate in Colombia and through of BAC Credomatic is the largest and most profitable regional group in Central America. It has four banks: Banco de Bogotá, Banco de Occidente, Banco Popular, and Banco AV Villas. Banco de Bogotá: it is the financial institution with the longest history in the country (founded in 1870), the second largest bank in Colombia by size of assets, and the largest company in the Aval Group due to its level of assets, deposits, and profits. Banco de Occidente: it is the fifth largest bank in Colombia by asset level and portfolio and the third largest bank in current accounts. Banco Popular: is a pioneer in the promissory note market and provides financial solutions for government entities in Colombia. AV Villas Bank: it has gone from being exclusively focused on housing loans, to be a consumer-oriented universal bank; it is the group's most active bank, in the use of nontraditional channels: mobile banking, nonbank correspondents, and virtual branches.

Another of the important groups is the Grupo Bancolombia, which denotes that 142 years have passed since its birth [16]; this group originated from the merger of the Bank of Colombia and the Industrial Colombiano Bank, later merged with Conavi (2005) and Confisura. Bancolombia is the largest private bank in the country due to the size of its equity and assets [17].

The Grupo Empresarial Bolívar is a conglomerate of companies that are coordinated and controlled through Sociedades Bolívar. Banco Davivienda belongs to this group, which originated in the Colombian Savings and Housing Corporation, Coldeahorro (founded in 1972) and the Superior Banks (merger in 2005) and Granbanco-Bancafe [18]. In mid-2017 and according to the information on its website [19], it is consistently positioned among the first three banks in the country, with a record number of customers for 2016 of 7,714,552 and 593 offices.

The Grupo Colpatria—Red Multibanca has a banking unit (bank and fiduciary), a construction unit, an investment unit, and Colfondos. In October 2011, Colpatria, the group's holding, announced the Bank's new partner: the multinational Scotiabank, with whom they signed a long-term strategic alliance; this multinational institution is one of the main financial institutions in North America and Canada [20].

The international financial groups with presence in Colombia are: BBVA, Citibank, GNB Sudameris, and Corpbanca.

Grupo BBVA: is composed of a Spanish banking entity with over a century and a half of experience, which after mergers and acquisitions both nationally and internationally, is currently the Grupo BBVA. According to [21], in 1996, it made a presence in Colombia through the purchase of 40% of Banco Ganadero; in 2004, it was renamed BBVA Colombia, and in 2006, it merged with Banco Granahorrar.

Grupo Citibank: the history of Citibank in the world began in the United States in 1812, when the City Bank of New York (today Citibank) was founded [22]. Citibank Colombia is a Citigroup franchise. The group is composed of the controlling company, Citibank Colombia S.A. and the subordinates [23]. According to [24], Citi in Colombia opened the first branch in 1916; since 1986, it has organized according to the model of Consumer Banking and Corporate Banking.

Grupo GNB Sudameris: in 1920, the bank was born as a Colombian mercantile company; after acquisitions and sales in 2004 Banco Sudameris acquires the majority shareholding of Banco Tequendama and Servibanca. At the beginning of 2014, the incorporation of the HSBC operation was formalized. The group currently consists of the Banco GNB Sudameris and 7 companies [25].

Grupo Corpbanca: Banco CorpBanca was created in 1997 from the merger of Banco Concepción Chileno and Banco Corp Group. Since then, it has positioned itself in the Chilean market as the fourth most important bank. For [26], in 2012 arrived at the Colombian market. Currently, Banco Corpbanca Colombia has four subsidiary companies.

Otherwise, at the December 2016 cutoff and using the SFC as an information source, it can be seen that within the Colombian financial system, there are 25 banks that represent 11% of the total of the entities in the sector, but in terms of participation in assets, banks represent 93%. For example, savings and housing cooperatives have a 78% stake in the entities, but only 2.1% participation in the sector's assets. Within the aforementioned 25 banks, there are 14 national (including the two cooperatives owned), 10 foreign, and one public.

According to [27], the banking sector closed in 2016 with total assets of $ 548 billion, an increase of 8.6% over the previous year and 126% in relation to the result for 2010, with an increasing trend in this period; the entities with the greatest assets at the end of 2016 are the Bancolombia, the Banco de Bogotá and Davivienda.

With respect to the gross portfolio, 2016 closed with $ 394 billion, representing an increase of 12% with respect to 2015 and 148% with respect to 2010, with an ever-increasing trend. By the end of 2016, the commercial portfolio of the banking sector participates with 58%, while the consumer portfolio with 27%, housing with 13%, and microcredit with only 3%; these shares are very similar to those of the immediately previous year.

Looking at the behavior of the liability, it is established that at the end of 2016, it is 475 billion with a growth of 9% for this last year, and between the years 2010 and 2016 with a growth of 125% with an ever-increasing trend. The ratio between the granted portfolio and the deposits of the public (savings, CDT, and current accounts) for the sector is 0.92, which represents that the sector for each peso that captures places only 0.92 pesos.

3. Methodology

3.1 Data envelopment analysis (DEA)

It is a nonparametric methodology of advanced linear programming, in which a double process of optimization is carried out, establishing the relative efficiency of Decision Making Unit—DMU, for [28] specify that this is done by generating an efficient frontier that locates the individual relative indices without having prior knowledge of the production function. More specifically, according to [29], DEA is used to evaluate the relative efficiency of a set of n DMUs, by posing linear programming problems for each unit according to the data of resource utilization or inputs and product or output generation. More broadly, according to [30], the efficiency of each DMU is defined as the relationship between weighted outputs and weighted inputs, so that it is obtained by solving a double problem of linear programming in order to determine the optimum set of weights that maximizes.

DEA has experienced a dynamic development, for [31] it gradually becoming a set of concepts and methodologies, which have materialized in a series of models. The first DEA model to be developed was the CRS (constant returns to scale) model that results in the categorical classification of each DMU [32]; after a few years, the VRS (variable returns to scale) model appears, through which not only constant returns can be worked [33], but also as [34] clarifies these returns can be incremental and decremental.

3.2 CRS model

Consider a set composed of n DMU denoted as DMU_j (j = 1, ..., n), which uses resources x_{ij} (i = 1, ..., m) and generate s outputs y_{rj} (r = 1, ..., s), part of that the multipliers v_i, u_r associated with i inputs and r outputs respectively are known. So, specifically, if the DMU_0 is under study, this model is giving the solution to the problem of fractional programming for the measure of efficiency of that DMU_0 as well [31]:

$$e_0 = \max \sum_r u_r y_{ro} / \sum_i v_i x_{io} \tag{1}$$

Subject to:

$$\sum_r u_r y_{rj} - \sum_i v_i x_{ij} \le 0, \text{for all } j$$

$$u_r, v_i \ge \varepsilon, \text{for all } r, i$$

where ε is a nonarquimidian value designated strictly positive.

The theory of fractional programming expressed in [35] is applied and the following changes of variables are made:

$$\mu_r = t u_r y, v_i = t v_i$$

where:

$$t = \left(\sum_i v_i x_{i0}\right)^{-1}.$$

The initial problem can be transformed into the following linear programming model:

$$e_0 = \max \sum_r \mu_r y_{ro} \tag{2}$$

Subject to:

$$\sum_i v_i x_{i0} = 1$$

$$\sum_r \mu_r y_{rj} - \sum_i v_i x_{ij} \le 0, \text{for all } j$$

$$\mu_r, v_i \ge \varepsilon, \text{for all } r, i$$

3.3 VRS model

According to [31] and all the above parameters, for this model we have a mathematical approach:

$$e_0^* = \max\left[\sum_r u_r y_{r0} - u_o\right] / \sum_i v_i x_{i0} \tag{3}$$

Subject to:

$$\sum_r u_r y_{rj} - u_o - \sum_i v_i x_{ij} \le 0 \text{ j } = 1, \ldots, n$$

$$u_r \ge \varepsilon, v_i \ge \varepsilon, \text{for all } i, r$$

$$u_o \text{ not restricted in sign}$$

With its equivalent in linear programming:

$$e_0^* = \max \sum_r \mu_r y_{ro} - \mu_o \tag{4}$$

Subject to:

$$\sum_i v_i x_{io} = 1$$

$$\sum_r \mu_r y_{rj} - \mu_o - \sum_i v_i x_{ij} \le 0, j = 1, \ldots, n$$

$$\mu_r \ge \varepsilon, v_i \ge \varepsilon, \text{for all } i, r$$

$$\mu_o, \text{unrestricted}$$

The conventional measurement of DEA is based on the hypothesis that resources or inputs should be minimized, and products or outputs should be maximized according to [36]. Additionally, for each of these basic models, there is the orientation to the entrances and the orientation to the exits, depending on whether you want to prioritize the maximum decrease of the inputs keeping the outputs constant or the total maximization of the outputs with the constant inputs. One of the strengths of DEA is that a single efficiency result (%) is obtained for each unit, in a multi-input and multi-output context.

4. Review of current literature

The study of relative efficiency through DEA applied to banks is one of the most recurrent issues, so that some of the most recently published research is presented below. Ref. [37] measured the efficiency of the offices of a state bank in India by applying a DEA model based on slack. The diffuse DEA models are used in those occasions in which it is considered that the accuracy of the data is not the best; for this reason, [38] used this type of model to analyze the banking sector in India.

The CRS and VRS models were used by [39] in their research to measure the efficiency of commercial banks in Slovenia, Poland, Austria, Hungary, Slovenia, and Czech Republic. These two models were also applied by [40] to study the banks of Côte d'Ivoire in West Africa, for the period 2008–2010. The VRS model was also used by [41] to study 79 bank branches in Canada, only this time they used the orientation to the inputs.

Islamic banks were studied by [42] comparing them with traditional banks using the Meta-Frontier Analysis (MFA) model. The Nash negotiation game model was combined with the centralized two-stage DEA model for [43] to study banks in China. Through an additive efficiency decomposition approach in DEA [44] evaluated the management and investment efficiencies of ICTs in Taiwan's banks for the 2007–2011 period. The Iranian banks were studied by [45] through the CRS model with output orientation. The research of [46] proposes a model with a multi-stage procedure that integrates robust methods, cluster analysis, and DEA to identify and study the efficiency of management in the different branches of the banks.

The main banks in Cambodia were studied by [47] through the DEA panel model for 13 years. [48] studied Taiwanese banks developing a new model based on DEA gaps to decompose their different components. Research from [49] focused on measuring the efficiency of marketing as a measure of performance after the merger, and this was investigated through DEA applied to 20 merger and acquisition agreements within the US commercial banking.

[50] studied the relative efficiency of 23 Colombian commercial banks for a period of 10 years, with the CRS and VRS models oriented to inputs and outputs. [51] investigated the efficiency of Colombian banking from the year 2000 until 2012; they applied the VRS model with orientation to the outputs. On the other hand, [52] compared the relative efficiency of the real sector with that of the financial sector of the Colombian economy for 2014 using the VRS model oriented to the outputs.

5. Specific methodological design

Information source: Superintendencia Financiera de Colombia.

Delimitation of the DMU: given that the Colombian banking sector is studied, the DMUs are the banks that year after year, and from 2002 and until 2016, reported their financial statements to the SFC. Reaffirming the statement by [53], DEA is generally interpreted using the notion of production technology generated by the set of observed units. For the study period, 5 government banks reported to the SFC, because they were from the government; they withdrew from the database, accepting what was expressed by [54] as to which institutions of the Government alone are comparable to each other. Additionally, a private bank that was liquidated in 1999 also reported its financial statements, but because it was not comparable, because it was not fully operational, it was also removed from the database. **Table 1** shows the number of banks that are part of the investigation for each year.

Determining the specific DEA model, as recommended by [55], since there is no evidence of constant returns to scale, we choose to use the DEA VRS model, and having a particular interest to evaluate how to obtain best results, we work with orientation to the outputs.

Delimitation of inputs and outputs: given that DEA is a nonparametric boundary model, in which it is not necessary to previously establish the production function, and that the determinant variables of the model are the resources used (inputs) and

Year	# of banks
2002	24
2003	24
2004	24
2005	19
2006	15
2007	15
2008	17
2009	17
2010	18
2011	22
2012	22
2013	23
2014	21
2015	24
2016	24

Source: self-made.

Table 1.
Number of banks that are part of the study.

Inputs	Outputs
Current assets	Operating income
Property, plant & equipment	Net income
Noncurrent liabilities	
Equity	

Source: self-made.

Table 2.
Input and output variables.

what is obtained from the process of transformation of them (outputs), the variables used in this investigation are shown in **Table 2**.

Taking into account the inputs and outputs chosen for this model and going to [56], it can be established that this model is what they call an intermediation model, which consists in measuring how the entity operates based on the monetary assets it gathers (inputs), making loans and investments (outputs).

These same input and output variables have been used by [52] and by (Rodríguez-Lozano) [57] to study the insurance brokerage companies in the Colombian financial environment through DEA indicators, and also by [58] to determine relative efficiency in two subsectors of the Colombian economy from 1993 to 2002 and [59] to determine the measurement of relative efficiency in three subsectors of the Colombian economy from 1993 to 1999.

6. Results

Table 3 shows the result of the efficient units, in percentage terms, for each year. **Table 4** shows the average efficiency levels for each year.

Year	# Units	%
2002	6	25%
2003	9	38%
2004	9	38%
2005	8	42%
2006	9	60%
2007	10	67%
2008	11	65%
2009	10	59%
2010	11	61%
2011	11	50%
2012	14	64%
2013	14	61%
2014	15	71%
2015	13	54%
2016	15	63%

Source: self-made.

Table 3.
Efficient units per year.

The study period begins with the lowest index, as previously stated is a period of recovery of the Colombian financial sector after its crisis; the highest index is in 2007, which coincides with the time before it started the first global financial crisis of this century; as of 2014 (with the second best average), the consequences of the global economic recession begin to be evident.

Year	Average efficiency (%)
2002	70
2003	79
2004	78
2005	88
2006	89
2007	95
2008	87
2009	88
2010	89
2011	90
2012	92
2013	89
2014	92
2015	86
2016	86

Source: self-made.

Table 4.
Average efficiency levels per year.

Figure 1 shows the behavior of the units, by efficiency range for the entire study period.

The results show that between the years 2002 and 2007, the percentage of efficient banks was in a clear rise, an increase of 42% points, despite the fact that the number of banks decreased; even 2007 was the year in which there were fewer banks for the 15 years studied. This confirms that since 2002, the sector experienced improvements. Additionally, in 2005, 79% of the banks were between 100 and 80% efficiency, ratifying the good results.

In 2007, the first financial crisis of the twenty-first century began and the impact of this crisis is evident from the results of 2008 and until 2011; there is a decrease of 17% points for efficient banks. At the same time, the efficiency range between 79.9 and 60 increased until reaching its maximum level in 2010 (33%). These results are consistent with those found by [60], in the sense that global financial crises impact the behavior of banks. As of 2011, a slow recovery process begins for the efficient, until 2014.

The Colombian economy ended 2015 with a rate slightly above 3%, despite the stagnation experienced in Latin America and the low global growth [61]; for the financial sector, the growth was 4.3% and according to [62], the balance of the banking sector was satisfactory. In terms of relative efficiency, it can be established that the share of the efficiency decreased at the same time as the holdings of the banks with low ranks increased. The efficient banking institutions had a fall of 17% points in a single year and for this same year the ranges 99.9–60% and lower to 40% increased their participation by 11 and 8% points, respectively. This means that more banks were unconcerned about managing their resources efficiently prioritizing other types of behavior as a process of deceleration of the beginning of world economy.

For the year 2016, in spite of continuing the deceleration of the economy, the Colombian banking sector reported good results, for example increased profits by 17% over the immediately previous year. In the same way, the participation of efficient banks improved to 63%, and the rest of the ranks were concentrated around a 10% share; leaving in this way very clear the difference of participation between the efficient banks and those that are not, a difference that began to take shape from the year 2006 and that from 2014 is stabilized by the grouping of the inefficient in participations below 20%.

6.1 Analysis by groups

Figure 2 shows the behavior in terms of relative efficiency of the 8 financial groups mentioned previously.

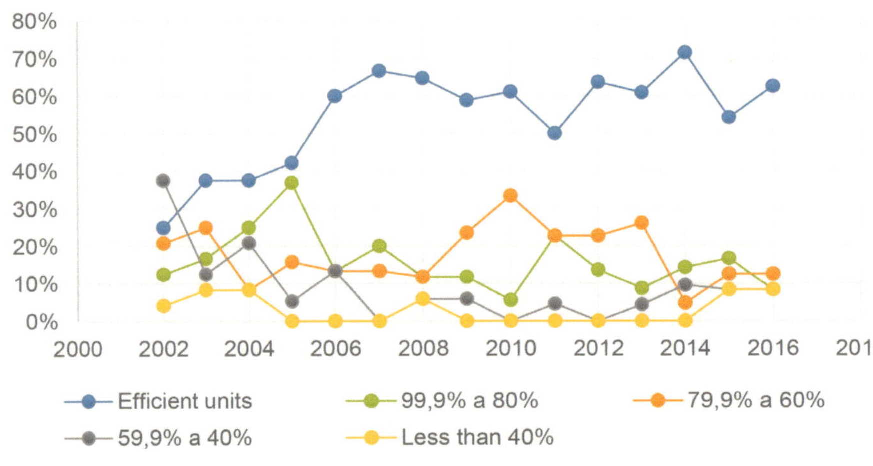

Figure 1.
Units by efficiency range. Source: self-made.

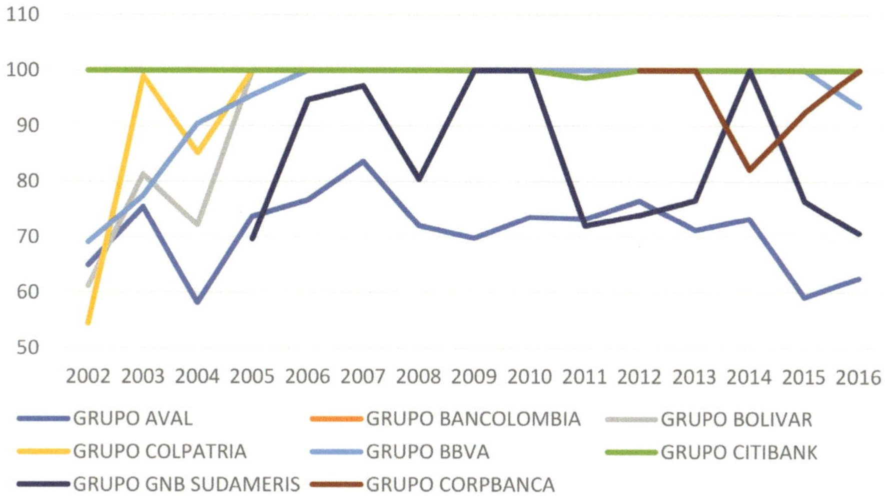

Figure 2.
Efficiency (%) of financial groups. Source: self-made.

The one that is considered the largest financial group in Colombia, Grupo Aval, is definitely the most inefficient. On the other hand, the only efficient group in all 15 years of the study is Grupo Bancolombia. Both the Grupo Bolívar (Banco Davivienda) and the Grupo Colpatria (Banco Colpatria) began the study period with relatively low rates (61 and 54% respectively) but they were improving to the point that since 2005 they are efficient and they maintained the same until 2016.

Regarding international groups, Citibank is the one with the best results, because only in 2011, it is not efficient, but its index is 99%. The behavior of the Grupo BBVA is very similar to that of the Grupo Bolívar and Grupo Colpatria, but this group is not efficient for 2016; although it is very close, its index is 93%. The Grupo Corbanca in the 5 years of presence has rates above 80% and even in 3 years it is efficient. The Grupo GNB started in 2005 with an index of 70%, improved for the next 2 years, fell, and then became efficient, and from 2011 it repeats the cycle ending in 2016 with an indicator close to 70%. The Grupo CorpBanca Group shows better results, since in three of the 5 years of presence it is efficient, and its inefficiency is not more than 20% since the indexes of those 2 years are higher than 80%.

In **Figure 3**, the behavior of the 4 banks of the Grupo AVAL is presented; there it is evident that although the one of Banco de Bogota has indexes above 80% and is efficient in 9 of the 15 years of the study, this is not enough because the remaining three banks mark downwards since they have rates that are around 60% and in the last 3 years the Banco Popular and the Banco AV Villas have very low rates.

Figure 4 shows the behavior of efficient banks according to whether they are national or not.

Foreign banks have a greater participation in 90% of the years; only for 2016, the participation of efficient national banks exceeds the participation of foreigners, although by very little.

Now, of all the banks that are part of the study, only 24% are efficient during all the years that each of them is present in the study period: Bancolombia: 15 years, Procredit: 9 years, Finandina: 6 years, Santander de Negocios: 4 years, Bankboston: 3 years, Megabaco: 4 years, Mundo Mujer: 2 years, Multibank: 2 years, and Standard Chartered: 3 years. They are not within this group because only in 1 year they were not efficient, although this indicator is above 95%: Coomeva, WWB, Bancompartir, and Citibank.

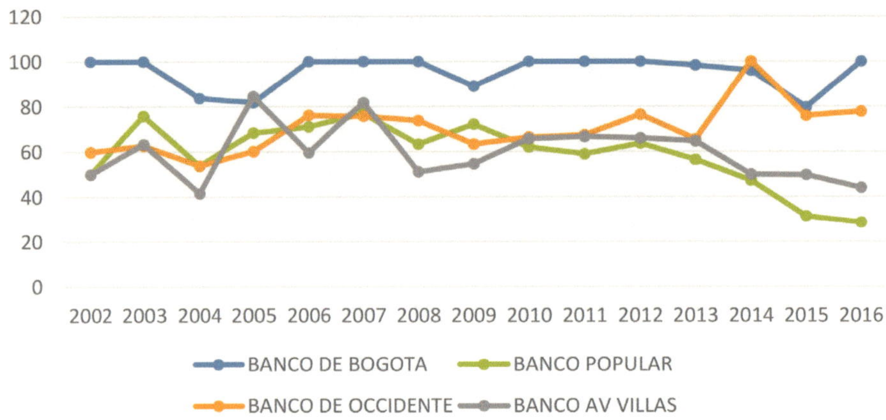

Figure 3.
Efficiency (%) of banks of the grupo aval. Source: self-made.

Figure 4.
Behavior of efficient banks. Source: self-made.

7. Conclusions

The financial sector is important in any economy, since it is the one that brings together those agents that have monetary resources to spare with the agents that need those resources. In Colombia, this sector is one of those that have marked the growth of the economy for several years; even more, at a time of global economic slowdown, it is the one that has definitely had very good profits. An example of this is the year 2016 in which despite the global recession this sector obtained an increase of 24.8% in its profits.

The Colombian banking sector emerges from the crises of the twentieth century, preparing itself to tackle the twenty-first century, through an organization that prioritizes the formation of financial groups of both national and foreign origin; this means that purchases and mergers have been the order of the day. In the last 15 years, the period of time of this investigation, there has been a great movement regarding the number of banks in the sector in each year. Although this period began with 24 banks in 2002, discounting the banks owned by the government and closing 2016 with this same number of banks do not mean that there was no movement in the intervening years.

The Santander Colombia, Scotiabank, HSBC Colombia, Helm, Union Colombiano, Tequendama, Bansuperior, Megabanco, Granahorrar, Colmena, and Conavi banks disappeared from the sector, either because they were purchased or because they merged. The Standard Chartered bank closed voluntarily and the bank Bankboston also settled voluntarily. The following banks entered: Corpbanca, Procredit Colombia, Bancamia, WWB, Coomeva, Finandina, Falabella, Pichincha Coopcentral, Santander de Negocios, Mundo Mujer, Multibank, and Bancompartir.

To study this sector, the DEA methodology is perfectly adjusted since it is not necessary to predetermine the production function and a set of multiple entries and multiple outputs can be worked out to obtain a single indicator per bank and per period studied.

Within this investigation, it has been possible to establish that of the 13 banks that entered the sector during the study period, and that were previously mentioned, 62% have obtained such good results that at least 90% of the years of existence have been efficient and when they are not, the indicator is above 90%.

Overall, it can be determined that the average relative efficiency over the last 15 years of the Colombian banking sector is 86%, despite having raffled the first global financial crisis of this century and living a severe economic recession worldwide.

The highest percentage of efficient banks is in 2014 (71%), for 2016 37% of banks are inefficient.

It was also established that one does not need to be the most important bank, nor the largest one in terms of assets, nor have they obtained the most voluminous profits, nor belong to the most powerful financial group, and nor have years of evolution, to be efficient. Additionally, not being a bank that is part of a very important financial group, it is efficient. Otherwise, so be a robust financial group, this does not mean that as a group, be efficient.

The results also show that there is a large gap between the percentage of efficient and inefficient banks. This gap began in 2006 with 47% points of difference, which increases for 2008 and ends in the last year of the study with more than 50% points.

On the other hand, the foreign banks have been more representative in the total of efficiency than the national banks; this situation only has a very slight change for the year 2016.

8. Recommendations

This research was carried out using public information to which there is free access, but it is evident that for a better approximation to the measurement of relative efficiency it would be very good to have access to information that in Colombia is considered as private. In other words, the directors of the different banks agree to provide the information required to improve this study.

On the other hand, bank managers should take advantage of the conclusions reached to improve their performance, and if that were the case, the performance of the whole group.

Author details

Gloria Rodriguez-Lozano
Faculty of Economics Sciences, National University of Colombia, Bogotá, Colombia

*Address all correspondence to: girodriguezl@unal.edu.co

IntechOpen

© 2019 The Author(s). Licensee IntechOpen. This chapter is distributed under the terms of the Creative Commons Attribution License (http://creativecommons.org/licenses/by/3.0), which permits unrestricted use, distribution, and reproduction in any medium, provided the original work is properly cited.

References

[1] BID. Progreso Económico y Social en América Latina. Informe 2005. Buenos Aires: El Ateneo; 2004. 323 p

[2] Parlamento Europeo. A Su Servicio [Internet]. 2017. Available from: http://www.europarl.europa.eu/atyourservice/es/displayFtu.html?ftuId=FTU_4.2.4.html

[3] Superintendencia de Bancos e Instituciones Financieras—SBIF Chile [Internet]. 2017. Available from: http://www.bancafacil.cl/bancafacil/servlet/Contenido?indice=1.2&idPublicacion=3000000000000076&idCategoria=2

[4] Secretaría de Hacienda y Crédito Público. Boletín Estadístico Banca Múltiple. Méjico, D.F.: Comisión Nacional Bancaria y de Valores; 2016. 208 p

[5] Uribe J. Nota Editorial: El sistema financiero colombiano: Estructura y evolución reciente. Revista del Banco de la República. 2013;**LXXXVI**(1023):5-17

[6] Coltefinanciera. Educación Financiera [Internet]. 2017. Available from: http://www.coltefinanciera.com.co/educacion-financiera/sistema-financiero/385-como-esta-estructurado-el-sistema-financiero-en-colombia

[7] SFC. Superintendencia Financiera de Colombia [Internet]. 2017. Available from: https://www.superfinanciera.gov.co/jsp/loader.jsf?lServicio=Publicaciones&lTipo=publicaciones&lFuncion=loadContenidoPublicacion&id=10083710

[8] Secretaría Senado. Leyes Desde 1992, Vigencia Expresa y Control de Constitucional [Internet]. 2017. Available from: http://www.secretariasenado.gov.co/senado/basedoc/estatuto_organico_sistema_financiero.html

[9] Banco de la República. La Historia del Banco [Internet]. 2017. Available from: http://www.banrep.gov.co/sites/default/files/paginas/historia-banco-sept.pdf

[10] Meisel Roca A. El Banco de la República: Antecedentes, Evolución y Estructura. Bogotá: Banco de la República; 1990. 750 p

[11] Arango M. Evolución y Crisis del Sistema Financiero Colombiano, Serie Estudios y Perspectivas. Bogotá: CEPAL; 2006. 101 p

[12] Galán J, Veiga H, Wiper M. Dynamic effects in inefficiency: Evidence from the Colombian banking sector. European Journal of Operational Research. 2015;**240**:562-571

[13] Moncada A, Dopacio C. Information technology: Potential in the Colombian banking sector. In: Soft Comput. in Manag. and Bus. Econ. STUDFUZZ. Vol. 287. 2012. pp. 77-95

[14] Escuela Nacional Sindical. Sector Financiero y Bancario Colombiano. Medellín: Escuela Nacional Sindical; 2015. 222 p

[15] Grupo Aval. Nuestras Compañías [Internet]. 2017. Available from: https://www.grupoaval.com/wps/portal/grupo-aval/aval/nuestras-companias

[16] Grupo Bancolombia. Grupo Bancolombia [Internet]. 2017. Available from: https://www.grupobancolombia.com/wps/portal/acerca-de/informacion-corporativa/quienes-somos/

[17] Grupo Sura. Grupo Sura [Internet]. 2017. Available from: https://www.gruposura.com/Inversiones/grupo-bancolombia/Paginas/default.aspx

[18] Davivienda. Banco Davivienda [Internet]. 2017. Available from: https://www.davivienda.com/wps/portal/inversionistasespanol/

inversionistas/AcercaBanco/quienes_somos/davivienda/!ut/p/b1/04_SjzQzszCysDAzMdWP0I_KSyzLTE8syczPS8wB8aPM4sNMDQK9vN0NDfw9PdwMPF18QkzMAwONvEMMgQoigQoMcABHA0L6w_Wj8CpxNYAqwGOFn0d-bqp-blSOpaeuoyIAOw

[19] Banco Davivienda. Informe de Gestión 2016. Bogotá: Davivienda; 2017. 141 p

[20] Colpatria. Colpatria Multibanca [Internet]. 2017. Available from: https://www.colpatria.com/Acerca-de/banco-colpatria/informacion-institucional/nuestra-organizacion

[21] BBVA. Banco Bilbao Vizcaya Argentaria Colombia [Internet]. 2017. Available from: https://www.bbva.com.co/meta/historia/

[22] Citibank. Citi España [Internet]. 2017. Available from: http://www.citibank.com/spain/citi/historia.htm

[23] Citibank Colombia. Citi [Internet]. 2017. Available from: https://www.citibank.com.co/Citivalores/quienes_somos/index.htm

[24] Citigroup. Citi Institucional [Internet]. 2017. Available from: https://www.citibank.com.co/institucional/historia.htm#

[25] Banco GNB Sudameris. Banco GNB Sudameris [Internet]. 2017. Available from: https://www.gnbsudameris.com.co/quienes-somos

[26] Corpbanca. Presentación Institucional. Bogotá: Corpbanca; 2015. 25 p

[27] Asobancaria. Informe de Tipificación la Banca Colombiana en 2016. Bogotá: Asobancaria; 2017. 146 p

[28] Yang M, Li Y, Chen Y, Liang L. An equilibrium efficiency frontier data envelopment analysis approach for evaluating decision-making units with fixed-sum outputs. European Journal of Operational Research. 2014;**239**(2):479-489

[29] Wei Q, Yang H. A data envelopment analysis (DEA) evaluation method based on sample decision making units. International Journal of Information Technology and Decision Making. 2010;**9**(4):601-624

[30] Dotoi M, Epicoco N, Falagario M, Sciancalepore F. A cross-efficiency fuzzy data envelopment analysis technique for performance evaluation of decision making units under uncertainty. Computers & Industrial Engineering. 2015;**79**:103-114

[31] Rodríguez G. Indicadores DEA (Data Envelopment Analysis) de eficiencia y productividad para las actividades de extensión universitaria. Aplicación en la Universidad Nacional de Colombia. Bogotá: Centro Editorial Facultad de Ciencias Económicas Universidad Nacional de Colombia; 2011, 2011. 262 p

[32] Iyer K, Banerjee P. Measuring and benchmarking managerial efficiency of project execution schedule performance. International Journal of Project Management. 2016;**34**(2):219-236

[33] Ghasemi M, Ignatius J, Lozano S, Emrouznejad A, Hatami-Marbini A. A fuzzy expected value approach under generalized data envelopment analysis. Knowledge-Based Systems. 2015;**89**:148-159

[34] Lau K. Measuring distribution efficiency of a retail network through data envelopment analysis. International Journal of Production Economics. 2013;**146**(2):598-611

[35] Charnes A, Cooper W. Programming with linear fractional

functional. Naval Research Logistics Quarterly. 1962;**9**:67-88

[36] Mahdiloo M, Tavana M, Saen R, Noorizadeh A. A game theoretic approach to mideling undesirable outputs and efficiency decomposition in data envelopment analysis. Applied Mathematics and Computation. 2014;**244**:479-492

[37] Puri J, Yadav S. A concept of fuzzy imput mix-efficiency in fuzzy DEA and its application. Expert Systems with Applications. 2013;**40**(5):1437-1450

[38] Puri J, Yadav S. A fuzzy DEA model with undesirable fuzzy outputs and its application to the banking sector in India. Expert Systems with Applications. 2014;**41**(14):6419-6432

[39] Svitálková Z. Comparision and evaluation of Bank efficiency in selected countries in EU. Procedia Economics and Finance. 2014;**12**:644-653

[40] Samuel G, Hongzhong Z, Thierry B. Technical efficiency assessment using data envelopment analysis: An application to the banking sector of Côte d'Ivoire. Procedia-Social and Behavioral Sciences. 2016;**235**:198-207

[41] Ghahraman A, Prior D. A learning ladder to ward efficiency: Proposing network-based stepwise benchmark selection. Omega. 2016;**63**:83-93

[42] Johnes J, Izzeldin M, Pappas V. A comparison of performance of Islamic and conventional banks. Journal of Economic Behavior and Organization. 2014;**103**:93-107

[43] Zhou Z, Sun L, Yang W, Liu W, Ma C, Bargaining Game A. Model for efficiency decomposition in the centralized model. Computers & Industrial Engineering. 2013;**4**(1):103-108

[44] Liu J, Lu W, Kweh Q, Wang C. Exploring the benchmarks of the Taiwanese investment trust corporations: Management and investment efficiency perspectives. European Journal of Operational Research. 2016;**248**:607-618

[45] Aghayi N, Maleki B. Efficiency measurement of DMUs with undesirable outputs under uncertainty based on the directional distance function: Application on bank industry. Energy. 2016;**112**:376-387

[46] Herrera-Restrepo O, Triantis K, Seaver W, Paradi J. Bank branch operational performance: A robus tmultivariate and clustering approach. Expert Systems with Applications. 2016;**50**:107-119

[47] Okuda H, Aiba D. Determinants of operational efficiency and total factor productivity change of major Cambodian financial institutions: A data envelopment analysis during 2006-13. Emerging Markets Finance and Trade. 2016;**52**:1455-1471

[48] Juo J, Fu T, Yu M, Lin Y. Non-radial profit performance: An application to Taiwanese banks. Omega. 2016;**65**:111-121

[49] Rahman M, Lambkin M, Hussain D. Value creation and appropriation following M & A: A data envelopment analysis. Journal of Business Research. 2016;**69**:5628-5635

[50] Sarmiento M, Cepeda A, Mutis H, Pérez J. Nueva evidencia sobre la eficiencia de la Banca Colombiana: Una medición con modelos de frontera no-paramétricos. Archivos de Economía. 2014;**392**:1-49

[51] Piraquete J, Piñeros J, Mondragón L. Eficiencia de los establecimientos bancarios (EB): una aproximación

mediante modelos DEA. Borradores de Economía. 2013;**798**:1-39

[52] Rodríguez-Lozano G, Sarmiento-Muñoz M. La eficiencia relativa del sector real vs. la del sector financiero de la economía colombiana. Apuntes del CENES. 2017;**36**(64):113-140

[53] Podinovski V, Bouzdine-Chameeva T. Consistent weight restrictions in data envelopment analysis. European Journal of Operational Research. 2015;**244**:201-209

[54] Charnes A, Cooper W, Golany B, Seiford L, Stutz J. Foundations of data envelopment analysis and Pareto-Koopmans empirical production functions. Journal of Econometrics. 1985;**30**:91-107

[55] Banker R, Charnes A, Cooper W. Some models for estimating technical and scale inefficiencies in data envelopment analysis. Management Science. 1984;**30**(9):1078-1092

[56] Paradi J, Rouatt S, Zhu H. Two-stage evaluation of bank branch efficiency using data envelopment analysis. Omega. 2011;**39**:99-109

[57] Rodríguez-Lozano G. Indicadores de eficiencia DEA (data envelopment analysis) para las sociedades corredoras de seguros en el medio financiero colombiano. In: Facultad vde Ciencias Económicas Universidad Nacional de Colombia, editor. La administración en el siglo XXI: Herencia e innovación en conceptos y herramientas para las ciencias de gestión. EditorialUN; 2016. pp. 173-183

[58] Rodríguez-Lozano G. Medición de la eficiencia relativa a en dos subsectores de la economía colombiana desde 1993 a 2002 utilizando Data Envelopment Analysis (DEA). INNOVAR. Revista de Ciencias Administrativas y Sociales. 2003;**22**:121-144

[59] Rodríguez-Lozano G. Medición de la eficiencia relativa en tres subsectores de la economía colombiana desde 1993 a 1999 utilizando Data Envelopment Analysis (DEA). INNOVAR. Revista de Ciencias Administrativas y Sociales. 2003;**21**:145-181

[60] Feng G, Wang C. Why European banks are less profitable than U.S. banks: A decomposition approach. Journal of Banking & Finance. 2018;**90**:1-16

[61] Andi. Colombia: Balance 2015 y Perspectivas 2016. Bogotá; 2016. 77 p

[62] Asobancaria. La Banca Colombiana en 2015. Bogotá: Asobancaria; 2016. 146 p

Chapter 8

Sustainability in Indebtedness: A Proposal for a Treaty-Based Framework in Sovereign Debt Restructuring

Maximo Paulino T. Sison III

Abstract

The debate concerning reforms in sovereign debt restructuring (SDR) ranges from those which maximize flexibility (e.g. adoption of clauses in debt contracts) to those which maximize uniformity and predictability (e.g. enacting a fixed framework similar to a domestic bankruptcy regime). This paper proposes that the principles of SDR in the United Nations General Assembly Resolution 69/319 be broadly codified into a treaty. This includes the principle of sustainability which emphasizes "inclusive growth and sustainable development" of stakeholders in SDR. In the current unsystematized regime of SDR, this proposal seeks to bridge the value of flexibility in *ex post* negotiations and the values of uniformity and predictability in *ex ante* rules under fixed bankruptcy regimes.

Keywords: sovereign debt, bankruptcy, sustainability, inclusive growth, reform, resolution, restructuring

1. Introduction

The challenge of resolving a sovereign debt crisis becomes problematic when it has international dimensions. A State with purely domestic indebtedness has the flexibility to enact measures to address a debt problem that is entirely within its jurisdiction. It can, for example, enact laws that reduce the face value of its debts or lengthen their maturities to prevent the possibility of default. In this case, the State's creditors would normally be without any recourse because debt contracts are created, operationalized and sustained within the legal framework of the Debtor State.

Thus, the restructuring measure enacted by the State to preserve its fiscal health and the general well-being of the country would prevail over the creditors' rights to receive the "full value" of the debt. In short, a purely domestic sovereign debt would permit unilateral restructuring by the Debtor State. In the positivistic language of Justice Oliver Wendell Holmes, "there can be no legal right as against the authority that makes the law on which the right depends..." [1]. In economic terms, however, the creditors' lack of legal recourse is effectively part of sharing the burden (like everyone else) to address the collective problem of a sovereign debt crisis.

Once sovereign debt has international dimensions, the Debtor State has to contend with legal systems other than its own. Access to international debt markets

IntechOpen

made the resolution of sovereign debt crises an international affair. For their protection, creditors often demand that debt contracts contain provisions that introduce elements of foreign legal systems to them—e.g. provisions on governing law and the choice of a tribunal other than the domestic courts of the Debtor State. As in the Argentinian crisis that started in the early 2000s, Argentina had to contend with creditor holdouts carrying injunctions secured from New York courts (see generally [2]).

Thus, in the case of sovereign debt crises with international dimensions, their resolution involves the interaction between at least two legal systems. The protection given by a foreign legal system has a deterrent effect on unilateral debt restructuring by the concerned State. However, this situation also creates the opportunity for delays and holdout behavior by giving creditors protection contractually drawn from a legal system outside that of the Debtor State. This becomes a collective action problem that ultimately hinders the resolution of a sovereign debt crisis. The International Monetary Fund (IMF) has long acknowledged that "While private creditors as a group may recognize that support for rapid restructuring is in their own interest, they may hesitate to agree to a restructuring out of concern that other creditors may hold out and press for full payment on the original terms after the agreement has been reached" ([3], p. 12).

Proposals for a collective and multilateral proceeding for the resolution of sovereign debt crises have been fairly widespread in the literature (see [4], p. 87). These proposals usually draw from domestic bankruptcy rules which, among others, support a global stay on debt collection efforts, the application of the automatic acceleration principle, priority rules and variants of creditor equality treatments [4, 5]. However, the resolution of a sovereign debt crisis would not be as simple as, to paraphrase the legal positivist H.L.A. Hart, a "bankruptcy regime writ large."[1] [6, 7]. There are significant characteristics of sovereign debt restructuring (SDR) that are not found in domestic bankruptcy regimes such as the non-availability of an option to liquidate the Debtor State and the critical role of macroeconomic policy and economic growth in the restructuring process.

Given that the current resolution efforts primarily occur within a conflict of law regime, this paper argues that a minimum level of multilateralization is needed to universalize broad norms that channel collective behavior during a sovereign debt crisis. These norms should consider "inclusive economic growth and sustainable development" within the Debtor State as rational and viable strategy in SDR.

More concretely, this paper proposes that efforts should be exerted so that the principles of sovereignty, good faith, transparency, impartiality, equitable treatment, sovereign immunity, legitimacy, majority structuring and sustainability under the United Nations (UN) General Assembly Resolution 69/319 adopted on 10 September 2015 ("Basic Principles on Sovereign Debt Restructuring Processes") may be broadly codified into a treaty. Such codification may include changes in the articulation of these principles to accommodate the positions of States such as those in the European Union (EU) (see [8]). Significantly, the principle of sustainability should emphasize the concept of "inclusive growth and sustainable development" within the Debtor State as part of the SDR framework. This proposal is being made in the context of the lack of support from countries which have the "major financial centers from which most of the sovereign debt has been issued" ([9], p. 47).

While certainly not an easy feat, the foregoing follows the so-called Incremental Approach that "complement, rather than replace, existing mechanisms, including contractual approaches and the activities of the International Financial Institutions

1 In criticizing John Austin's Command Theory of law, H.L.A. Hart described it as a "gunman situation writ large."

or the Paris Club, and guide their operation" ([10], p. 38). The thrust of this proposal is to broadly formalize the norms in UN General Assembly Resolution 69/319 that would give States the flexibility to negotiate and adopt specific measures in resolving a sovereign debt crisis. As such, this would not involve an unqualified "surrender of sovereignty" which has been part of the reason for States' reluctance to enter into a "hard law" approach in SDR.

In what follows, Part II of this essay describes the current conflicts of law regime where SDR occurs. Part III sets out the proposal to codify the principles embodied in UN General Assembly Resolution 69/319 into a treaty. Part IV discusses the principle of sustainability which emphasizes the concept of "inclusive growth and sustainable development" as a rational and viable strategy of SDR. Part V concludes.

2. The conflict of law regime of sovereign debt restructuring

The international dimension of SDR does not arise primarily from international law as set out in Article 38 of the Statute of the International Court of Justice, but by contractual provisions that incorporate laws other than the legal system of the Debtor State. In general, there are no "top-down" international norms that govern SDR in the form of treaties, custom or principles. Instead, there is a "bottom-up" normative system that arises from the number of debt contracts issued by Debtor States that contain provisions which introduce elements of a foreign legal system to them. While there are international laws that may apply to SDR such as the principle of sovereign immunity or the Hague Convention on the Recognition and Enforcement of Foreign Judgments [11], these international norms are limited to the extent that they do not dictate how SDR is implemented as a whole.

The legal environment where SDR presently operates is a regime of "conflict of laws" or "private international law." Conflict of laws does not principally concern itself with international law. Rather, it is the application of domestic law whenever a particular jurisdiction is "faced with a claim that contains a foreign element" ([12], p. 2). In other words, a conflict of law approach does not depend on the application of international law among States. Instead, it entails the operation of a particular domestic law in cases involving the interaction between at least two legal systems. Thus, "[t]he *raison d'être* of private international law is the existence in the world of a number of separate municipal systems of law—a number of separate legal units—that differ...from each other in the rules by which they regulate the various legal relations arising in daily life" ([12], p. 4).

As in the current regime of SDR, the rules that generally govern SDR are strictly not international law, but a State's domestic rules that apply when foreign laws that are incorporated in a debt contracts. The practice of SDR is therefore situated within a conflict of laws regime which recognizes that while a State may enact rules in the treatment of foreign laws, "[a] sovereign is supreme within [its] own territory...[It] can, if [it] chooses, refuse to consider any law but [its] own" ([12], p. 4). As discussed below, while the principle of State sovereignty is conceded in theory, its practical application in SDR is qualified by the set of debt contracts that a State has entered into in accessing capital.

The provisions that are often stipulated to introduce elements of a foreign legal system in debt contracts are (a) dispute settlement clauses that confer jurisdiction to a tribunal other than the domestic courts of the Debtor State; and (b) governing law clauses (see [13]). Incidentally, these contractual provisions mirror the scope of private international law which is "always concerned one or more of three questions, namely," (a) the jurisdiction of the domestic court; (b) recognition and enforcement of foreign judgments; and (c) the choice of law ([12], p. 7). Governing law clauses in a debt instrument may include the specific choice of law provision designating, for instance, New York or English or Japanese Law; or a provision that references an international norm such as a particular treaty or the principle of State sovereignty. Moreover, the adjudicatory body constituted by the debt instrument applies the choice of law or the referenced international norm pursuant to the contract.

During a sovereign debt crisis, the Debtor State may unilaterally restructure its debts pursuant to its sovereign right to set its own macroeconomic policy. In this context, foreign creditors demand that such clauses be contained in debt instruments to effectively act as restraints against unilateral restructuring by a Debtor State. These clauses therefore imply a waiver of the Debtor State's sovereignty. By introducing elements of foreign laws in debt instruments, these clauses "internationalize" what would otherwise be purely domestic contracts that may be amended by the Debtor State. It has been said that "[m]any countries do not regard foreign creditors with great sympathy, especially when the country is bankrupt and the citizens are throwing stones in the street" ([13], p. 4). Thus, for instance, New York or English Law "continue to dominate sovereign and quasi-sovereign lending in large parts of the world[,] including many emerging markets" [14]. New York and English Law are often the choices of law in debt instruments because of their "insulating effect" against "legislative changes in the borrower's country" ([13], p. 4). It is also a factor that the United States and the United Kingdom are often the home States of creditors who hold sovereign bonds. On the other hand, Debtor States normally agree to these clauses in order to access capital from foreign creditors.

The present regime of SDR has also been described as "contractual" given that debt contracts are the primary sources of rights and obligations with respect to sovereign debt. In the case of purely domestic indebtedness, a State has wide leeway in dealing with its debts using its own legal system. However, when a State borrows capital from foreign markets, debt contracts trigger the interaction between two legal systems and, to a more limited extent, between the Debtor State's legal system and international law. This precisely characterizes a conflict of law regime which is principally concerned with the "existence in the world of a number of separate municipal systems of law" ([12], p. 4). As a consequence, without an overarching framework, SDR is necessarily based on the Debtor State's *ad hoc* negotiations with its creditors based on those contracts. The *ad hoc* nature of these negotiations is reinforced by the fact that the distressed Debtor State has to deal with a diverse set of creditors including hedge funds and institutional investors.

Thus, the myriad of sovereign debt contracts is the core operational legal framework in the resolution of a sovereign debt crisis. The application of the international principle of State sovereignty is, to a large extent, determined through the State's express and implied waivers of such principle contained in the debt contracts. In most cases, international law is applied in SDR if it is incorporated in these contracts. While sovereign debt contracts may be covered by an investment treaty as in *Abaclat v. Argentina* [15], this has not evolved into a consistent international norm in SDR. In the first place, whether debt contracts are considered protected under an investment treaty would depend on the specific definition of "investment" under such treaty. Second, such expansive interpretation of "investment" has been criticized as a departure from the common understanding of "investment" as long-term commitment of capital that contributes to the economic development of the host State (see [16], pp. 515–517). In any event, there are currently more than 3000 investment treaties in the world with varying definitions of investment [17]. This situation in international investment law is hardly one that reflects a consistent international norm in SDR.

According to Guzman and Stiglitz, the contractual regime of SDR is essentially a "*non-system*...characterized by bargaining based on decentralized and non-binding market-based instruments centered on collective action clauses and competing codes of conduct" ([18], p. 3). There is widespread literature which strongly criticizes this contractual regime as "disorderly, inefficient, and overly costly" especially with respect to creditor collective action problems, in particular debt runs, holdouts and litigation (Citations omitted. [14], p. 88). Thus, there have been calls for a "hard law" approach or a multilateral legal framework to address these problems (for an overview, see [14], pp. 87–96). On the other hand, there are those skeptical of these proposals because of fears of rigidity of rules and "regulatory overkill" ([14], p. 88). Thus, an *ex ante* SDR framework consisting of mandatory rules carries the risk of stifling the flexibility that is present in *ex post* negotiations. Thus, "[i]nstead of creating a statutory framework 'top-down,' it could suffice to alter the documentation of bond and loan contracts to regulate the restructuring process in a more efficient way" ([14], p. 88). This would be the contractual regime of SDR where the prevailing norms of conduct are based on the terms and conditions stipulated in contracts entered into by the parties.

As discussed below, one way to bridge these opposing views is to formalize norms that are broad enough to allow parties the flexibility to adopt certain measures in the resolution of sovereign debt crises. At the same time, the formalization of these norms should be effective enough to influence collective behavior in SDR. The next part will discuss this proposal through a treaty-based codification of the norms contained in UN General Assembly Resolution 69/319.

3. Establishing a treaty-based normative framework

When UN General Assembly Resolution 69/319 ("Basic Principles on Sovereign Debt Restructuring Processes") was adopted 10 September 2015, 136 States voted in favor, six States voted against, and 41 States abstained ([18], p. 4; [10], p. 37). Among those which opposed the resolution were the US and the UK, "the two major jurisdictions for sovereign debt issuances by emerging economies, as well as Canada, Germany, Israel and Japan" ([18], p. 4).

UN General Assembly Resolution 69/319 articulates nine principles in SDR:

1. The *principle of State sovereignty* which is the Debtor State's right "to design its macroeconomic policy, including restructuring its sovereign debt..."

2. *Good faith* which is the "engagement in a constructive sovereign debt restructuring workout negotiations and other stages..."

3. *Transparency* which is "accountability of the actors concerned..."

4. *Impartiality* which means "independence" and the prevention of "undue influence" and "conflicts of interest."

5. *Equitable treatment* which is a State duty to refrain from arbitrary discrimination among its creditors.

6. *Sovereign immunity* of States from jurisdiction and execution which includes courts' obligation to "restrictively interpret" it in favor of State immunity.

7. *Legitimacy* entails "inclusiveness and the rule of law" in SDR "at all levels."

8. *Sustainability* "implies that sovereign debt restructuring workouts are completed in a timely and efficient manner and lead to a stable debt situation in the debtor State..."

9. *Majority restructuring* "implies that sovereign debt restructuring agreements that are approved by a qualified majority of the creditors of a State are not to be...impeded by other States or a non-representative minority of creditors..."

The thrust of this paper's proposal to codify UN General Assembly Resolution 69/319 into a treaty is to formalize the foregoing principles without necessarily adopting the resolution *in toto*. Its main objective is to introduce these norms as a multilateral framework in SDR without, at the moment, imposing any specific measure to implement them. If the voting of UN General Assembly Resolution 69/319 is any indication, only the 136 States which voted in favor of the resolution would probably regard its codification in a positive light. Thus, to improve chances of success, a proposal for codification should court the "no" and "abstain" votes by introducing changes that accommodate their positions.

It is not surprising that the influential developed economies voted against UN General Assembly Resolution 69/319 given that their citizens are mostly the creditors of distressed Debtor States and that they have an interest in protecting the sovereign debt market within their jurisdictions. One of their declared principal objections is that "the IMF as the primary forum to discuss sovereign debt restructuring issues" instead of the UN General Assembly [8]. Equally important are their objections against the specific manner by which the nine principles in SDR were articulated in the resolution.

The European Union, for instance, has declared its objections to a number of statements in the resolution:

> *Paragraph 4 requests "all institutions and actors involved in sovereign debt restructuring," including "at regional level," to "refrain from exercising any undue influence over the process and other stakeholders or engaging in actions that would give rise to conflicts of interest." Such a statement fits poorly with both the EU institutional setting and its practical situation, where possible discussions related to the stock of debt of a Member State take place primarily at regional level, against a background in which the Member States themselves are often the main creditors (directly or via the financial assistance mechanisms that they have established).*
>
> *Paragraph 5 states that "creditors have the right to receive the same proportionate treatment in accordance with their credit and its characteristics" and that "no creditors or creditor groups should be excluded ex ante from the sovereign debt restructuring process." Such a statement denies the customary preferred creditor status recognized to the International Financial Institutions (IMF, ESM...) when lending to a Sovereign in distress, with possible major negative implications on their ability to fulfill their primary mission.*
>
> *Paragraph 9 states that "sovereign debt restructuring agreements that are approved by a qualified majority of the creditors of a State are not to be affected, jeopardized or otherwise impeded by other States." This statement is very problematic for issuances under foreign jurisdiction (the overwhelmingly dominant case for developing and emerging countries). Issuing under a foreign jurisdiction does, by definition, involve accepting the competence of the courts of another State. This is a point of major importance for the EU, as a very large part of the world's sovereign issuances under foreign jurisdiction are made under the jurisdiction of one of its Member States (UK) [19].*

In codifying the SDR principles into a treaty, it is worthwhile to consider these objections which relate to the specific wording of UN General Assembly Resolution 69/319. In general, it is easier to accommodate concerns about specific wordings than the more fundamental objections premised on the market-based approach to SDR. This is driven by a pragmatic consideration of securing State support by postponing the imposition of concrete SDR measures and, at the same time, advancing the main objective of introducing a multilateral normative framework for SDR.

The broad codification of the norms contained in UN General Assembly Resolution 69/319 would be a compromise between a "soft law" approach (usually in the form of guidelines that may be adopted by States) and a "hard law" approach in the form of enforceable specific SDR measures. While the proposal takes the form of a treaty, the SDR norms to be articulated therein should be deliberately broad to give States the flexibility to implement them within specific contexts. As such, this would not be an unqualified "surrender of sovereignty" on the part of State parties that may be part of their reluctance towards UN General Assembly Resolution 69/319 (consider, for instance, the 41 abstaining votes). A degree of flexibility is needed in SDR especially in view of the rapidly changing nature of the sovereign debt market (see generally [20]).

This proposal is preliminary: it is designed to set the grounds for more concrete and "hard law" SDR measures in the future. It follows the so-called Incremental Approach that "complement, rather than replace, existing mechanisms, including contractual approaches and the activities of the International Financial Institutions or the Paris Club, and guide their operation" ([10], p. 38). Despite not codifying concrete SDR measures, a treaty-based multilateral normative framework would hopefully exert a "compliance pull" on the current SDR regime towards a shared understanding of these norms. A multilateral normative framework in SDR would reinforce the "use [of] the interpretive space between the factual and the normative, between apology and utopia, in order to highlight and strengthen trends in current practice that support debt sustainability" ([10], pp. 38–39).

The broad codification of the norms in UN General Assembly Resolution 69/319 would be a significant advancement in itself towards establishing a multilateral normative framework in SDR. Even from the perspective of market-based proponents, the efficiency and equity problems in the present regime of SDR is serious enough to warrant reform. It is the kind of reforms needed that is the subject of intense debate which straddles between, on one hand, the value of uniform and predetermined rules and, on the other, that of flexibility. This paper's proposal seeks to bridge those values by supporting a predetermined normative framework in SDR while retaining the flexibility needed to undertake the restructuring process.

As a political matter, there is no doubt that the support of influential developed countries like the US and the UK would considerably help the success of the proposed treaty. However, even if there are limited States which become parties to the treaty, it cannot be discounted that some actors might prefer to deal with States with a reasonably predictable SDR framework as forwarded by the proposal.

A treaty-based normative framework in SDR would also mean that domestic courts of State parties would be obligated to apply the norms embodied in the treaty. Howse observes that even with international arbitration, "domestic courts are the ultimate mechanism [in SDR], as is illustrated by international investment law." ([21], p. 244). This is because it is often the case that most of the Debtor State's assets are located within its jurisdiction. Moreover, it is not unlikely for treaty-based SDR norms to influence international arbitration (based on debt contracts) given that State parties to the arbitration now have a treaty obligation to follow the SDR norms. Thus, while treaty-based SDR norms are deliberately broad, there will be some measure of enforcement of those norms especially at the domestic level.

In sum, this treaty-based normative framework in SDR may set the course for a series of actions which may lead to more concrete and enforceable measures in SDR. This is the import of the Incremental Approach that "highlight[s] and strengthen[s] trends in current [and future] practice" ([10], p. 38) and which is designed to continuously build upon the broad normative framework set out at the start.

4. The principle of sustainability

The principles or norms provided in UN General Assembly Resolution 69/319 may be seen as addressing two aspects of SDR: first is the conduct of the relevant actors in the restructuring process; and second is the objective of the restructuring process itself. The principles of sovereignty, good faith, transparency, impartiality, equitable treatment, sovereign immunity, legitimacy and majority structuring primarily relate to the behavior of States, creditors, tribunals and other relevant actors. These principles are meant to address the collective action problem in SDR—that is, the failure to coordinate among the Debtor State and its creditors even if all would be better off if they coordinate.

For instance, the end of Argentina's 15-year stand-off with the majority of its holdout creditors in April 2016 has been attributed to "good faith negotiations" ([22], p. 1). In lifting the injunction against Argentina's payment of its restructured bonds, the New York court judgment "indicated...that the election of Argentina's new government with its willingness to negotiate in good faith...was pivotal in [the] decision..." ([22], p. 2; [23]). In this case, "good faith" was used to characterize the conduct of negotiations between Argentina and its holdout creditors in reaching a settlement that allowed the recovery of 75% of the creditors' original claim. This would be an example of an application of the principle of good faith to address a collective action problem in SDR. Moreover, the principles of sovereignty, transparency, impartiality, equitable treatment, sovereign immunity, legitimacy and majority structuring would have, in varying degrees, played a part in reaching the Argentine settlement as it is often necessary that these principles operate in concert to address a collective action problem.

On the other hand, the principle of sustainability would relate more to the objective of SDR. Guzman and Stiglitz, for instance, stated that "[t]he ultimate goal of sovereign debt restructuring is to restore the sustainability of public debt *with high probability*" ([18], p. 3). The articulation of the principle of sustainability in UN General Assembly Resolution 69/319, however, is noteworthy for its progressiveness:

> *Sustainability implies that sovereign debt restructuring workouts are completed in a timely and efficient manner and lead to a stable debt situation in the debtor State, preserving at the outset creditors' rights while promoting sustained and inclusive economic growth and sustainable development, minimizing economic and social costs, warranting the stability of the international financial system and respecting human rights.*

The foregoing is not only concerned with the debt sustainability within the Debtor State, but also "inclusive economic growth and sustainable development... [which] respect[] human rights." This particular articulation is holistic in the sense that it considers the long-term social and economic welfare of the Debtor State which may very include the well-being of its citizens. Significantly, the mention of "human rights" may make SDR resonate with other international norms such as those embodied in the Universal Declaration of Human Rights [24] and the International Covenant on Civil and Political Rights [25].

The IMF has defined debt sustainability "as a situation in which a borrower is expected to be able to continue servicing its debts without an unrealistically large future correction to the balance of income and expenditure" ([26], p. 4). Compared to the UN General Assembly Resolution 69/319, the IMF's definition of "debt sustainability" has been described as being "in purely financial terms" ([10], p. 25).

The IMF's definition consists of two components: (1) that the Debtor State "cannot indefinitely accumulate debt faster than [its] capacity to service these debts" ([14], p. 71); and (2) that the service of debts should not require "an unrealistically large future correction to the balance of income and expenditure." The first relates to a financial aspect of sustainability whose objective is to prevent an unbounded accumulation of debts. On the other hand, the second component "implies that there are social and political limits to adjustment[s]" ([14], p. 71) needed to achieve debt sustainability. Thus, "[n]ot all fiscal adjustment paths are realistic, because political and other constraints will influence a country's willingness to pay (as opposed to ability to pay)" ([14], p. 71).

The recognition by the IMF of social and political limits to adjustment policies in SDR may be seen as going beyond a purely financial framework of sustainability. Although this recognition may be limited insofar as it affects a State's willingness to pay its debts, the IMF's definition is significant to the extent that it implies an assessment of the social and political situation of the debtor State in the context of a sovereign debt crisis. Thus, it may not be entirely accurate to say that the IMF's definition of "debt sustainability" is incompatible with the principle of sustainability in UN General Assembly Resolution 69/319. The IMF's willingness to recognize the social and political aspects of SDR has some kinship with UN General Assembly Resolution 69/319 in its emphasis on "inclusive economic growth and sustainable development." In fact, recent debt sustainability analysis of the IMF factors economic growth within the Debtor State as a contributing variable in assessing sustainability ([14], p. 75).

In short, there may be possible points of contact between the IMF's definition of "debt sustainability" and the principle of sustainability in UN General Assembly Resolution 69/319. These points of contact in the understanding of sustainability may play a role in gathering support for a proposal to include "inclusive economic growth and sustainable development" as part of a codified principle of sustainability. Indeed, there is significance in going beyond a purely financial assessment of whether a State's payment of its existing debts may lead to an unbounded accumulation of future debts.

Fostering "inclusive economic growth" within the Debtor State should be seen as rational and viable strategy in SDR because it addresses the following characteristics that is not found in domestic bankruptcy regimes:

1. Unlike corporations that are the subject of bankruptcy proceedings, there is no option of liquidation or dissolution of a Debtor State because the likelihood of such an event happening as result of a debt crisis is minimal. As such, the debt crisis is necessarily temporary. In most instances, the question to be addressed is not the survival of the State but the duration of the crisis given the measures that may be implemented to resolve it.

2. The assets of the Debtor State that may be used to service its debts are dynamic given a functioning economy that utilizes fiscal policies like raising revenue through taxation.

3. A distressed Debtor State is also accountable to its citizens who are beneficiaries of its social welfare programs such as "pensioners, those depending on the

government for health benefits or education, etc." ([27], p. 9). Guzman and Stiglitz have called these beneficiaries as the State's informal creditors whose "benefits are part of the social contract" ([27], p. 9).

The lack of an option for "State dissolution" should shift an SDR framework towards economic growth rather than the common pool of assets (see [28]) problem in bankruptcy cases. This is a paradigm shift from a static understanding of a limited pool of assets in domestic bankruptcy proceedings to a dynamic one which accounts for the possibility of growth (or conversely, diminution) based on the macroeconomic conditions of the Debtor State. Thus, "inclusive economic growth" within the Debtor State shortens the duration of a sovereign debt crisis and expands the available pool of State assets that may be used to service its debts. The expansion of the State's asset base also prevents the unbounded accumulation of future debts.

Such endogenous growth implies a multiplier effect in the economy that increases a Debtor State's tax revenues, consumption, investments and expenditures. In the context of SDR, "inclusive economic growth" ultimately means the increased likelihood of the Debtor State to pay its debts. In this way, it benefits everyone: the Debtor State, the creditors and other stakeholders especially those who derive benefit from the Debtor State's social contract.

The project of fostering "inclusive economic growth" within the Debtor State would be seen in contradistinction to the imposition of austerity measures purportedly to set aside money to pay the State's creditors. Guzman and Stiglitz observes that "austerity policies are normally counterproductive *even from a creditors' perspective*" ([27], p. 8). Austerity measures may, in fact, decrease the State's ability to pay its debts by preventing it from spending money domestically to fix its economy.

There is also a question of social justice involved in prioritizing foreign creditors over the Debtor State's social welfare beneficiaries. Limited State assets are being used to pay wealthy investors (which include the so-called "vulture funds") over the State's citizens who may be invoking rights more closely related to value of human life—*e.g.* rights to public health and a sustainable environment. This arrangement in effect violates a hierarchy of values in international law which upholds the primacy of human life over property. Thus, the U.N. Commission on Human Rights issued a Resolution which "affirms that the exercise of the basic rights of the people of debtor countries to food, housing, clothing, employment, education, health services and a healthy environment cannot be subordinated to the implementation of structural adjustment policies and economic reforms arising from the debt" [29].

The treaty codification of "inclusive economic growth and sustainable development" as part of the principle of sustainability helps SDR move beyond a purely financial framework of sustainability. While it may be desirable to adopt more targeted measures to improve the human rights situation in SDR, an emphasis on the Debtor State's endogenous economic growth also addresses this aspect by improving the collective welfare of the relevant actors involved. In particular, an improved economic situation helps the Debtor State fulfill its social contract with its citizens through an increased ability to pay.

5. Conclusion

As shown above, there are not only efficiency and equity problems in the present regime of SDR but also fundamental questions of social justice. These problems are serious enough to be simply ignored. The literature in SDR reveals that there is hardly any dispute that reforms are needed to address these problems. There is dispute, however, in the *kind of reforms* needed in SDR which ranges from the adoption of clauses in debt contracts (to maximize flexibility) to enacting a fixed framework similar to a domestic bankruptcy regime (to maximize uniformity and predictability). If anything, a pragmatic approach to reform in SDR would have to be between these two ends.

This paper's proposal to establish a treaty-based normative framework in SDR seeks to bridge the flexibility in *ex post* negotiations and the uniformity in *ex ante* rules. This proposal is not as ambitious as it sounds. By broadly codifying the norms in UN General Assembly Resolution 69/319, the proposal follows the Incremental Approach for reforms in SDR in the way it is *backward-looking*, because it builds on existing "soft law" approaches and other mechanisms and practices in SDR (for instance, it recognizes the role of *ad hoc* negotiations); and *forward-looking*, because it hopes to set the ground for more concrete and effective SDR measures in the future. As a whole, the approach of the proposal is to complement and improve the current conflicts of law regime of SDR.

Despite the proposal's declared strategy to accommodate the various positions of States, one thing that stands out is its call to codify the principle of sustainability that emphasizes "inclusive economic growth and sustainable development." This comes from the recognition that the articulation and implementation of this principle is a more effective way of undertaking SDR than a purely financial framework of preventing the State's unbounded accumulation of debts. It is premised on the idea that while there are no systematized rules, SDR is nevertheless situated in an international order that upholds a complexus of norms which includes human rights. Thus, this proposal is a call for a movement from SDR's current focus on the protection of debt markets to the broader economic, social and political context of the international order.

Author details

Maximo Paulino T. Sison III
University of the Philippines, Diliman, Quezon City, Philippines

*Address all correspondence to: mowiesison@gmail.com

IntechOpen

© 2019 The Author(s). Licensee IntechOpen. This chapter is distributed under the terms of the Creative Commons Attribution License (http://creativecommons.org/licenses/by/3.0), which permits unrestricted use, distribution, and reproduction in any medium, provided the original work is properly cited.

References

[1] Kawananokoa v. Polyblank. 205 U.S. 349; 1907

[2] Blackman J. The evolution of modern sovereign debt litigation: Vultures, alter egos, and other legal fauna. Law and Contemporary Problems. 2010;**73**:47

[3] International Monetary Fund. Sovereign Debt Restructuring—Recent Developments and Implications for the Funds Legal and Policy Framework (26 April 2013). Available at: <https://www.imf.org/external/np/pp/eng/2013/042613.pdf> [Accessed: 26 October 2017]

[4] Li Y. Question the unquestionable beauty of a collective proceeding for all sovereign debt claims. International Insolvency Review. 2013;**22**:85

[5] Bolton P. Toward a statutory approach to sovereign debt restructuring: Lessons from corporate bankruptcy practice around the world. IMF Staff Papers. 2003;**50**:41

[6] Hart HLA. Positivism and the separation of law and morals. Harvard Law Review. 1957;**71**:593-603

[7] Hart HLA. The Concept of Law. 3rd ed. 2012. pp. 50-78

[8] European Union. EU Explanation of Vote – United Nations General Assembly: Sovereign Debt Restructuring Process. Statement on behalf of the Member States of the European Union, EUUN14-112EN. Sept. 9, 2014. Available from: <http://eu-un.europa.eu/eu-explanation-of-vote-united-nations-general-assembly-draft-resolution-sovereign-debt-restructuring/> [Accessed: 27 April 2017]

[9] Uy M, Zhou S. Sovereign debt of developing countries: Overview of trends and policy perspectives. In: Guzman M, Ocampo JA, Stiglitz JE, editors. Too Little, Too Late: The Quest to Resolve Sovereign Debt Crises. 2016

[10] Bohoslavsky JP, Goldmann M. An incremental approach to sovereign debt restructuring: Sovereign debt sustainability as a principle of public international law. Yale Journal of International Law Online. 2016;**41**:13

[11] Hague Convention on the Recognition and Enforcement of Foreign Judgments, 1144 UNTS 249 [Adopted: 1 February 1971]

[12] North P, Fawcett JJ. Cheshire and North's Private International Law. 13th ed. UK: LexisNexis; 2004

[13] Allen & Overy Global Intelligence Unit. State Insolvency – What Bondholders and Other Creditors Should Know; 2010. Available from: <http://www.allenovery.com/archive/Documents/Legacy/64484.pdf> [Accessed: 25 April 2017]

[14] Das U, Papaioannou M, Trebesch C. Sovereign debt restructurings: Concepts, literature survey, and stylized facts. IMF Working Paper WP/12/203. August 2012

[15] Abaclat and Others v. The Argentine Republic, ICSID Case No. ARB/07/5. Decision on Jurisdiction and Admissibility. 4 August 2011

[16] Beess J, Chrostin C. Sovereign debt restructuring and mass claims arbitration before the ICSID. The abaclat case. Harvard International Law Journal. 2012;**53**:505

[17] Investment Policy Hub of the United Nations Conference on Trade and Development. Available from: <http://investmentpolicyhub.unctad.org/IIA> [Accessed: 26 April 2017]

[18] Guzman M, Stiglitz J. A soft law mechanism for sovereign debt restructuring based on UN principles. International Policy Analysis. 2016

[19] Council of the European Union. EU common position on the UN draft resolution A/69/L.84 on basic principles on sovereign debt restructuring processes, 11705/15. (7 September 2015). Available from: <http://data.consilium.europa.eu/doc/document/ST-11705-2015-INIT/en/pdf> [Accessed: 27 April 2017]

[20] Anderson P, Silva A, Velandia-Rubiano A. Public debt management in emerging market economies: Has this time been different? World Bank Policy Research Working Paper. August 2010

[21] Howse R. Toward a framework for sovereign debt restructuring: What can public international law contribute? In: Guzman M, Ocampo JA, Stiglitz JE, editors. Too Little, Too Late: The Quest to Resolve Sovereign Debt Crises. 2016

[22] Erichsen K, Wilkinson A. Sovereign debt restructuring: What can we learn from Argentina, Greece and Ukraine? Journal of International Banking and Financial Law. 2016;7:1

[23] Financial Times. US Appeals Court Clears Way for Argentina Bond Sale. 13 April 2016. Available from: <https://www.ft.com/content/5039a628-01a0-11e6-99cb-83242733f755> [Accessed: 27 April 2017]

[24] Universal Declaration of Human Rights. General Assembly Resolution 217 A(III) [Adopted: 10 December 1948]

[25] International Covenant on Civil and Political Rights. 999 UNTS 171 [Adopted: 16 December 1966, entered into force 23 March 1976]

[26] International Monetary Fund. Assessing sustainability. IMF Policy Paper. 28 May 2002

[27] Guzman M, Stiglitz JE. Creating a framework for sovereign debt restructuring that works. In: Guzman M, Ocampo JA, Stiglitz JE, editors. Too Little, Too Late: The Quest to Resolve Sovereign Debt Crises. 2016

[28] Block-Lieb S. Fishing in muddy waters: Clarifying the common pool analogy as applied to the standard for commencement of a bankruptcy case, 42. American University Law Review. 1993:337

[29] U.N. Commission on Human Rights Res. 1999/22. Effects on the full enjoyment of human rights of the economic adjustment policies arising from foreign debt and, in particular, on the implementation of the Declaration on the Right to Development, E/CN.4/RES/1991/13. April 23, 1999